Humber College Library
3199 Lakeshore Blvd. West
Toronto, ON   M8V 1K8

Advanced Sciences and Technologies for Security Applications

Series editor

Anthony J. Masys, Associate Professor, Director of Global Disaster Management, Humanitarian Assistance and Homeland Security, University of South Florida, Tampa, USA

Advisory Board

Gisela Bichler, California State University, San Bernardino, CA, USA
Thirimachos Bourlai, WVU - Statler College of Engineering and Mineral Resources, Morgantown, WV, USA
Chris Johnson, University of Glasgow, UK
Panagiotis Karampelas, Hellenic Air Force Academy, Attica, Greece
Christian Leuprecht, Royal Military College of Canada, Kingston, ON, Canada
Edward C. Morse, University of California, Berkeley, CA, USA
David Skillicorn, Queen's University, Kingston, ON, Canada
Yoshiki Yamagata, National Institute for Environmental Studies, Tsukuba, Japan

The series Advanced Sciences and Technologies for Security Applications comprises interdisciplinary research covering the theory, foundations and domain-specific topics pertaining to security. Publications within the series are peer-reviewed monographs and edited works in the areas of:

- biological and chemical threat recognition and detection (e.g., biosensors, aerosols, forensics)
- crisis and disaster management
- terrorism
- cyber security and secure information systems (e.g., encryption, optical and photonic systems)
- traditional and non-traditional security
- energy, food and resource security
- economic security and securitization (including associated infrastructures)
- transnational crime
- human security and health security
- social, political and psychological aspects of security
- recognition and identification (e.g., optical imaging, biometrics, authentication and verification)
- smart surveillance systems
- applications of theoretical frameworks and methodologies (e.g., grounded theory, complexity, network sciences, modelling and simulation)

Together, the high-quality contributions to this series provide a cross-disciplinary overview of forefront research endeavours aiming to make the world a safer place.

More information about this series at http://www.springer.com/series/5540

Hamid Jahankhani
Editor

Cyber Criminology

HUMBER LIBRARIES LAKESHORE CAMPUS
3199 Lakeshore Blvd West
TORONTO, ON.   M8V 1K8

Editor
Hamid Jahankhani
QAHE and Northumbria University London
London, UK

ISSN 1613-5113　　　　　　ISSN 2363-9466　(electronic)
Advanced Sciences and Technologies for Security Applications
ISBN 978-3-319-97180-3　　　　ISBN 978-3-319-97181-0　(eBook)
https://doi.org/10.1007/978-3-319-97181-0

Library of Congress Control Number: 2018960872

© Springer Nature Switzerland AG 2018
This work is subject to copyright. All rights are reserved by the Publisher, whether the whole or part of the material is concerned, specifically the rights of translation, reprinting, reuse of illustrations, recitation, broadcasting, reproduction on microfilms or in any other physical way, and transmission or information storage and retrieval, electronic adaptation, computer software, or by similar or dissimilar methodology now known or hereafter developed.
The use of general descriptive names, registered names, trademarks, service marks, etc. in this publication does not imply, even in the absence of a specific statement, that such names are exempt from the relevant protective laws and regulations and therefore free for general use.
The publisher, the authors and the editors are safe to assume that the advice and information in this book are believed to be true and accurate at the date of publication. Neither the publisher nor the authors or the editors give a warranty, express or implied, with respect to the material contained herein or for any errors or omissions that may have been made. The publisher remains neutral with regard to jurisdictional claims in published maps and institutional affiliations.

This Springer imprint is published by the registered company Springer Nature Switzerland AG
The registered company address is: Gewerbestrasse 11, 6330 Cham, Switzerland

Foreword

This book could not be more timely. Every day we learn of new developments in artificial intelligence. The Internet of Things (IoT) is becoming a kind of parallel universe. The skills of scientists and inventors have the capacity at their best to enhance and even extend lives, to provide new abilities for people with disabilities, to make us all more inventive, to process in moments information that previously challenged the best mathematical and statistical skills and to develop and satiate our natural curiosity.

At the same time, we know that the same skills in advanced sciences and technologies, if misused, will be the instruments of crime and even oppression. Most users of the Internet experience weekly, if not daily, attacks on their privacy and financial integrity, even on their very identity. Data is mined and misused. Sexually motivated grooming, bullying, victimisation and terrorist radicalisation become ever more methodical and concealed. Like-minded criminals congregate on the Dark Web, using difficult to detect pseudonyms and acronyms, and often impenetrable security, to achieve their purposes.

The benefits of artificial intelligence and the Internet of Things can be reaped beneficially only if sown securely; and the online world has not yet the power or creator motivation to secure itself.

Nowhere have the challenges been demonstrated more clearly than in the efforts of the State to counter terrorism propagated on the Internet. Such has been the impact of radicalising websites, whether for violent Islamism or right-wing extremism, that the authorities are now removing tens of thousands of such sites every month. Although that battle is being won, this is happening by attrition, with the net loss of such sites occurring at a worryingly slow pace.

Cybercrime is having an even greater impact than terrorism on the general public. Daily attacks are made on electricity and other energy suppliers, banks, law firms and accountants, private companies, medical records and other caches of evidence of human activity. Even keeping pace with the operational range of cybercrime is a hugely expensive endeavour.

This book provides instructive guidance for readers interested in tackling these huge, contemporary problems. It explains the criminological context of cybercrime. It demonstrates the mental and physical components that are required for readers to understand cybercrime. It deals with the psychology of cyber criminals, analysing the motives which move them and the methodologies that they adopt. It sets out the intelligence networks that are used to bring together information about crime falling into this exponentially growing category. It explains the power of the State to intervene in private data for the detection of crime and the protection of the public. Also, it teaches readers of the legal protections of confidentiality, the extent of those protections and the extent and limits of data protection legislation.

The categories of crime described in the book are very wide. They are every bit as psychologically complex as offences of, for example, murder and manslaughter. The extensive written or graphic evidential material comprising the actus reus of such crimes tells us much about the nature of the criminals who are undermining the benefits of the electronic world by abusing AI and IOT.

The available evidence often is of a kind analogous to that used by psychiatrists and psychologists in analysing mental health, motive and loss of control in crimes of violence. The book explains how such analysis can be used in understanding and thereby detecting the perpetrators of crimes against confidentiality and safety on the Internet. The text will prove instructive to police and other regulatory authorities in their pursuit of cybercrime. It will also be especially valuable for the increasing number of organisations willing to take private prosecutions against perpetrators, in cases in which the State does not act because of resource limitations. Deductive, psychologically trained reasoning should enable the detection of many criminals in this range.

For AI and IoT to benefit society, it needs to be policed. That policing must be conducted in a strictly ethical context, proportionate and in the overall public interest. The ethical base must be founded on high-quality training, education and awareness for all those who carry out the policing – just as ethical parameters should be set out during education and training for those who are learning how to use the advanced sciences and technologies under discussion.

Further, it is just as important for safety to be a watchword in this virtual world as it is in the physical world – as when we teach our children across the road or to be safe in their teenage lives.

The potential for technology to cause or contribute to serious mental illness, lack of confidence and economic failure cannot be exaggerated; just as its potential to create great happiness and economic and professional success knows almost no bounds.

This book will provide professionals, teachers and students alike with an excellent reference guide for the multi-faceted issues which will come their way in dealing with advanced technologies in the years to come. Just as there are standard works on criminal law, family law and the law of tort, equally there will have to be standard reference volumes on lawful and unlawful activities in the virtual world. I believe that this volume is one of the first of such works and promises great benefit.

In addition, it provides an understanding of the potential of the criminal and civil jurisdiction in protecting the public from the kinds of crime under contemplation. There is bound to be an ever-increasing number of cases brought before courts, as dividing lines are set concerning the acceptability or otherwise of questioned behaviours.

This is new territory for the judiciary too. Some judges are technically very proficient, whilst others less so. Non-specialist judges, including lay magistrates, will have to be able to deal with these issues. All would be well advised to read this important work. It will provide them with the full necessary background and answers to many of the specific problems that they will encounter.

In the years to come, we will be grateful for the impetus given in this area of the law by Prof. Jahankhani and his colleagues who have contributed to the widely ranging chapters in this work.

June 2018   Lord Alex Carlile of Berriew CBE QC

Contents

Part I Cyber Criminology and Psychology

Crime and Social Media: Legal Responses to Offensive Online
Communications and Abuse 3
Oriola Sallavaci

Explaining Why Cybercrime Occurs: Criminological
and Psychological Theories 25
Loretta J. Stalans and Christopher M. Donner

Cyber Aggression and Cyberbullying: Widening the Net 47
John M. Hyland, Pauline K. Hyland, and Lucie Corcoran

Part II Cyber-Threat Landscape

Policies, Innovative Self-Adaptive Techniques and Understanding
Psychology of Cybersecurity to Counter Adversarial Attacks in
Network and Cyber Environments... 71
Reza Montasari, Amin Hosseinian-Far, and Richard Hill

The Dark Web 95
Peter Lars Dordal

Tor Black Markets: Economics, Characterization and Investigation
Technique 119
Gianluigi Me and Liberato Pesticcio

A New Scalable Botnet Detection Method in the Frequency Domain 141
Giovanni Bottazzi, Giuseppe F. Italiano, and Giuseppe G. Rutigliano

Part III Cybercrime Detection

Predicting the Cyber Attackers; A Comparison of Different
Classification Techniques 169
Sina Pournouri, Shahrzad Zargari, and Babak Akhgar

Crime Data Mining, Threat Analysis and Prediction 183
Maryam Farsi, Alireza Daneshkhah, Amin Hosseinian-Far,
Omid Chatrabgoun, and Reza Montasari

SMERF: Social Media, Ethics and Risk Framework 203
Ian Mitchell, Tracey Cockerton, Sukhvinder Hara, and Carl Evans

Understanding the Cyber-Victimisation of People with Long Term Conditions and the Need for Collaborative Forensics-Enabled Disease Management Programmes... 227
Zhraa A. Alhaboby, Doaa Alhaboby, Haider M. Al-Khateeb,
Gregory Epiphaniou, Dhouha Kbaier Ben Ismail, Hamid Jahankhani,
and Prashant Pillai

An Investigator's Christmas Carol: Past, Present, and Future Law Enforcement Agency Data Mining Practices................................. 251
James A. Sherer, Nichole L. Sterling, Laszlo Burger, Meribeth Banaschik,
and Amie Taal

DaP∀: Deconstruct and Preserve for All: A Procedure for the Preservation of Digital Evidence on Solid State Drives and Traditional Storage Media 275
Ian Mitchell, Josué Ferriera, Tharmila Anandaraja, and Sukhvinder Hara

Part IV Education, Training and Awareness in Cybercrime Prevention

An Examination into the Effect of Early Education on Cyber Security Awareness Within the U.K. 291
Timothy Brittan, Hamid Jahankhani, and John McCarthy

An Examination into the Level of Training, Education and Awareness Among Frontline Police Officers in Tackling Cybercrime Within the Metropolitan Police Service........................ 307
Homan Forouzan, Hamid Jahankhani, and John McCarthy

Combating Cyber Victimisation: Cybercrime Prevention................... 325
Abdelrahman Abdalla Al-Ali, Amer Nimrat, and Chafika Benzaid

Information Security Landscape in Vietnam: Insights from Two Research Surveys... 341
Mathews Nkhoma, Duy Dang Pham Thien, Tram Le Hoai,
and Clara Nkhoma

Part I
Cyber Criminology and Psychology

Crime and Social Media: Legal Responses to Offensive Online Communications and Abuse

Oriola Sallavaci

1 Introduction

Social media is defined as "websites and applications that enable users to create and share content or to participate in social networking" (The Law Society 2015). It commonly refers to the use of electronic devices to create, share and exchange information, pictures, videos via virtual communities and networks (CPS guidelines n.d.-a). Some of the most popular social networking platforms include Facebook; Twitter; LinkedIn; YouTube; WhatsApp; Snapchat; Instagram and Pinterest. Facebook and Twitter are among the oldest and were founded in 2004 and 2006 respectively. Approximately 2 billion internet users are using social networks and these figures are expected to grow further as mobile device usage and mobile social networks increasingly gain traction (The Statistics Portal). Taken together, social media platforms are likely to contain several millions of daily communications.

The strong and rapid emergence of social media platforms over the past decade has significantly facilitated the contact and exchange of information between people across geographical, political and economic borders. At the same time it has opened up avenues to new threats and offensive online behaviour. Such behaviour includes inter alia (House of Lords 2014, p. 7):

- *Cyber bullying* – which refers to bullying and harassing behaviour conducted using the social media or other electronic means;
- *Trolling* – which refers to the intentional disruption of an online forum, by causing offence or starting an argument;

O. Sallavaci (✉)
University of Essex, Colchester, UK
e-mail: osallavaci@essex.ac.uk

© Springer Nature Switzerland AG 2018
H. Jahankhani (ed.), *Cyber Criminology*, Advanced Sciences and Technologies for Security Applications, https://doi.org/10.1007/978-3-319-97181-0_1

- *Virtual mobbing* – whereby a number of individuals use social media or messaging to make comments to or about another individual, usually because they are opposed to that person's opinions;
- *Revenge pornography* – which involves the electronic publication or distribution of sexually explicit material (principally images) without consent, usually following the breakup of a couple, the material having originally been provided consensually for private use.

In addition to these apparently modern offences there are other 'traditional' offences, which involve the use of words or images that can also be committed via social media. Harassment, malicious communications, stalking, threatening violence, incitement are among these traditional crimes which have existed and have been prohibited for a long time before the emergence of social media platforms. It can however be argued that the commission of these offences has been facilitated by the use of technology and the widespread use of social media, acquiring new dimensions that require careful legal and policy considerations. As a commentator puts it "online abuse is underpinned by entrenched power differentials on the basis of gender, age and other factors and 'crosses over' with offline harms such as domestic violence, bullying and sexual harassment. Social media has come to saturate social life to such an extent that the distinction between 'online' and 'offline' abuse has become increasingly obsolete, requiring a nuanced understanding of the role of new media technologies in abuse, crime and justice responses" (Salter 2017, p. 13).

The need to have in place legislation that clearly and adequately provides for the prohibition and punishment of online offences committed through social media is paramount. In England and Wales offensive online communications include a range of offences which are categorised as follows (CPS guidelines n.d.-a):

1. Credible threats of violence to the person or damage to property:
    - Threat to Kill (Offences Against the Person Act 1861, s 16)
    - Putting another in fear of violence; Stalking involving fear of Violence or serious alarm or distress (Protection From Harassment Act 1997, s 4 and s4A respectively)
    - Sending of an electronic communication which involves threat (Malicious Communications Act 1988, s 1)
    - Sending of messages of a "menacing character" via public telecommunications network (Communications Act 2003, s 127)

2. Communications targeting specific individuals:
    - Harassment and stalking (Protection from Harassment Act 1997, s 2; s4 and s4A)
    - Offence of controlling or coercive behaviour (Serious Crime Act 2015, s76)
    - Disclosing private sexual images without consent (revenge pornography) (Criminal Justice and Courts Act 2015, s33)
    - Other offences involving communications targeting specific individuals, such as offences under the Sexual Offences Act 2003 or Blackmail.

3. Breach of court order, e.g. as to anonymity. This can include:
    - Juror misconduct offences under the Juries Act 1974 (sections 20A-G);
    - Contempts under the Contempt of Court Act 1981;
    - An offence under section 5 of the Sexual Offences (Amendment) Act 1992 (identification of a victim of a sexual offence);
    - Breaches of a restraining order; or
    - Breaches of bail.
4. Communications which are grossly offensive, indecent, obscene or false:
    - Electronic communications which are indecent or grossly offensive, convey a threat false, provided that there is an intention to cause distress or anxiety to the victim (Malicious Communications Act 1988, s 1)
    - Electronic communications which are grossly offensive or indecent, obscene or menacing, or false, for the purpose of causing annoyance, inconvenience or needless anxiety to another (Communications Act 2003, s 127)

Almost all these offences pre-date the invention of social media. This chapter will focus on the legal aspects of offensive online communications including cyberbullying, revenge pornography and other related offences. These types of offensive and abusive behaviour have spread considerably in the recent years, acquiring new dimensions and posing new challenges for the public, legal community, law enforcement and policymaking. It will be argued that the current legal framework is complex. The legislation dealing with offensive online communications is in need of clarification and simplification. This is a necessary step that must go hand in hand with reforms in the area of law enforcement and preventative measures aimed at raising public awareness and education.

2 Cyberbullying, Cyber-Harassment and Cyberstalking

Cyberbullying, *cyber-harassment* and *cyberstalking* are terms that cover a variety of forms of behaviour that display similar features. Sometime the terms are used interchangeably and at other times they are distinguished (Gillespie 2016, p. 257). Bullying and harassment could be considered to be different to stalking even though there is some overlap between them. Bullying and harassment involve individualised negative behaviour whereby someone acts in an aggressive or hostile manner in order to intimidate the victim. This includes a variety of types of behaviour such as: *flaming* (the posting of provocative or abusive posts); *outing* (the posting or misuse of personal information) and/or the distribution of malware. (Gillespie 2016, p. 258) Cyberstalking could involve: communicating with the victim (both passive and aggressive forms); publishing information about the victim (similar to outing); targeting the victim's computer (especially to gain personal data); placing the victim under surveillance including cyber-surveillance). Apart from the final factor there are similarities between cyberstalking and cyberbullying in terms of how the offences are committed. (Gillespie 2016, p. 261) This chapter focuses on bulling, harassment and stalking via online communications. Hacking and distribution of malware have received attention elsewhere (Sallavaci 2017).

At the time of writing, there is no specific criminal offence of bullying or cyberbullying. There are a wide range of offences within the categories 1, 2 and 4 presented above which are used to prosecute bullying conducted online e.g. via social media. One such category includes communications which may constitute threats of violence to the person (CPS guidelines n.d.-a). If the online communication includes a threat to kill, it may be prosecuted under s16 of the Offences Against the Person Act 1861. Other threats of violence to the person may fall to be considered under the provisions of the Protection from Harassment Act 1997, namely section 4 (putting another in fear of violence) or 4A (stalking involving fear of violence or serious alarm or distress), if they constitute a course of conduct which amounts to harassment or stalking – see below.

Threats of violence to the person or damage to property may also fall to be considered under section 1 of the Malicious Communications Act 1988, which prohibits the sending of an electronic communication which conveys a threat, or section 127 of the Communications Act 2003 which prohibits the sending of messages of a "menacing character" by means of a public telecommunications network. According to *Chambers v DPP* [2012] EWH2 2157 (Admin): "... a message which does not create fear or apprehension in those to whom it is communicated, or may reasonably be expected to see it, falls outside [section 127(i)(a)], for the simple reason that the message lacks menace" (Paragraph 30).

Offensive communications sent via social media that target a specific individual or individuals may fall to be considered under: Sections 2, 2A, 4 or 4A of the Protection from Harassment Act 1997 if they constitute an offence of harassment or stalking; or Section 76 of the Serious Crime Act 2015 if they constitute an offence of *controlling or coercive behaviour*.

Harassment can include repeated attempts to impose unwanted communications or contact upon an individual in a manner that could be expected to cause distress or fear in any reasonable person (CPS guidelines). It can include harassment by two or more defendants against an individual or harassment against more than one individual (S.1A (a) Protection from Harassment Act 1997). There is no legal definition of *cyberstalking*, nor is there any specific legislation to address the behaviour. Generally, cyberstalking is described as a threatening behaviour or unwanted advances directed at another, using forms of online communications (CPS guidelines). Cyberstalking and online harassment are often combined with other forms of 'traditional' stalking or harassment, such as being followed or receiving unsolicited phone calls or letters. Examples of offensive behaviour may include (see s. 2A (3) of the Protection from Harassment Act 1997): threatening or obscene emails or text messages; live chat harassment or "flaming"; "baiting", or humiliating peers online by labelling them as sexually promiscuous; leaving improper messages on online forums or message boards; unwanted indirect contact with a person that may be threatening or menacing, such as posting images of that person's children or workplace on a social media site, without any reference to the person's name or account; posting "photoshopped" images of persons on social media platforms; sending unsolicited emails; spamming, where the offender sends the victim multiple junk emails; hacking into social media accounts and then monitoring and controlling the accounts; distribution of malware; cyber identity theft etc.(CPS guidelines).

Whether any of these cyber activities amount to an offence will depend on the context and particular circumstances of the action in question. The Protection from Harassment Act 1997 requires the prosecution to prove that the defendant pursued *a course of conduct* which amounted to harassment or stalking. The Act states that a "course of conduct" must involve conduct on at least two occasions. The conduct in question must form a sequence of events and must not be two distant incidents (Lau v DPP [2000] 1 FLR 799; R v Hills (2000) Times 20-Dec-2000). Each individual act forming part of a course of conduct need not be of sufficient gravity to be a crime in itself; however, the fewer the incidents, the more serious each is likely to have to be for the course of conduct to amount to harassment (Jones v DPP [2011] 1 W.L.R. 833). Where an individual receives unwanted communications from another person via social media in addition to other off-line unwanted behaviour, all the behaviour should be considered together in the round in determining whether or not a course of conduct is made out (CPS guidelines n.d.-a).

Communications sent via social media may alone, or together with other behaviour, amount to an offence of *Controlling or coercive behaviour in an intimate or family relationship* under section 76 of the Serious Crime Act 2015. This offence only applies to offenders and victims who are personally connected: in an intimate personal relationship; or they live together and they have previously been in an intimate personal relationship; or they live together and are family members (s76 (2)). The controlling or coercive behaviour in question must be repeated or continuous, it must have a serious effect on the victim, and the offender must know or ought to know that the behaviour will have such an effect. According to s76 (4), "serious effect" is one that either causes the victim to fear, on at least two occasions, that violence will be used against them, or it causes the victim serious alarm or distress that has a substantial adverse effect on their usual day-to-day activities.

According to CPS the patterns of behaviour associated with coercive or controlling behaviour might include: isolating a person from their friends and family, which may involve limiting their access to and use of social media; depriving them of their basic needs; monitoring their time; taking control over where they can go, who they can see, what to wear and when they can sleep. It could also include control of finances, such as only allowing a person a punitive allowance, or preventing them from having access to transport or from working. Controlling or coercive behaviour does not only occur in the home. For instance, the offender may track and monitor the whereabouts of the victim by communications with the victim via social media, texts, email, and/or by the use of spyware and software.

If the offender and victim are no longer in a relationship and no longer live together, or are not family members, the offences of harassment or stalking may apply if the offender is continuing to exert controlling or coercive behaviour beyond the marriage, relationship or period of co-habitation.

Communications which are *grossly offensive, indecent, obscene or false* will usually fall to be considered either under section 1 of the Malicious Communications Act 1988 or under section 127 of the Communications Act 2003. These provisions also prohibit communications conveying a threat (s.1 of the 1988 Act] or which are of a menacing character (s.127 of the 2003 Act) discussed above. It need be noted that some indecent or obscene communications may more appropriately be prosecuted under other legislation, which may contain more severe penalties, rather than as a communications offence. For instance, in *R v GS* [2012] EWCA Crim 398, the defendant was charged with publishing an obscene article contrary to section 2(1) of the Obscene Publications Act 1959, relating to an explicit internet relay chat or conversation with one other person, concerning fantasy incestuous, sadistic paedophile sex acts on young and very young children.

Section 1 of the Malicious Communications Act 1988 prohibits the sending of an electronic communication which is indecent, grossly offensive, or which is false, or which the sender believes to be false if, the purpose or one of the purposes of the sender is to cause distress or anxiety to the recipient. The offence is committed when the communication is sent; there is no legal requirement for the communication to reach the intended recipient. According to *Connolly v DPP* [2007] 1 ALL ER 1012 the terms "indecent or grossly offensive" were said to be ordinary English words. Section 32 of the Criminal Justice and Courts Act 2015 amended section 1 making the offence an either-way offence and increased the maximum penalty to 2 years' imprisonment for offences committed on or after 13 April 2015. This amendment allowed more time for investigation, and a more serious penalty available in appropriate cases.

Section 127 of the Communications Act 2003 makes it an offence to send or cause to be sent through a "public electronic communications network" a message or other matter that is "grossly offensive" or of an "indecent or obscene character". The same section also provides that it is an offence to send or cause to be sent a false message "for the purpose of causing annoyance, inconvenience or needless anxiety to another". The defendant must either intend the message to be grossly offensive, indecent or obscene or at least be aware that it was so. This can be inferred from the terms of the message or from the defendant's knowledge of the likely recipient (*DPP v Collins* [2006] UKHL 40). The offence is committed by sending the message. There is no requirement that any person sees the message or be offended by it. The s127 offence is summary-only, with a maximum penalty of 6 months' imprisonment. Prosecutions may be brought up to 3 years from commission of the offence, as long as this is also within 6 months of the prosecutor having knowledge of sufficient evidence to justify proceedings (s.51 of the Criminal Justice and Courts Act 2015).

According to *Chambers v DPP* [2012] EWHC 2157 (Admin), a message sent by Twitter is a message sent via a "public electronic communications network" as it is accessible to all who have access to the internet. The same principle applies to any such communications sent via social media platforms. However, section 127 of the Communications Act 2003 does not apply to anything done in the course of providing a programme service within the meaning of the Broadcasting Act 1990.

Those who encourage others to commit a communications offence may be charged with encouraging an offence under the Serious Crime Act 2007: for instance, encouragement to tweet or re-tweet ("RT") a grossly offensive message; or the creation of a derogatory hashtag; or making available personal information (doxing/doxxing), so that individuals can more easily be targeted by others. Such encouragement may sometimes lead to a campaign of harassment or "virtual mobbing" or "dog-piling", whereby a number of individuals use social media or messaging to disparage another person, usually because they are opposed to that person's opinions (CPS guidelines n.d.-a).

There is a high threshold that must be met at the evidential stage as per the Code for the Crown Prosecutors. Even if the high evidential threshold is met, in many cases a prosecution is unlikely to be required in the public interest (CPS guidelines n.d.-a). According to *Chambers v DPP* [2012] EWHC 2157 (Admin) "Satirical, or iconoclastic, or rude comment, the expression of unpopular or unfashionable opinion about serious or trivial matters, banter or humour, even if distasteful to some or painful to those subjected to it should and no doubt will continue at their customary level, quite undiminished by [section 127 of the Communications Act 2003]."

Section 1 of the Malicious Communications Act 1988 and section 127 of the Communications Act 2003 prohibit the sending of a communication that is *grossly offensive*. This is problematic area of law and the legislation has been criticized for lacking clarity and certainty, as discussed further below. According to CPS a communication sent has to be more than simply offensive to be contrary to the criminal law. Just because the content expressed in the communication is in bad taste, controversial or unpopular, and may cause offence to individuals or a specific community, this is not in itself sufficient reason to engage the criminal law. As per *DPP v Collins* [2006] UKHL 40: "There can be no yardstick of gross offensiveness otherwise than by the application of reasonably enlightened, but not perfectionist, contemporary standards to the particular message sent in its particular context. The test is whether a message is couched in terms liable to cause gross offence to those to whom it relates" (Para 9).

According to CPS prosecutors should only proceed with cases under section 1 of the Malicious Communications Act 1988 and section 127 of the Communications Act 2003 where they are satisfied there is sufficient evidence that the communication in question is *more than*:

- Offensive, shocking or disturbing; or
- Satirical, iconoclastic or rude comment; or
- The expression of unpopular or unfashionable opinion about serious or trivial matters, or banter or humour, even if distasteful to some or painful to those subjected to it (CPS guidelines n.d.-a).

The next step to be considered is whether a prosecution is required in the public interest. Given that every day several millions of communications are sent via social media, the application of section 1 of the Malicious Communications Act 1988 and section 127 of the Communications Act 2003 to such comments creates the potential that a very large number of cases could be prosecuted before the courts. In these circumstances there is the potential for a chilling effect on free speech. Both section 1 of the Malicious Communications Act 1988 and section 127 of the Communications Act 2003 will often engage Article 10 of the European Convention on Human Rights. These provisions must be interpreted consistently with the free speech principles in Article 10, which provide that: "Everyone has the right to freedom of expression. This right shall include the freedom to hold opinions and to receive and impart information and ideas without interference by public authority and regardless of frontiers ..."

Article 10 protects not only speech which is well-received and popular, but also speech which is offensive, shocking or disturbing. According to *Sunday Times v UK* (No 2) [1992] 14 EHRR 229 "Freedom of expression constitutes one of the essential foundations of a democratic society ... it is applicable not only to 'information' or 'ideas' that are favourably received or regarded as inoffensive or as a matter of indifference, but also as to those that offend, shock or disturb ...". In addition, there is only limited scope for prosecution in relation to political speech or debate on questions of public interest (*Sener v Turkey* [2003] 37 EHRR 34). Freedom of expression and the right to receive and impart information are not absolute rights. They may be restricted but only where a restriction can be shown to be both *necessary* and *proportionate*. These exceptions, however, must be narrowly interpreted and the necessity for any restrictions convincingly established (*Sunday Times v UK* (No 2); *Goodwin v UK* [1996] 22 EHRR 123). Accordingly, no prosecution will be brought under section 1 of the Malicious Communications Act 1988 or section 127 of the Communications Act 2003 (Category 4 cases) unless it can be shown on its own facts and merits to be both *necessary* and *proportionate* (CPS guidelines n.d.-a).

3 Revenge Pornography, Sexting, Sextortion and Related Offences

Revenge pornography involves the distribution of sexually explicit images or videos of individuals without their consent and with the purpose of causing embarrassment or distress. The images are sometimes accompanied by personal information about the subject, including their full name, address and links to their social media profiles (Ministry of Justice 2014). The offence applies both online and offline and to images which are shared electronically or in a more traditional way. It includes the uploading of images on the internet, sharing by text and e-mail, or showing someone a physical or electronic image.

There are subtle differences between 'non-consensual pornography' and 'revenge pornography'. Non-consensual pornography is a broader term, that encompasses a number of offences including Revenge Pornography (Criminal Justice and Courts Act 2017, s 33) voyeurism (Sexual Offences Act 2003, s67), hacking to obtain materials (Computer Misuse Act 1990, s 1,2,3) and other offences if the person depicted is under 18 (Protection of Children Act1978, s1; Criminal Justice Act 1988, s160). Revenge Pornography is a more specific term, "usually following the breakup of a couple, the electronic publication or distribution of sexually explicit material (principally images) of one or both of the couple, the material having originally been provided consensually for private use" (House of Lords 2014).

The sharing of private communications during or after the breakdown of a relationship is not a new phenomenon in itself but it has become more widespread in the past decade. The causing of serious harm after the collapse of trust between previously consenting individuals is not unusual, especially considering the sheer amount of unauthorised 'celebrity' sex tapes. In the early 2000's these videos and images were distributed among many websites and gained attention across message boards predating the emergence of today's social media platforms. The eruption of social media has not only fuelled the obsession with the 'celebrity culture' but has opened up possibilities for breaches of the same nature that could affect almost anyone.

Prior to April 2015 in the UK a range of existing laws were used to prosecute cases of revenge porn. This legislation is still used for offences committed prior to that date. Sending explicit or nude images of this kind may, depending on the circumstances, be an offence under the Communications Act 2003 or the Malicious Communications Act 1988. Behaviour of this kind, if repeated, may also amount to an offence of harassment under the Protection from Harassment Act 1997 as discussed above. Section 33 of the Criminal Justice and Courts Act 2015 created a specific offence for this practice and those found guilty of the crime could face a sentence of up to 2 years in prison. It came in force on 13 April 2015 and does not have retrospective effect.

The new offence criminalises the sharing of private, sexual photographs or films, where what is shown would not usually be seen in public (s34). Sexual material not only covers images that show the genitals but also anything that a reasonable person would consider to be sexual, so this could be a picture of someone who is engaged in sexual behaviour or posing in a sexually provocative way (s 35). The available defences for the offence under the Act and (5) are where the defendant "reasonably believed the publication to be necessary for the prevention, detection or investigation of crime", or such publication is in the public interest or previously disclosed for reward by consent (CJCA 2015 s 33 (3), (4)).

CJCA 2015, specifically Section 33 was introduced with the aim of addressing the growing concerns associated with technological advances and the increasing use of social media. The criminalisation of acts such as revenge porn was considered as one of the ways to deal with these challenges (Phippen and Agate 2015). The new legislation takes into account the societal changes by ensuring that it applies to material distributed both online and offline unlike the previously existing legislation that failed to acknowledge or reflect the changes in time. It has been observed that a number of statutes passed before the invention of the internet (e.g. Children and Young Persons Act 1933) refer to publications in terms only of print media;

electronic communications and social media are not being provided for (House of Lords 2014, para. 47). As argued further below, this state of affairs is not satisfactory and need be addressed by policymakers.

Although the legislation has been largely welcomed, it has been argued that the offence is not as far reaching as it could have been. The element of "intention to cause distress" is arguably weakened by Section 33(8) according to which intention to cause distress cannot be found "merely because that was a natural and probable consequence of the disclosure". A person will only be guilty of the offence if the reason for disclosing the photograph, or one of reasons, is to cause distress to a person depicted in the photograph or film. On the same basis, anyone who re-tweets or forwards without consent, a private sexual photograph or film would only be committing an offence if the purpose, or one of the purposes was to cause distress to the individual depicted in the photograph or film who had not consented to the disclosure. Anyone who sends the message for any other reason would not be committing the offence (CPS guidelines n.d.-b). It has been argued that due to this limitation the offence is "not harsh enough" (Nimmo 2015) and "an opportunity missed"(Pegg 2015). It results in the offence to be a limited and restrictive tool, rather than encompassing more circumstances such as where intention to distribute the material was not to cause distress, but instead motivated for financial gain or sexual purpose (Pegg 2015) or "for a laugh" (Phippen and Agate 2015, p. 85). These motives do not lessen the harm caused to the victim and are likely to fall outside the remit of the intention required by the offence.

Despite the *mens rea* issues, the type of material prescribed by the S.33 offence is broader than that under the Malicious Communications Act 2003, s1 due to the wider definitions as compared with the stricter requirements for the content of the latter (as discussed above). From a technical perspective, the offence is drafted so that it only applies to material which looks photographic and which originates from an original photograph or film recording. This is because the harm intended to be tackled by the offence is *the misuse of intimate photographs or films*. The offence will still apply to an image which appears photographic and originated from a photograph or film even if the original has been altered in some way or where two or more photographed or filmed images are combined. However the offence does not apply if it is only because of the alteration or combination that the film or photograph has become private and sexual or if the intended victim is only depicted in a sexual way as a result of the alteration or combination. For example, a person who has non-consensually disclosed a private and sexual photograph of his or her former partner in order to cause that person distress will not be able to avoid liability for the offence by digitally changing the colour of the intended victim's hair. However, a person who simply transposes the head of a former partner onto a sexual photograph of another person will not commit the offence. Images which are completely computer generated but made to look like a photograph or film are not covered by the offence (CPS guidelines n.d.-b).

There is a significant overlap between different offences in this area of law. Despite the specific legislation, cases involving 'revenge pornography' may also fall to be considered under stalking and harassment offences discussed above (S2,

S2a, S4, S4a of the Protection from Harassment Act 1997) and the offences of sending a communication that is grossly offensive, indecent, obscene, menacing or false (S127 of the Communications Act 2003 or S1 Malicious Communications Act 1988). Where the images have been obtained through computer hacking, S1 of the Misuse of Computers Act 1990 – *unauthorised access to computer material* – would be the relevant offence (Sallavaci 2017).

Where the images may have been taken when the victim was under 18, offences under section 1 of the Protection of Children Act 1978 (taking, distributing, possessing or publishing indecent photographs of a child) or under section 160 of the Criminal Justice Act 1988 (possession of an indecent photograph of a child) may have been committed. Specific issues arise in cases of "sexting" that involve images taken of persons under 18. *Sexting* commonly refers to the sharing of illicit images, videos or other content between two or more persons. *Sexting* can cover a broad range of activities, from the consensual sharing of an image between two children of a similar age in a relationship, to instances of children being exploited, groomed, and bullied into sharing images, which in turn may be shared with peers or adults without their consent. An image may have been generated by an individual as a result of a request from another; an image may have been generated by an individual and sent to a recipient who has not asked for it; an image may have been redistributed by a recipient to further third parties online or offline. Within the broader *sexting* context therefore, there could be a variety of acts and motives that should warrant different types of responses by law enforcement (Phippen and Agate 2015, p. 5).

In terms of prosecution, one factor that may warrant particular consideration is the involvement of younger or immature perpetrators. Children may not appreciate the potential harm and seriousness of their communications and as such the age and maturity of suspects should be given significant weight (CPS guidelines n.d.-b). According to the Association of Chief Police Officers (ACPO), with regard to the images self-generated by children, the consequences of applying the current legislation are far reaching. A prosecution for any of related offences means that an offender is placed on the sex offenders register for a duration that is commensurate with the sentence they receive. Even though the sentencing and time limits are generally reduced for those younger than 18, this can still mean in some cases a considerable time spent on the register. According to ACPO, first time offenders should not usually face prosecution for such activities, instead an investigation to ensure that the young person is not at any risk and the use of established education programmes should be utilised (see below). Nevertheless, in some cases, e.g. persistent offenders, a more robust approach may be called for, such as the use of reprimands. It is recommended that prosecution options are avoided, in particular the use legislation that would attract sex offender registration (ACPO – Lead position).

According to CPS, whilst it would not usually be in the public interest to prosecute the consensual sharing of an image between two children of a similar age in a relationship, a prosecution may be appropriate in other scenarios, such as those involving exploitation, grooming or bullying (CPS guidelines n.d.-b). In addition to the offences outlined above, consideration may be given to the offence of *Causing or inciting a child to engage in sexual activity* under section 8 (child under 13) or section 10 (child) of the Sexual Offences Act 2003 (SOA) – see below.

Section 15A of the SOA 2003, *Sexual communication with a child*, may be used to prosecute cases of sexting between an adult and a person under 16, where the conduct took place on or after 3 April 2017. This offence is committed where an adult intentionally communicates with another person who s/he does not believe to be over 16, for purposes of obtaining sexual gratification. The communication must be sexual i.e. any part of it relates to sexual activity or a reasonable person would, in all circumstances consider it to be sexual. According to the Ministry of Justice "ordinary social or educational interactions between children and adults or communications between young people themselves are not caught by the offence" (Ministry of Justice 2015).

Where intimate images or other communications are used to coerce victims into sexual activity, or in an effort to do so, other offences under the Sexual Offences Act 2003 could be considered, such as:

- Section 4, Causing sexual activity without consent, if coercion of an adult has resulted in sexual activity.
- Sections 8 (child under 13) and 10 (child), Causing or inciting a child to engage in sexual activity: 'causing' activity if coercion has resulted in sexual activity; and 'inciting' such activity if it has not.
- Section 15 – Meeting a child following sexual grooming.
- Section 62 – Committing an offence with intent to commit a sexual offence, if no activity has taken place but there is clear evidence that an offence was intended to lead to a further sexual offence.

Where intimate images or other communications are used to threaten and make demands from a person, the offence of *Blackmail* may apply. For example, so called "webcam blackmail", where victims are lured into taking off their clothes in front of their webcam, and sometimes performing sexual acts, on social networking or online dating sites, allowing the offender to record a video. A threat is subsequently made to publish the video, perhaps with false allegations of paedophilia, unless money is paid. These acts of online blackmail are known as *sextortion* (Interpol – online safety).

According to Interpol, sextortion is often conducted by sophisticated organized criminal networks operating out of business-like locations similar to call centres. While there is no one method by which criminals target their victims, many individuals are targeted through websites including social media, dating, webcam or adult pornography sites. Criminals often target hundreds of individuals around the world simultaneously, in an attempt to increase their chances of finding a victim. (Interpol- online safety) In England and Wales the offences committed under such circumstances are that of blackmail or attempted blackmail besides any other offence under Sexual Offences Act 2003 such as the ones indicated above (CPS guidelines n.d.-a).

4 Tackling Offensive Online Communications and Abuse: Issues and Concerns

4.1 Is the Legislation Fit for Purpose?

From a legal perspective, the above review demonstrates that the current legal framework dealing with offensive online communications and abuse is complex. There is need to consider whether it is capable of dealing with offensive internet communications effectively and whether there is scope for simplifying the law in this difficult area. There is considerable overlap between existing offences as shown above. For example, Part 1 of the Malicious Communications Act 1988 makes it an offence to send a communication which is "indecent or grossly offensive" with the intention of causing "distress or anxiety"; section 127 of the Communications Act 2003 applies to threats and statements known to be false, but also contains areas of overlap with the 1988 Act. In addition to the 1988 and 2003 Acts, online abuse may be caught by several other provisions. The scope and inter-relationship between these provisions covering inter alia harassment, stalking, public order offences and revenge porn is unclear (The Law Commission 2018).

One of the main criticisms is the ambiguity of the existing legislation. One prime example is the confusion surrounding the broad definition of "grossly offensive" in the 1988 and 2003 Acts, which may fall foul of the principle of legal certainty. It is inherently difficult to judge between what is offensive (but legal) and grossly offensive (and illegal). Context and circumstances are highly relevant for prosecuting decisions being made whilst giving due consideration to the freedom of expression. Despite the guidance offered by the CPS, decisions on prosecuting remain highly subjective. This confusion is increased by the scarcity of legal argument available due to the frequency of guilty pleas in cases of this nature (Law Commission 2018). Even when a case is brought before a jury, the line between 'offensive' and 'grossly offensive' can be highly subjective and depend on the jury members' personal interpretations. There is an obvious need for clearer and more precise statutory provisions.

Another example is the definition of 'public communications network' in section 127 which still requires clarification. According to *DPP v Collins* [2006] UKHL 40, the purpose of section 127(1) (a) is not to protect people against the receipt of offensive messages which is covered by the Malicious Communications Act 1988. Instead, section 127 (1)(a) was designed to prohibit the use of a service provided and funded by the public for the benefit of the public for the transmission of communications which contravene the basic standards of our society. The Communications Act 2003 was drawn up before the popularisation of social networking, and could not have foreseen how pervasive social networking would become in a short space of time. The original intent was to prevent the waste of public services funded by public money. Social media platforms such as Twitter and Facebook are "public" in the sense that they are free to use and open to view unless specified otherwise, however they are not public services but profit-making companies funded by investors and advertising (see the defence's argument in *Chambers v DPP* [2012] EWHC 2157 (QB)) Despite the decision in *Chambers* to include social media platforms within the s127 provision, there remains ambiguity over what constitutes a "public communications network" that needs clarification in the legislation. Moreover, it is not clear whether the current legislation requires proof of fault or of intention to prosecute online communications (The Law Commission 2018).

The criminal law in this area is almost entirely enacted before the invention of social media and recent technological developments. One of the challenges that the legal community and policy makers face is to ensure that the legislation on 'offline offences' is capable of being used to combat the electronic versions of these offences. This has led to proposals for legislative changes (Gillespie 2016, p. 257). An update of the existing legislation would be welcome, so as those statutes pre-dating the invention of the internet, refer to publications not only in terms of print media but also the online one. As the House of Lords recognised in their 2014 review, there are aspects of the existing legislation that could 'appropriately be adjusted and certain gaps which might be filled' (HL 2014, para 94). According to the Law Commission 'there is need to update definitions in the law which technology has rendered obsolete or confused, such as the meaning of "sender"' (Law Commission 2018). With regard to sentencing, calls have been made to increase the severity of sentences available for the punishment of these online offences (HL 2014, para 49) as well as updating the Sentencing Guidelines so as to clearly refer to communications via internet as it is arguably unreasonable to sentence people under guidelines which do not relate to the nature of their offence (see Magistrates Court Sentencing guidelines on s 127).

According to the House of Lords "the starting point is that what is not an offence off-line should not be an offence online". In their 2014 review it was concluded that the existing legislation is generally appropriate for the prosecution of offences committed using the social media (House of Lords 2014, para 94). The House of Lords deemed it was not necessary to create a new set of offences specifically for acts committed using the social media and other information technology. With regard to cyberbullying for instance, since there is no specific criminal offence of bullying (offline) the current range of offences, particularly those under the Protection from Harassment Act 1997 and Malicious Communications Act 1988, was found sufficient to prosecute bullying conducted using social media. In a similar fashion, although "trolling" causes offence, the House of Lords did not "see a need to create a specific and more severely punished offence for this behaviour" (House of Lords 2014; para 32).

Research shows that in 2017 28% of UK internet users were on the receiving end of trolling, harassment or cyberbullying (The Law Commission 2018). There is a clear public interest in tackling online abuse in all forms including those that do not correspond to 'offline' or 'traditional' offences. This must be done through clear and predictable legal provisions that keep up to date with changes in society. Updating and consolidating the legislation is highly desirable. At the time of writing the Law Commission has been commissioned by the UK Government to undertake an analysis of the laws around offensive online communications. This is part of the UKs Government reform plans to make the UK the safest place online in the world (HM Government 2017; Gov.uk –press release).

It is paramount to take into consideration that the context in which interactive social media dialogue takes place is quite different to the context in which other forms of communications take place. Access is ubiquitous and instantaneous. The use of technology and social media platforms facilitate a much higher volume of crime and the consequences could become more serious given the widespread circulation of the information. Communications intended for a few may reach millions. Online abuse could escalate fast as multiple offenders could be instantaneously involved. There is a difference in how subjects get involved in offensive and abusive behaviour which happens more easily online than offline (see below). Online abuse could lead to extremely distressing and often devastating personal consequences for victims. The internet never 'forgets' (despite 'The right to be forgotten' – see art 17 General Data Protection Regulation 2016/679) as images and comments may be easily distributed and stored by subjects even after their removal from a particular website or social media platform. A range of related issues including the anonymity of social media users, jurisdictional and evidence collection challenges, make the prosecution of online crime particularly difficult. For all these reasons and more, online abuse requires careful and special strategic consideration which should aim not only punishment but also prevention. The strategy must focus not only on criminalisation and updating the legislation but also on its enforcement including training, raising public awareness and education. To these issues the attention now turns.

4.2 Enforcement Challenges

4.2.1 Anonymity

One of the greatest challenges of combating online crime is the identification of perpetrators. The internet readily facilitates its users doing so anonymously. Even though it is possible to identify the computer used to post a statement (based on its unique "internet protocol address"), it is not necessarily possible to identify who used that computer to do so. This is in part because many website operators facilitate the anonymous use of their service. There is no consistent attitude taken by website operators: some require the use of real names (Facebook, although users' identities are not actively confirmed); some allow anonymity but challenge impersonation (Twitter); others allow absolute anonymity (House of Lords 2014).

There are two conflicting aspects to anonymity. Anonymity is of great value in ensuring freedom of speech especially for human rights workers, dissidents and journalists working in conflict areas as it enables them to publish information and opinion without placing themselves at risk. (House of Lords 2014) However, there is a less positive side to anonymity related to a lack of apparent accountability and immediate confrontation that facilitates offensive behaviour, notably in the forms of cyber bullying and trolling. Being anonymous online provides people the opportunity to act in ways that they would not if exposing their identity (Rosewarne 2016, p. 90–91). The Internet is conceived as a place separate and distinct from real life. There is the idea that cyberspace is a world of its own and for some people the entire online experience is construed as life in another dimension. Different rules apply which provides part of the explanation for the Internet serving as an instigator in online bullying (Rosewarne 2016, p. 91).

An example is 'ask.fm' a Latvian-based social networking site where users can ask each other questions with the (popular) option of anonymity. The site, popular with British teenagers, is sadly infamous for the bullying conducted using it and for the consequences of that bullying. In 2012, Erin Gallagher committed suicide at the age of 13 naming 'ask.fm' in her suicide note and stating that she could not cope with the bullying. Anthony Stubbs committed suicide in 2013; his girlfriend received abuse on 'ask.fm'. There are further similar incidents relating to the same and other websites (House of Lords 2014).

A potential solution to ensure that law enforcement agencies can properly investigate crime is to require the operators of websites and social media platforms which enable their users to post opinions or share images to establish the identity of people opening accounts to use their services, whether or not they subsequently allow those people to use their service anonymously. According to the House of Lords "if the behaviour which is currently criminal is to remain criminal and also capable of prosecution it would be proportionate to require the operators of websites first to establish the identity of people opening accounts but that it is also proportionate to allow people thereafter to use websites using pseudonyms or anonymously. There is little point in criminalising certain behaviour and at the same time legitimately making that same behaviour impossible to detect" (House of Lords 2014; para 94).

4.2.2 Jurisdictional Issues

The issue of anonymity is related to that of locating the perpetrator and evidence collection which in turn pose challenges for law enforcement as it requires cooperation by social media and website operators which is not always given. This highlights jurisdictional challenges given that online crime is 'crime without borders'. From the perspective of the offences discussed above, in the circumstances where material is posted on a website hosted abroad, the court would need to be satisfied that it was in substance an offence committed within the jurisdiction. For example, if the perpetrator was physically located in England or Wales it would be possible for the offence to be committed. According to *R v Smith* (Wallace Duncan) (No.4) [2004] EWCA Crim 631 [2004] QB 1418 an English court has jurisdiction to try a substantive offence if "substantial activities constituting [the] crime take place in England"; or "a substantial part of the crime was committed here". This approach "requires the crime to have a substantial connection with this jurisdiction".

In the case of revenge porn, the removal of the images uploaded to the internet would be the responsibility of the website or social media provider. The offence does not itself force website operators to take action in relation to the uploaded material. Where a forum is specifically provided for the dissemination of material, then the provider of the website could, depending on all the circumstances, be guilty of encouraging or assisting the commission of the offence even if they are based abroad – although there may be practical difficulties about prosecuting foreign companies (CPS guidelines n.d.-a).

Section 33(10) refers to Schedule 8 of the Act which makes special provision in relation to persons providing information society services. The Schedule reflects the requirement in the e-commerce directive that information services providers based in the EEA should not usually be prosecuted for any offences which might be committed by providing services in the country where they are established. In rare cases, where all the requirements of the offence are satisfied including the intention to cause distress to the victim, the Schedule does not stop an operator being guilty of the offence if it actively participates in the disclosure in question or fails to remove the material once it is aware of the criminal nature of its content.

According to the House of Lords 2014 (para 94) the only way to resolve questions of jurisdiction and access to communications data would be by international treaty. The question relates to wider issues of the law and public protection that go beyond criminal offences committed using social media and is politically contentious in most countries.

4.2.3 Police Training

A related issue concerning the enforcement of legislation is that of awareness and training. Policing agencies often lack the capacity or the motivation to investigate adult complaints of online abuse even where it takes clearly illegal forms such as death or rape threats (Slater 2017, p. 154). Users report lack of understanding from law enforcement. While many forms of online abuse are already covered by existing laws, these are frequently not enforced in practice. Research recently conducted in England and Wales highlights the confusion associated with revenge pornography legislation among police officers and staff, and the restricted nature of the legislation itself (Bond and Tyrrell 2018). The uncertainty relating to the legislation and misunderstandings of the socio-technicalities associated with revenge pornography may lead to miscommunications with victims and inconsistencies in police responses and the ability to manage revenge pornography referrals and cases effectively. A total of 94.7% of police officers and staff responded to the research that they had not received any formal training on how to conduct investigations into revenge pornography. Of the 41 individuals who replied that they had received training, for nearly half of these respondents, the training was delivered via an online tutorial (Bond and Tyrrell 2018). While this is only one example, there is clear need to improve training of law enforcement officers on the forms and impact of online abuse, and for investment of law enforcement resources into the investigation and prosecution of online abuse (Slater 2017, p. 154).

4.3 Raising Awareness and Education: Online Abuse, VAWG[1] and Young Offenders

The prevalence of online abuse and harassment and its impact on women and girls has been evident since the internet's popularisation in the 1990s. With the advent of social media, online abuse and harassment continues to be a highly gendered phenomenon (Slater 2017, p. 105). Yet, little attention has been given to understanding the ways in which new technologies are used to facilitate sexual violence, online abuse and harassment against women (Bond and Tyrrell 2018, p. 3). Such lack of understanding results in the inability of the criminal justice system to adequately respond to online offensive behaviour. The landscape in which VAWG offences are committed is changing. The use of the internet, social media platforms, emails, text messages, smartphone apps (such as WhatsApp; Snapchat), spyware and GPS (Global Positioning System) tracking software to commit VAWG offences is rising. Online activity is used to humiliate, control and threaten victims, as well as to plan and orchestrate acts of violence (CPS guidance n.d.-a).

Online violence and abuse are often sexist and misogynistic in nature, targeting women's multiple identities such as their race, religion or sexual orientation, and can include threats of physical and sexual violence (Amnesty International UK 2017). While it is acknowledged that online abuse is a similarly serious issue for men and boys (CPS guidelines n.d.-a; Government Equalities Office 2015) research suggests that online abuse "disproportionately affects women, both in terms of the number of women affected and the amount of social stigma attached" (Cooper 2016, p. 819).

The content of online abuse is inextricably linked to patterns of harassing and intrusive conduct, embedded within larger inequalities of power (Salter 2017, p. 127). The characteristic nature or context of VAWG offending is usually that the perpetrator exerts power and/or a controlling influence over the victim's life (CPS guidance n.d.-a). Gender power imbalances at the level of relationship and on social media are generally recognised; at the same time the locus of control and therefore of responsibility, is consistently located in girls and women (Salter 2017, p. 116). Societal attitudes to female victims of online abuse, such as revenge pornography for instance, are often dominated by victim blaming, in that the breach of privacy which arises from the non- consensual sharing of the images is deemed, in some way, to be the responsibility of the women who produced, or allowed to be produced, the images in the first place (Salter 2017, p. 116–117).

---

[1] Violence against women and girls.

Recently there have been calls to amend the legislation so as to give victims of revenge porn anonymity in a bid to reduce the number of discontinued prosecutions.[2] What is still missing from the debate around legislation to protect victims of sexting, revenge porn, cyberbullying and online abuse, is effective raising of public awareness, challenging of societal attitudes and education (Phippen and Agate 2015, p. 86). Most internet users do not consider the long term implications on themselves or others when making online comments or sharing content that could constitute an offence. Most people are "extremely ignorant about the laws" around online offences (Phippen and Agate 2015, p. 84). Awareness campaigns and educational initiatives should take place to increase the public awareness as to what constitutes as offending behaviour, its impact as well as to tackle discriminatory societal attitudes. An example of what is needed to raise public awareness is the campaign 'BE AWARE B4 YOU SHARE' which encourages victims coming forward and invites them to familiarise themselves with the offence of revenge porn (Ministry of Justice 2015). The slogan can apply equally to the perpetrators and victims.

As recent widespread developments of technology transform the social world they present new challenges. The behaviour and understanding of the 'millenials' is startlingly different from that of previous generations. As discussed above, part of the offensive or abusive behaviour can be attributed to technological advancements and its effect to "potentially normalise" online acts, where that same act committed offline is considered as unacceptable. In the case of sexting and revenge porn for instance, a factor that appears to drive the creation or distribution of self-taken images is children and young people's natural propensity to take risks and experiment with their developing sexuality (ACPO). This is linked to, and facilitated by, the global escalation in the use of the internet, multi-media devices and social networking sites. Children and young people may not realise that what they are doing is illegal or that it may be potentially harmful to them or others in the future. This is also the case with their involvement in cyberbullying.

Children and young people creating indecent images of themselves or engaging in other types of online offensive behaviour may be an indicator of other underlying vulnerabilities, and such children may be at risk in other ways. According to the ACPO Investigating Child Abuse Guidance (2009), any such minor offending behaviour by children and young people should result into a referral to children's social care so that any issues that are present can be dealt with at an early stage. A safeguarding approach should be at the heart of any intervention. As a commentator puts it "furnishing young people with multiple strategies to prevent online abuse and negotiate technologically mediated relationships is likely to be far more effective in reducing online abuse than punitive or shaming responses to young people's online practices" (Salter 2017, p. 157). Integrating social media into education curricula, focusing on what constitutes an offensive behaviour, sexual ethics and negotiation

---

[2]See https://www.telegraph.co.uk/news/2018/06/14/revenge-porn-allegations-dropped-third-cases-campaigners-call/accessed03/08/2018

of consent, "could be a step in the right direction and recognizes the embeddedness of social media in peer and intimate relations" (Salter 2017, p. 157).

The ability to bring offenders to justice is beneficial in reducing further crime and punishing those that break the law. Having in place a legal framework fit for purpose is indisputably important. This however should be part of a broader strategy that assists our technology-reliant society and an increase in awareness and education to support a more cohesive and safer approach to the use of social media. It is better for the online offences not to occur in the first place, rather than to have offenders to bring to justice.

References

ACPO CPAI. *Lead's position on young people who post self-taken indecent images*. http://www.cardinalallen.co.uk/documents/safeguarding/safeguarding-acpo-lead-position-on-self-taken-images.pdf. Accessed 09 Apr 2018.

Amnesty International. (2017). *Social media can be a dangerous place for UK women* [Report briefing] https://www.amnesty.org.uk/files/Resources/OVAW%20poll%20report.pdf. Accessed 09 Apr 2018.

Bond, E., & Tyrrell, K. (2018). Understanding revenge pornography: A national survey of police officers and staff in England and Wales. *Journal of Interpersonal Violence*, first published online February 2018, https://doi.org/10.1177/0886260518760011.

Cooper, P. W. (2016). The right to be virtually clothed. *Washington Law Review, 91*, 817–846.

CPS (Crown Prosecution Service). (n.d.-a). *Guidelines on prosecuting cases involving communications sent via social media*. https://www.cps.gov.uk/legal-guidance/social-media-guidelines-prosecuting-cases-involving-communications-sent-social-media. Accessed 09 Apr 2018.

CPS (Crown Prosecution Service). (n.d.-b). *Revenge pornography – Guidelines on prosecuting the offence of disclosing private sexual photographs and films*. https://www.cps.gov.uk/legal-guidance/revenge-pornography-guidelines-prosecuting-offence-disclosing-private-sexual. Accessed 09 Apr 2018

CPS. *The code for the crown prosecutors*. https://www.cps.gov.uk/publication/code-crown-prosecutors. Accessed 09 Apr 2018.

Gillespie, A. (2016). *Cybercrime: Key issues and debate*. Oxford: Routledge ISBN 978-0-415-71220-0.

Gov.uk. Press release. https://www.gov.uk/government/news/government-outlines-next-steps-to-make-the-uk-the-safest-place-to-be-online. Accessed 09 Apr 2018.

Government Equalities Office. (2015). *Hundreds of victims of revenge porn seek support from helpline [Press release]*. https://www.gov.uk/government/news/hundreds-of-victims-of-revenge-porn-seek-support-from-helpline. Accessed 09 Apr 2018.

HM Government Internet Safety Strategy – Green paper. (2017, October) https://assets.publishing.service.gov.uk/government/uploads/system/uploads/attachment_data/file/650949/Internet_Safety_Strategy_green_paper.pdf. Accessed 09 Apr 2018.

House of Lords Select Committee on Communications. (2014) *Social media and criminal offences* (1st Report of Session 2014–15, July 2014). London: The Stationery Office Limited.

Interpol. *Online safety*.https://www.interpol.int/Crime-areas/Cybercrime/Online-safety/Sextortion. Accessed 09 Apr 2018.

Ministry of Justice. (2014). *Factsheet – Serious crime act 2015: Offence of sexual communication with a child*. https://www.gov.uk/government/uploads/system/uploads/attachment_data/file/416003/Fact_sheet_-_Offence_of_sexual_communication_with_a_child.pdf. Accessed 09 Apr 2018.

Ministry of Justice. (2015). Revenge porn: Be aware b4 you share available at https://www.gov.uk/government/publications/revenge-porn-be-aware-b4-you-share. Accessed 09 Apr 2018

Nimmo, J. (2015, September 1). Revenge porn: Opinion divided on the new law. *BBC England*.http://www.bbc.co.uk/news/uk-england-33807243. Accessed 09 Apr 2018

Pegg, S. (2015). *Wrong on 'revenge porn' (Law Gazette, 23 February 2015)*.http://www.lawgazette.co.uk/analysis/comment-and-opinion/wrong-on-revengeporn/5046957.article. Accessed 09 Apr 2018.

Phippen, A., & Agate, J. (2015). New social media offences under the criminal justice and courts act and serious crime act: The cultural context. *Entertainment Law Review, 26*(3), 82–87.

Rosewarne, L. (2016). *Cyberbullies, cyberactivists, cyberpredators: Film, TV and internet stereotypes*. Santa Barbara: Praeger ISBN: 9781440834400.

Sallavaci, O. (2017). Combating cyber dependent crimes: The legal framework in the UK. *Communications in Computer and Information Science, 630*, 53–66. https://doi.org/10.1007/978-3-319-51064-4_5.

Salter, M. (2017). *Crime, justice and social media*. Oxford: Routledge ISBN: 978-1-138-91967-9.

The Law Commission. (2018). *Online communications project*. https://www.lawcom.gov.uk/online-communications/. Accessed 09 Apr 2018.

The Law Society. (2015). *'Social media' practice notes*. http://www.lawsociety.org.uk/support-services/advice/practice-notes/social-media/. Accessed 09 Apr 2018.

The Statistics Portal. https://www.statista.com/statistics/272014/global-social-networks-ranked-by-number-of-users/. Accessed 09 Apr 2018.

Explaining Why Cybercrime Occurs: Criminological and Psychological Theories

Loretta J. Stalans and Christopher M. Donner

1 Introduction

Long before the internet and related technology were invented, criminological and psychological theories provided explanations for why people committed crime in the real world. From these theories, a voluminous amount of empirical research has expanded our knowledge about why people commit crime in the real world (Akers et al. 2016). Research on cybercrime is a relatively new field of inquiry and has focused on testing whether well-established theories about criminal offending in the real world can explain the crimes that people commit utilizing the internet and related technologies in the virtual world.

To what extent does the internet attract a unique population of persons who commit cybercrimes, but do not commit crimes in the real world? If only offenders who commit crimes in the real world also are committing crimes on the internet, cybercrimes are simply crimes occurring in the real world, but with new tools. For example, digital piracy is the download, streaming, or producing of copyrighted material without paying the required fees or without permission from the owners. It is theft of intellectual property that before the internet was accomplished using tape-recorders, copy machines, and typewriters. Some scholars (e.g., Grabosky 2001) contend that basic motivations to commit crime (e.g., greed, pleasure, control and thrill) are ubiquitous; thus, traditional theories would still be pertinent because computers and the internet merely act as a new avenue to engage in the same

---

L. J. Stalans (✉)
Department of Criminal Justice and Criminology, Psychology, Loyola University Chicago, Chicago, IL, USA
e-mail: Lstalan@luc.edu

C. M. Donner
Department of Criminal Justice and Criminology, Loyola University Chicago, Chicago, IL, USA

antisocial behaviors. Moreover, because many criminological theories are "general" in conceptualization, they should be able to explain a wide scope of deviant behaviors.

Other scholars (e.g., Wall 1998) believe that some real-world crimes have direct analogies to cybercrimes (e.g., fraud), but there are also certain cybercrimes (e.g., hacking, spreading malware) that may not be able to be explained as well by traditional theories because such offenses are dependent on acquiring knowledge about the operation of computer/internet technology. Moreover, research on the perceived and actual features of the internet and related technology has begun to explore how these features are associated with the perpetration of cybercrime (e.g., Barlett and Gentile 2012; Lowry et al. 2016; Stalans and Finn 2016c). Most studies on understanding cybercrime, however, have not tested new or integrated theories, but have tested whether well-established criminological or psychological theories also explain why people commit cybercrimes. The aim of this research is to use valid evidence-based knowledge to inform policies and practices that can reduce the occurrence of cybercrimes. We review the extant literature regarding the applicability—and empirical validity—of several traditional criminological and psychological theories as they relate to cybercrime.

2 Rational Choice Theories: Deterrence Theory and Routine Activity Theory

The rational choice framework was born out of the classical school of criminology (Beccaria 1764), emphasizing rational thought and choice as major influences on human behavior. This perspective asserts that people freely choose to seek pleasure in a rational way that considers whether the benefits of a behavior outweigh the possible negative consequences that might result from the behavior. For example, before individuals illegally download copyrighted music or books or commit acts of piracy, they might consider whether the savings for stealing these items outweighs the possible consequences from the criminal justice system if caught. They may also consider the possible consequences from their social networks. Two of the most prominent rational choice theories in criminology are deterrence theory and routine activity theory.

2.1 Deterrence Theory

Beccaria (1764) argued that crime in society reflected ineffective law rather than the presence of evil, which was contrary to some of the early origins of criminological thought based on religion and spirituality. Beccaria theorized that the effectiveness of criminal laws depends on how punishments are administered. To make the costs of committing crime outweigh the benefits, Beccaria asserted that punishments need to be certain, severe, and swift. Certainty meant offenders had a high chance of being caught. Severity meant that the punishment was sufficiently severe enough to deter would be offenders, but not so severe that it went beyond the harm done and what would deter most potential offenders. Celerity meant that the punishment would be delivered in a timely manner soon after the crime was committed.

These basic tenets of deterrence are the foundation of many criminal justice systems whereby laws and formal sanctions work to keep people from acting on their hedonistic intentions. According to Brenner (2012), nearly all countries have criminal laws on their books regarding a range of cybercrimes, such as hacking, malware, cyberstalking, and digital piracy. Creation and enforcement of these laws are expected to effectively deter criminal behavior if formal punishments are administered in a certain, severe, and swift manner (Hollinger and Lanza-Kaduce 1988; McQuade 2006; O'Neill 2000). A voluminous body of research over the last 50 years has demonstrated a modest deterrent effect for a variety of traditional crimes and across a range of methodological contexts; the certainty of detection rather than severity or celerity of punishment has been most consistently associated with the modest deterrent effect (for reviews, see Nagin 2013; Pratt et al. 2008).

Few empirical studies have examined the influence of deterrence on cybercrime. Lack of knowledge about what actions constitute cybercrimes and the severity of punishment for these crimes might hamper deterrence. Similar to the public's lack of awareness of the severity of punishment for crimes in the real world (e.g., Roberts and Stalans 1997), most people lack awareness about the criminality of many online behaviors and the punishment associated with specific cybercrimes. For example, Irdeto (2017) conducted a survey of more than 25,000 adults across 30 countries in 2016 and found that only 41% were unaware that streaming or downloading pirated content for personal use was illegal and 30% were unaware that producing or sharing pirated video content was illegal. Most respondents in Russia were unaware of the unlawfulness of digital piracy, with only 13% knowing that producing or sharing counterfeit copies of videos was illegal.

Showing a modest potential for criminal prosecutions to reduce digital piracy, Bachmann (2007) examined whether the Recording Industry Association of America (RIAA) campaign to increase public awareness about the severe criminal penalties for downloading and sharing copyrighted music reduced the prevalence of digital piracy. He used three national surveys conducted by the Pew Center with one conducted in early 2003 before the RIAA campaign, one conducted in late 2003 after the campaign began, and one conducted in 2005. He found that the RIAA campaign had no discernible effect on illegal filesharing; however, the illegal downloading was halved at the time of the campaign in 2003, but showed deterrence was short-lived with illegal downloading increasing from 14.5% to 21% by 2005. Moreover, only one quarter of those who stopped downloading reported that they were afraid to be sued or prosecuted whereas the majority reported stopping for practical reasons including fear of viruses or malware, poor quality of the illegal material, and the slowness of the downloads.

Kigerl (2009, 2015) examined the effectiveness of the United States' CAN SPAM Act on reducing spam email. The data were collected from a purposive sample of millions of spam e-mails downloaded from the Untroubled Software website. His 2009 study found that the law had no effect on the amount or nature of the spam email. Moreover, his more recent 2015 study suggested that a decrease in the frequency of sending spam emails was accompanied by a *decreased compliance* with e-mail header requirements in an effort to evade detection of violating the CAN SPAM Act. Thus, instead of supporting deterrence theory, Kigerl (2015) study supports the notion of restrictive deterrence.

Gibbs (1975) coined the term, restrictive deterrence, to convey that individuals limit the frequency or volume of their offending based on the belief that their luck might eventually run out. Thus, individuals do not cease their offending; the threat of punishment merely makes them think more rationally of how to avoid detection. Research on active offenders has discovered a wide range of evasive strategies that persistent offenders use to reduce the likelihood of detection. These evasive strategies can include displacing their cybercrime to less risky websites or computers, changing the nature of their offending to reduce the severity if caught, and using technology in ways that reduce the chance of detection (e.g., Stalans and Finn 2016b). Restrictive deterrence also was further shown in that many pimps reported refraining from the lucrative sex trade of minors on the internet due to the severe federal prison sentences associated with this crime (Stalans and Finn 2016b).

Research on hacking offenses also shows evidence of restrictive deterrence and limited effectiveness of surveillance techniques such as warning banners. Maimon and colleagues (Maimon et al. 2014; Wilson et al. 2015) have found that warning banners have limited effectiveness at reducing the progression, frequency, and duration of computer intrusions in a controlled, simulated computing environment. Specifically, Maimon et al. (2014) found that the warnings had no effect on terminating the hack, though it did reduce the duration of the attack. Wilson et al. (2015) conducted a randomized control trial and found that surveillance banners reduced the probability of hacking commands being entered into the system only during an individual's first hacking event and only for hacking attacks lasting longer than 50 s. Overall, it appears that the threat of detection and the severity of formal sanctions has only a modest and circumscribed effect on reducing cybercrime, which is similar to the research findings from traditional crime outcomes (e.g., Nagin 2013).

2.2 *Routine Activity Theory*

Routine activity theory also assumes that offenders are rational and hedonistic. While the importance of opportunity is implied within deterrence theory, Cohen's and Felson's (1979) routine activity theory actively highlights the role of opportunity with a noticeable focus on how 'direct-contact' (i.e. offender-victim) criminal opportunities arise. Simply put, Cohen and Felson argue that opportunities arise when there is a convergence in time and space of (1) a motivated offender, (2) a suitable target, and (3) a lack of capable guardianship. Targets can be people or property, and targets that are more suitable are those that have some value to the offender, are portable, are visible, and are accessible (Felson 1998). Guardianship, on the other hand, serves to protect the target and can take various forms such as security cameras, locks, neighborhood watch, traveling in groups, and carrying a weapon.

Although some have been critical of the application of routine activity theory to the cyber-world (e.g., Yar 2005), others have argued that the internet is conducive to the convergence of motivated offenders and suitable targets in the absence of capable guardianship (e.g., Grabosky and Smith 2001; Holt and Bossler 2013). Yar (2005) noted that suitable targets are those that have value, are less resistant to attack (inertia), and are visible and accessible, but that value and inertia are difficult to translate in cyberspace. There are plenty of motivated offenders on the internet who learn of vulnerable and suitable targets through exchanges in chatrooms or social media and may discover inertia and valuable targets through hacking weak firewalls on computer networks containing financial accounts or other unprotected and desirable confidential information. Measuring value and inertia in survey studies, however, does pose a challenge. Rather than converging in physical time and space, cyber-criminals and cyber-victims meet through network devices and internet connections (Holt and Bossler 2013). Moreover, it is possible for these offenders and targets to come into contact with one another in the absence of cyber-guardianship, such as antivirus of malware detection software, weak firewalls, or password protections.

Unlike deterrence theory, there is a sizable body of research using routine activity theory to explain both traditional crime (e.g., Henson et al. 2017; Mustaine and Tewksbury 1999) and cybercrime (for a review, see Leukfeldt and Yar 2016). From a review of the prior eleven studies applying routine activities theory to specific cybercrime and a secondary analysis of 9161 Dutch citizens, Leukfeldt and Yar (2016) operationalized value as financial value in their analysis and did not have a measure for inertia; value was not a significant predictor of victimization from six types of cybercrimes. Leukfeldt and Yar (2016) found that visibility played a role across a wide range of cybercrime; in their study, visibility was operationalized through twelve measures, and more than half of these measures predicted victimization from the cybercrimes of malware attacks, consumer fraud, and receiving threats through cyberspace. Conversely, fewer visibility measures predicted victimization from hacking or cyberstalking, and only more frequent targeted browsing were related to identity theft victimization. The amount of time spent on-line increased the risk of consumer fraud and malware attacks, which is consistent with other research (Pratt et al. 2010; Reyns 2013; Van Wilsem 2013). Time spent on directed communication such as email, MSN or Skype increased interpersonal crimes of stalking and cyberthreat as well as consumer fraud. Other studies have found that more time spent online, particularly in chatrooms, social network sites, and email, (e.g., Bossler et al. 2012; Holt and Bossler 2008; Hinduja and Patchin 2008), and risky online behaviors such as giving passwords to friends or sharing information with strangers (see for a review Chen et al. 2017), increased the likelihood of cyberbullying and cyber-harassment because it differentially expands exposure to motivated offenders.

Capable guardianship is expected to reduce the opportunity for victimization and offending, but empirical support is mixed. For example, having antivirus software has been found to be unrelated (and in some case, positively related) to cyber-victimization across a range of cybercrimes (see Leukfeldt and Yar 2016; Ngo and Paternoster 2011). However, Holt and Turner (2012), using data from students, faculty, and staff at a large university, found that those who updated their protective software programs (e.g., antivirus, Spybot) for a victimization incident were less likely to be repeat victims. Leukfeldt and Yar (2016) found that more awareness of online risks reduced the likelihood of victimization from stalking or hacking. Lastly, computer skills, which have been used as a proxy for personal guardianship, have also generally been found to be unrelated to harassment victimization (e.g., Holt and Bossler 2008); however, having more computer skills was a significant predictor of a general measure of cybercrime for those who were both victims and perpetrators of cybercrime (Kranenbarg et al. 2017). This body of research, overall, has provided limited support for using routine activity theory to explain cyber-offending/victimization, with visibility having the most empirical support across a wide range of cybercrimes.

3 Self-Control Theory

Gottfredson's and Hirschi's (1990) general theory of crime focuses on the concept of self-control: defined as the personal ability to avoid behaviors whose long-term costs exceed the immediate rewards. The theorists suggest that those with low self-control are impulsive, adventure-seeking, self-centered, have a low tolerance for frustration, have a lack of diligence, and have an inability to defer gratification. According to this theory, self-control is acquired through early socialization, particularly effective parenting (Gottfredson and Hirschi 1990). To instill self-control in their children, parents must be to be able to effectively supervise their children, recognize deviant behavior when it occurs, and consistently punish said deviant behavior. Moreover, after adolescence, self-control (or low self-control) will remain relatively stable over an individual's life-course, although there are bodies of research that both support (e.g., Beaver and Wright 2007) and challenge (e.g., Mitchell and Mackenzie 2006) the stability hypothesis. Consistent with life-course research indicating very little evidence for offense specialization (for a review, see DeLisi and Piquero 2011), a substantial self-control literature has also routinely found that individuals with low self-control engage in a wide variety of criminal/deviant behaviors (for meta-analyses, see Pratt and Cullen 2000; Vazsonyi et al. 2017).

Hirschi, in 2004, re-conceptualized self-control as the tendency to consider the full range of potential costs of a behavior. This revision moves the focus away from viewing self-control as a personality trait to rational choice decision-making, which is more consistent with the original intent of the theory. Hirschi posits that self-control refers to an internal set of inhibitors that influence the choices people make, and those with low self-control do not fully consider the formal and informal consequences before acting. Moreover, Hirschi (2004) brings self-control theory full circle with his earlier social control theory (1969) by suggesting that social bonds with family, friends, and work, school, and religious institutions are, in fact, the central inhibitors one considers before engaging in deviant behavior.

Though still relatively young in age, self-control theory has been abundantly tested, both on traditional crime (for reviews, see Pratt and Cullen 2000; Vazsonyi et al. 2017) as well as on cybercrime. Digital piracy is one type of cybercrime that has been the subject of numerous empirical tests within this theoretical context. Higgins and colleagues (e.g., Donner et al. 2014; Higgins 2005; Higgins et al. 2007; Marcum et al. 2011) have extensively studied digital piracy within college samples, and their research consistently finds a significant relationship between low self-control and pirating behavior. The Donner et al. (2014) study, which surveyed 488 undergraduate students from a southern U.S. state, found that low self-control predicted greater involvement in digital piracy as well as greater involvement in other forms of cybercrime such as cyber-harassment and unauthorized computer usage. Moreover, a 2008 study from Higgins et al. confirms the importance of both Gottfredson's and Hirschi's (1990) version of self-control theory in conjunction with Hirschi's (2004) version of the theory as self-control measures of each were related to digital piracy in the expected directions. Furthermore, Moon et al. (2010), in a longitudinal study of 2751 South Korean middle school students, found that low self-control was related to committing digital piracy and hacking. Taken together, these findings support the theory's generality hypothesis as low self-control has shown to be consistently predictive of several forms of cybercrime.

As it relates to engaging in cyberbullying and cyber-harassment, low self-control is, again, an important explanatory variable, including in a sample of teenagers from the Czech Republic (Bayraktar et al. 2015) and in a sample of middle school and high school students (Holt et al. 2012). Finally, a large, cross-cultural examination from Vazsonyi et al. (2012) demonstrated similar—and supportive—results. Using random samples of at least 1000 adolescents from 25 European countries, the authors found significant effects of low self-control on cyberbullying perpetration. Interestingly, though cyberbullying engagement varied noticeably across countries, there were only modest cross-cultural differences in the relationship between low self-control and cyberbullying behavior. Limited empirical attention has examined how personal and environmental characteristics modify the relationship between self-control and perpetration of cybercrime.

4 General Strain Theory

According to Agnew's (1992) general strain theory, there are three primary sources of strain: failure to achieve a positively valued goal, loss of positively-valued stimuli and the introduction of noxious stimuli. Agnew contends that when faced with strain, people experience negative emotions (e.g., anger, depression, anxiety, and fear). These negative emotions then, in the absence of pro-social coping mechanisms, lead people to commit crime. According to Agnew, strain is more likely to result in crime if a strain affects personally important areas, when proper coping skills and resources are absent, when conventional social support is absent, and when predispositions to engage in crime are present (e.g., those who are low self-control, those with weak social bonds, those with exposure to criminal role models). General strain theory is applicable to cybercrime in a number of contexts. For example, those who are financially strained may resort to cyber-theft (e.g., digital piracy) or phishing schemes. Those who may be strained in an interpersonal relationship may engage in cyber-harassment or revenge pornography. Moreover, those who may be strained through being fired from a job may resort to unleashing a virus in their former employer's computer system.

Though there has been a considerable amount of research examining—and validating—the impact of general strain theory on traditional types of crime (for a review, see Agnew 2006), the research testing the theory's effect on cybercrime is less pronounced. Using a large multi-school sample of middle school students in the United States, Patchin and Hinduja (2011) found that strain and anger/frustration were directly related to cyberbullying behavior, which is consistent with the theory. This test, however, provides some inconsistent results as well because anger/frustration did not fully mediate the direct effect of strain on cyberbullying. Additionally, research from Jang et al. (2014) also provides mixed results. In analyzing data among 3238 South Korean adolescents, the authors found that four types of strain (bullying victimization, parental strain, school strain, and financial strain) were all related to engagement in cyberbullying even while controlling for low self-control and deviant peers. However, this study did not test for the mediating effects of negative emotions, such as anger or anxiety. Substantially more research in this area is needed to assess the applicability of strain theory.

5 Social Learning Theory and Related Concepts and Theories

Social learning theory has its roots in the field of psychology. Skinner's (1938) idea of operant conditioning suggested rewarding consequences (i.e. positive reinforcements) reinforced behaviors whereas negative consequences decreased behaviors. Bandura (1973) showed that people also learn aggression vicariously through role models, and these learned behaviors could be maintained through vicarious observations of others being rewarded or through a formation of pleasure or pride from the action. According to social learning theories in psychology, individuals learn to commit deviant acts through social interactions, including both direct communication that reinforce their deviance or through observing role models that are rewarded for their deviance.

Social learning theory argues that individuals learn what behaviors are rewarding through directly performing behaviors and receiving more rewards than punishments. Committing similar crimes in the real world has consistently and moderately been associated with a range of cybercrime including piracy, cyberbullying, hacking, and cyber-fraud (Holt and Bossler 2014). Individuals who commit similar crimes in the real world have learned through direct experience the rewards and lower consequences for criminal behavior and have a much higher chance of committing these crimes on the internet. For example, two meta-analyses found that bullying in the real world was moderately associated with perpetrating cyberbullying (Chen et al. 2017; Kowalski et al. 2014). Relatedly, research suggests that about 90% of those who perpetrate cyberbullying commit bullying in the real world (Raskauskas and Stoltz 2007).

Akers is most associated with applying social learning concepts in psychology to explain why people commit crimes (e.g., Akers 1985). There are four main components of Akers' social learning theory: differential association, holding definitions favorable to committing crime, imitation and modeling and differential reinforcement. Differential association refers to social interactions with others who provide rationalizations, motives, and attitudes for committing—or refraining from—cybercrime. Favorable definitions for committing cybercrimes are attitudes learned from social interactions and contribute to the initiation and continuation of offending. Favorable attitudes indicate that the deviant act is not wrong and include rationalizations for why the cybercrime is not harmful or why they are not responsible for the harm (Hinduja and Ingram 2009). Imitation/modeling refers to observing others engaged in conventional or unconventional behaviors (e.g., cybercrimes) and then imitating that behavior. Finally, differential reinforcement includes both positive reinforcements (i.e. rewards) and negative reinforcement (i.e. unpleasant consequences). Differential reinforcement encompasses the perceived certainty and severity of legal sanctions in deterrence theory; however, it is much broader and includes personal rewards such as satisfaction or pride as well as social approval or disapproval from significant others or strangers.

Gunter (2008) conducted one of the more robust tests of how different components of Akers' social learning theory predicted commission of digital piracy. Cross-sectional survey data were collected from 587 undergraduate students. Differential reinforcement, measured as perceptions of the certainty and severity of negative consequences, belief that the behavior was morally justified, number of friends who engaged in digital piracy, and parental approval of digital piracy were predictors in three separate models of unlawfully downloading software, music and movies. For each of these forms of digital piracy, parental approval and deviant peers directly increased perpetration, and had indirect effects through increasing the technical ability to commit piracy and the belief that it was not wrong. Moreover, associating with more deviant peers and parental approval decreased the perceived certainty of being caught and punished, but perceived certainty of detection and perceived severity of punishment were not related to self-reported digital piracy.

Many studies have found that associating with deviant peers is related to self-reported piracy in youth and undergraduate samples (e.g., Skinner and Freams 1997; Morris et al. 2009; Marcum et al. 2011). It also is one of the most robust predictors of many forms of cybercrimes including digital piracy (e.g., Burruss et al. 2012; Holt et al. 2012; Morris et al. 2009), hacking (e.g., Bossler and Burruss 2011; Holt et al. 2012; Marcum et al. 2014), and cyberbullying (see Holt et al. 2012). For youth samples, having a greater number of friends who commit cybercrimes compared to self-control has been a stronger predictor of self-reported participation in a wide range of cybercrimes, though both are significant predictors (Bossler and Holt 2009; Holt et al. 2012).

Higgins et al. (2007) suggested that individuals with low self-control seek deviant peers to learn the technical skills needed to perform digital piracy and these deviant peers also reinforce their attitudes favorable to committing digital piracy. Their study, using structural equation modeling, found that a model where low self-control effects were fully mediated through social learning was a better fit of the data than a model where low self-control had both indirect and direct effects. Supporting this fully mediated model, self-control was not related to cyberbullying (Li et al. 2016) and other forms of cybercrime such as piracy (Higgins and Makin 2004; Morris and Higgins 2009; Moon et al. 2010) after accounting for deviant peers and unfavorable definitions. Some studies, using less sophisticated regression models, find support for direct effects of self-control on engaging in cybercrime after accounting for unfavorable definitions, deviant associates, general strain, and neutralizations (e.g., Hinduja 2008; Holt et al. 2012; Marcum et al. 2011). Moreover, self-control has inconsistent direct effects on hacking after controlling for peer association and grade point average, with it having direct effect on hacking into Facebook accounts or websites but having no direct effect for hacking into an email account (Marcum et al. 2014). These studies, however, demonstrate the potential advancement of the field's knowledge about perpetration of cybercrime through integrating constructs from different theories.

5.1 Sykes and Matza's Theory of Neutralization

Sykes and Matza (1957) argued that individuals hold beliefs supporting moral values found in criminal laws and must engage in cognitive activity to neutralize their guilt before they are able to commit crimes. They outlined five techniques of neutralization that temporarily lifted the constraints of moral beliefs and allowed individuals to drift into committing crimes. Denial of responsibility shifts the blame away from the offender and onto circumstances in the environment or onto third parties to deny or minimize responsibility. Denial of injury minimizes the harm that offending caused to others. Denial of victim involves claims that the victim is deserving of the harm or is partly responsible for the harm. In condemning of the condemners, individuals declare that the behavior really is not deviant or wrong or those who condemn the behavior do more wrongful actions. Finally, in appealing to higher loyalties, individuals claim that their actions are motivated by values that are more important. In the psychological field, these neutralizations are called techniques of moral disengagement (Bandura 1999).

Digital pirates, based on interview data and coding of web forums, often expressed neutralizations for engaging in digital piracy (Holt and Copes 2010; Harris and Dumas 2009; Moore and McMullan 2009). Morris (2010), using data from 785 college students, found that both neutralizations and association with deviant peers were significantly related to increased hacking and guessing passwords; moreover, though self-control was moderately related to both using neutralizations and having more deviant associates, it did not have a direct effect on hacking. Neutralizations are part of definitions favorable to committing crimes and are learned through social interaction and might maintain the offending. One aspect of neutralization theory has received little empirical attention; it is unclear whether these neutralizations occur before the commission of the crimes or serve as rationalizations after the commission of crimes (for preliminary findings see Higgins et al. 2015).

5.2 *Perceiving and Interpreting the Social Environment of Cyberspace and the Real World*

Individuals are not passive recipients of rewards and consequences, but actively learn about how social environments in cyberspace and the real world are related to rewards and 'negative consequences of illegal behavior'. As Giordano et al. (2015) noted in their life-course view of social learning, "this life-course view of social learning emphasizes the reciprocal relationship between social experiences and cognitive changes" (p. 336). The life-course view of social learning assumes individuals are active agents who navigate their environment and make decisions about their continual involvement with peers and family engaged in deviant behavior in the real and virtual world. Individuals also are more likely to learn which features or areas of the internet and associated social media technology provide the potential for more rewards from committing cybercrimes, allow moral disengagement and depersonalization of victims, and enhance the opportunity to associate with others engaged in specific forms of cybercrime.

Some researchers have discussed the features of cyberspace and related technology that might be perceived to facilitate the commission of cybercrimes (e.g., Barlett and Gentile 2012; Lowry et al. 2016; Seto 2013; Stalans and Finn 2016a). Table 1 defines and describes five dimensions of the internet and associated technology that might be related to increased prevalence of cybercrimes: perceived anonymity, depersonalization of targets or victims, amorphous geographical boundaries, ambi-

Table 1 Possible features of the internet environment facilitating cybercrime

Feature	Definition
Perceived Anonymity	Allows users' identity to be hidden and users perceive that their identity has a low chance of being revealed. Features that increase anonymity include IP masking services, having multiple accounts at the same IP address, google voice calling, creating fake email accounts, and using social media with fake identities
Depersonalization of targets/victims	Perpetrators often lack knowledge about the persons affected by the crimes and of the emotional, intellectual and material consequences to targets of cybercrime
Amorphous Geographical boundaries	Internet communication transcends regulatory and criminal laws of countries and makes it difficult to address cybercrimes that occur across national jurisdictions
Ambiguity of norms	Lack of consensus about what acts constitute certain cybercrimes as well as the varying definitions of legal and illegal acts across countries adds ambiguity about the code of conduct and what is harmful
Ease of affiliation	Social media and specialized websites for specific issues have proliferated and allows greater ease of finding and connecting with individuals who share similar interests, attitudes, and deviant lifestyles. Ease of affiliation, however, requires some knowledge of how to find and communicate on web forums or group chatrooms that host subcultures supportive of specific cybercrimes or cyber-deviance

guity of norms, and ease of affiliation with others engaged in specific cybercrimes or cyber-deviance. Perceived anonymity has been systematically conceptualized and integrated within social learning theory. The depersonalization of targets might facilitate moral disengagement and be associated with neutralizations that minimize how much the targets are harmed. Amorphous geographical boundaries, ambiguity of norms, and ease of affiliation facilitate the entry into oppositional subcultures, the contemplation of alternative self-identities and the creation of specialized knowledge.

Perceived anonymity has primarily been examined to understand cyberbullying. Lowry et al. (2016) proposed the social media cyberbullying model (SMCBM) to explain adult cyberbullying; the SMCBM model integrates anonymity into Akers' social learning model. Perceived anonymity was defined and measured as comprising five related concepts: inability of others to recognize them, limited proximity to observe their computer behavior, belief that social media features would keep their real identity hidden, lack of accountability for their actions, and confidence that the system would not malfunction or have features that could reveal their identity. Lowry and colleagues used data from 1003 adults who completed a survey on MTURK on adult cyberbullying and conducted sophisticated partial least squares regressions to examine how the effects of anonymity on cyberbullying were mediated through social learning theory. Individuals who perceived more anonymity had more moral disengagement, more neutralizations, and fewer perceived costs.

These beliefs, in turn, increased the frequency of adult cyberbullying, even after accounting for gender, hours spent on the computer, and association with deviant peers. Perceived anonymity was also associated with greater rewards, but this was unrelated to perpetrating cyberbullying. Barlett and Gentile (2012) found that differential reinforcement—measured as a scale of rewards or disapproval from friends as well as personal rewards—was related to cyberbullying among undergraduate students. Barlett and colleagues also have found that perceived anonymity is related to cyberbullying in samples of youth and undergraduates (Barlett et al. 2016, 2017; Barlett and Helmstetter 2017).

6 Subcultural Theories

Individuals who learn justifications to disengage their moral beliefs might become further enmeshed in deviant subcultures in the virtual and real world. Subcultural theories emphasize how societal structures can create oppositional subgroups whose values, attitudes, and behaviors conflict with the broader societal laws and values. Anderson's (1999) code of the street theory argues that the joblessness, racism, poverty, hopelessness and alienation, and mistrust of police and societal institutions contributed to the development of a subculture whose values conflicted with the wider conventional values (e.g., working hard, obtaining an education, complying with the law) that were also found in these disadvantaged neighborhoods. Anderson suggested that a "street" subculture supported using violence to address disrespect on the street and getting ahead through illegal behaviors. Eventually, in these neighborhoods, youths had to decide whether to believe in conventional values or in the "code of the street". Both qualitative studies (Anderson 1999) and quantitative studies (e.g., Stewart and Simon 2010) have found that adherence to street values is associated with high rates of violent crimes.

One study has empirically tested the concept of street code to cyberspace (Henson et al. 2017). Henson et al. (2017) argue that youths might share their street code values on social media platforms and specialized web forums. Street code values were adapted into code of the internet through changing Stewart and Simons (2010) quantitative scale to focus on an online code (e.g., "Appearing tough and aggressive is a good way to keep others from messing with you online"). Low self-control and higher fear of cyberbullying were related to a higher likelihood of adopting online street-oriented beliefs in an undergraduate sample. Other qualitative research has measured parental approval of street code values as self-reports of whether parents would approve or disapprove of the respondents' criminal behavior. Active pimps, running illicit prostitution businesses through online advertisements of sex workers, reported use of more indirect (psychological and economic concern) and direct physical or restraining coercive strategies if they had parents who supported street code values than parents who supported conventional values (Stalans and Finn 2016a).

Cyberspace features of amorphous geographical boundaries allow people to learn of behaviors such as prostitution or different copyright laws that create ambiguity about the appropriateness and wrongfulness of the behavior. Moreover, the many specialized website forums and 'how-to-do' websites for specific forms of deviance such as prostitution, digital pirating, hacking, and pimping provide easy affiliations and sharing of information for those interested in cyber-deviance. These features, discussed in Table 1, enhance the opportunities to learn about and participate in oppositional subcultures. Research has examined the subcultural values, norms, and practices of persistent digital pirates, hackers, and participants in the online-soliciting illicit commercial sex trade (e.g., Holt 2007; Holt and Copes 2010; Holt et al. 2017; Stalans and Finn 2016b).

Holt (2007) examined the subculture of hackers using interviews with 13 active hackers, coding of 365 posts to six public web forums for hackers, and observation data from the 2004 Defcon, the largest hacker convention in the United States. Holt (2007) identified five general 'normative orders', such as having a deep interest in technology and having a desire to demonstrate mastery in the ability to hack, establishing their identity within the subcategories of hackers, spending much effort to learn skills and complete successful acts, and having views about violating laws. Regardless of their support for illegal hacking, individuals shared information that others could use to perform illegal acts, and this sharing was often done with disclaimers that they did not support illegal hacking and neutralizations to minimize their responsibility. In part, these neutralizations supported the shared value of secrecy in the hackers subculture as all members were interested in avoiding legal sanctions, and practices such as 'spot the fed' at hacking conventions allowed members to develop knowledge about mannerism and interactions that differentiated true hackers from undercover cops. Hackers motivated by ideological agendas share these values of the hacker subculture, but selected targets for malicious hacking attacks based on religious and political agendas (Holt et al. 2017).

Research on both internet-solicited illegal sex trade and hacking suggests that individuals learn from specialized internet websites how to use technology and conduct their illicit behaviors in the real world to avoid arrest. For example, pimps reported using google voice, using "burner" (prepaid mobile) phones, changing advertising venues based on law enforcement focus, and attempting to disguise advertisements soliciting clients for illicit prostitution as legitimate businesses such as massage therapy. Moreover, the development of specialized knowledge about the behaviors of undercover cops compared to true participants is used in interactions to evade arrest (Stalans and Finn 2016b). Thus, the threat of uncovered stings on internet sites selling illicit prostitution services or drugs have limited effects on persistent offenders. Instead, these offenders consider detection and punishment, but invest their energy in finding ways to continue the illegal behavior.

7 Conclusions

People from a variety of socio-economic, intellectual, and cultural backgrounds participate in a wide range of cybercrimes for many different reasons. From this review of criminology and psychological theories and the associated empirical research on specific cybercrime, we can draw some broad conclusions. Social learning has been one of the most empirically tested theories for understanding cybercrime offending. Across studies, a greater number of deviant peers, beliefs that the crimes were not morally wrong, and providing a greater number of neutralizations consistently were related to a higher likelihood of engaging in cybercrime. Moreover, individuals with low self-control were more likely to associate with deviant peers, and have values and justifications that supported cyber-offending. These findings held across a wide variety of cybercrimes. Social media apps, web forums, chatrooms, and file-to-file transfers on the internet provide many educational opportunities to learn attitudes favorable to committing cybercrime, technological skills and strategies to avoid legal and social sanctions, and to affiliate with others who are supportive of committing cybercrime. Specialized web forums or exchanges on the deep web on specific crimes such as internet soliciting commercial sex trade or hacking allow the formation of subcultures with shared practices and norms.

Learning at the individual and group level facilitates the perpetration and continuation of cyber-offending. Some aspects of learning, however, have been understudied. For example, besides peers, few sources of approval or disapproval have been examined. Little is known whether digital bystanders can stop cyberbullying or cyber-harassment. Wong-Lo and Bullock (2014) argue that the perceived anonymity of cyberspace provides digital bystanders with more autonomy and discretion of how to respond when they observe cyberbullying or cyber-harassment: ignore it, spread and condone it publically or privately, or denounce it publically or privately. Digital bystanders might be "cyber-acquaintances, friends, or strangers" with little connection in the real world, but offer a potential means through which cybercrimes could be reduced. Moreover, digital upstanders are those who confront and address injustices; research on both the situational and personal characteristics that stimulate bystanders to address harmful behaviors is needed.

Research is also needed to examine the social interactions that occur or even how much youth, emerging adults, and adults confront others for flaming, harassment due to sexism, racism, or other biases. Moreover, little is known about how associations in the real world and online influence the formation of attitudes favorable or unfavorable to committing cybercrimes, except those studies that have coded website forums for specific crimes in the real or virtual world (e.g., Holt 2007; Holt and Copes 2010; Holt et al. 2017; Stalans and Finn 2016b). For example, few studies have examined how parental approval is related to engaging in cybercrime. Youth with parents that are unaware of their internet activity are more likely to engage in cyberbullying perpetration in a longitudinal survey study of 75 parent-child dyads (Barlett and Fennel 2016), and direct supervision did not seem to reduce cyberbullying. It might be that youth interpret parental lack of awareness as suggesting that there is no real harm in their virtual world behaviors. Vignette studies, computer experiments, and surveys could provide empirical data to address these issues and to create policies and interventions that might reduce cyber-offending. Researchers often note the unstructured and geographically unbounded effects of the internet, but empirical research has not examined why, when and how individuals gather information about laws in different countries and use this knowledge for further cyber-deviance.

Though components of social learning theory have received empirical support, the support for differential reinforcements and deterrence is limited. From a psychological perspective, the inconsistent and weak effects for rewards or deterrence is not surprising. Skinner's (1938) conditioning theory assumed that beyond basic needs authorities would need to learn what reinforcements were seen as rewarding, and then use these to reward or to provide costs for the unwanted behavior. Finding common rewards and costs might be difficult, though the frequency of internet use among those younger than 40 years of age suggests that limiting access to social media and internet could be an effective negative reinforcement for those who are not embedded in oppositional subcultures.

Several cybercrime scholars (e.g., Choi 2008; Higgins and Marcum 2011; Holt and Bossler 2014) have called for the integration of multiple theoretical perspectives in pursuit of trying to better explain the behavior. Similar to integration attempts to explain real world criminal behavior, theoretical integration in cybercrime would attempt to produce a more complete theoretical understanding of why people engage in cybercrime offenses. While running the risk of not being parsimonious, integrated theories offer a solution to the problem of viewing behavior from a single-lens perspective: human behavior—including criminal behavior—is multifaceted and complex, and it cannot be explained through a single viewpoint (Akers et al. 2016). Our review highlights that integrated models will need to include these consistent predictors of cyber-offending: low self-control, deviant peer associations, moral beliefs, neutralizations, past offending in the real world, visibility of targets (e.g., time spent on the computer and on social media). Prior offending in the real world also needs to be included to understand how real world behavior affects actions in the virtual world. Therefore, this chapter not only advocates for the continuation of research attempting to identify the causes and correlates of cybercrime, but also recommends creating—and testing—integrated theories based on the theoretical concepts that have already been identified as consistent predictors of cybercrime (e.g., social learning, low self-control, routine activities). Only then will we have a more thorough grasp on why people engage in such deviant behaviors as hacking, digitally piracy, cyberbullying, and cyber-solicitation.

References

Agnew, R. (1992). Foundation for a general strain theory of crime and delinquency. *Criminology, 30*, 47–87.
Agnew, R. (2006). General strain theory: Current status and directions for further researches. In F. T. Cullen, J. P. Wright, & K. R. Blevins (Eds.), *Taking stock: The status of criminological theory* (pp. 121–123). New Brunswick: Transaction Publishers.
Akers, R. L. (1985). *Deviant behavior: A social learning approach*. Belmont: Wadsworth.
Akers, R. L., Sellers, C. S., & Jennings, W. G. (2016). *Criminological theories: Introduction, evaluation, and application*. Oxford: Oxford University Press.
Anderson, E. (1999). *Code of the street: Decency, violence and the moral life of the Inner City*. New York: W. W. Norton and Company.
Bachmann, M. (2007). Lesson spurned? Reactions of online music pirates to legal prosecutions by the RIAA. *International Journal of Cyber Criminology, 1*(2), 213–227.
Bandura, A. (1973). *Aggression: A social learning analysis*. Englewood Cliffs: Prentice Hall.
Bandura, A. (1999). Moral disengagement in the perpetration of inhumanities. *Personality and Social Psychology Review, 3*(3), 193–209.
Barlett, C. P., & Fennel, M. (2016). Examining the relation between parental ignorance and youths' cyberbullying perpetration. *Psychology of Popular Media Culture, 7*(1), 444–449. https://doi.org/10.1016/j.chb.2017.02.009.
Barlett, C. P., & Gentile, D. A. (2012). Attacking others online: The formation of cyberbullying in late adolescence. *Psychology of Popular Media Culture, 1*(2), 123–135.
Barlett, C. P., & Helmstetter, K. M. (2017). Longitudinal relations between early online disinhibition and anonymity perceptions on later cyberbullying perpetration: A theoretical test on youth. Psychology of Popular Media Culture, Advance online publication. https://doi.org/10.1037/ppm0000149.
Barlett, C. P., Chamberlin, K., & Witkower, Z. (2017). Predicting cyberbullying perpetration in emerging adults: A theoretical test of the Barlette Gentile Cyberbullying Model. *Aggressive Behavior, 43*, 147–154.
Barlett, C. P., Gentile, D. A., & Chew, C. (2016). Predicting cyberbullying from anonymity. *Psychology of Popular Media Culture, 5*(2), 171–180. https://doi.org/10.1037/ppm0000055.
Bayraktar, F., Machackova, H., Dedkova, L., Cerna, A., & Sevcikova, A. (2015). Cyberbullying: The discriminant factors among cyberbullies, cybervictims, and cyberbully-victims in a Czech adolescent sample. *Journal of Interpersonal Violence, 30*(18), 3192–3216. https://doi.org/10.1177/088626051455006.
Beaver, K. M., & Wright, J. P. (2007). The stability of low self-control from kindergarten through first grade. *Journal of Crime and Justice, 30*(1), 63–86.
Beccaria, C. (1764). *On crimes and punishment* (H. Paolucci, Trans.). Indianapolis: Bobbs-Merrill.
Bossler, A. M., & Burruss, G. W. (2011). The general theory of crime and computer hacking: Low self-control hackers. In T. J. Holt & B. H. Schell (Eds.), *Corporate hacking and technology-driven crime* (pp. 38–67). Hershey: IGI Global.
Bossler, A. M., & Holt, T. J. (2009). On-line activities, guardianship, and malware infection: An examination of routine activities theory. International Journal of Cyber Criminology, *3(1)*, 400–420. Retrieved from https://doi.org/10.1177/1043986213507401.
Bossler, A. M., Holt, T. J., & May, D. C. (2012). Predicting online harassment victimization among a juvenile population. *Youth Society, 44*(4), 500–523. https://doi.org/10.1177/0044118X11407525.
Brenner, S. W. (2012). *Cybercrime and the law: Challenges, issues and outcomes*. Lebanon: Northeastern University Press.
Burruss, G. W., Bossler, A. M., & Holt, T. J. (2012). Assessing the mediation of a fuller social learning model on low self-control's influence on software piracy. *Crime & Delinquency, 59*(8), 1157–1184. https://doi.org/10.1177/0011128712437915.

Chen, L., Ho, S. S., & Lwin, M. O. (2017). A meta-analysis of factors predicting cyberbullying perpetration and victimization: From the social cognitive and media effects approach. *New Media & Society, 19*(8), 1194–1213.

Choi, K. S. (2008). Computer crime victimization and integrated theory: An empirical assessment. *International Journal of Cyber Criminology, 2*(1), 308.

Cohen, A. K., & Felson, M. (1979). Social change and crime rates: A routine activities approach. *American Sociological Review, 44*, 214–241.

DeLisi, M., & Piquero, A. R. (2011). New frontiers in criminal careers research, 2000–2011: A state-of-the-art review. *Journal of Criminal Justice, 39*, 289–301. https://doi.org/10.1016/j.jcrimjus.2011.05.001.

Donner, C. M., Marcum, C. D., Jennings, W. G., Higgins, G. E., & Banfield, J. (2014). Low self-control and cybercrime: Exploring the utility of the general theory of crime beyond digital piracy. *Computers in Human Behavior, 34*, 165–172. https://doi.org/10.1016/j.chb.2014.01.040.

Felson, M. (1998). *Crime & everyday life* (2nd ed.). Thousand Oaks: Pine Forge Press.

Gibbs, J. P. (1975). *Crime, Punishment, and Deterrence*. New York: Elsevier.

Giordano, P. C., Johnson, W. L., Manning, W. D., Longmore, M. A., & Minter, M. D. (2015). Intimate partner violence in young adulthood: Narratives of persistence and desistance. *Criminology, 53*(3), 330–365.

Gottfredson, M. R., & Hirschi, T. (1990). *A general theory of crime*. Stanford: Stanford University Press.

Grabosky, P. M. (2001). Virtual criminality: Old wine in new bottles? *Social & Legal Studies, 10*(2), 243–249. https://doi.org/10.1177/a017405.

Grabosky, P. N., & Smith, R. G. (2001). Digital crime in the twenty-first century. *Journal of Information Ethics, 10*(1), 8–26.

Gunter, W. D. (2008). Piracy on the high speeds: A test of social learning theory on digital piracy among college students. *International Journal of Criminal Justice Sciences, 3*(1), 54–68.

Harris, L. C., & Dumas, A. (2009). Online consumer misbehavior: An application of naturalization theory. *Marketing Theory, 9*(4), 379–402. https://doi.org/10.1177/1470593109346895.

Henson, B., Swartz, K., & Reyns, B. W. (2017). #Respect: Applying Anderson's code of the street to the online context. *Deviant Behavior, 38*(7), 768–780. https://doi.org/10.1080/01639625.2016.1197682.

Higgins, G. E. (2005). Can low self-control help understand the software piracy problem? *Deviant Behavior, 26*, 1–24.

Higgins, G. E., & Makin, D. A. (2004). Does social learning theory condition the effects of low self-control on college students' software piracy? *Journal of Economic Crime Management, 2*(2), 1–30.

Higgins, G. E., & Marcum, C. D. (2011). *Digital piracy: An integrated theoretical approach*. Raleigh: Carolina Academic Press.

Higgins, G. E., Fell, B. D., & Wilson, A. L. (2006). Digital piracy: Assessing the contributions of an integrated self-control theory and social learning theory using structural equation modeling. *Criminal Justice Studies, 19*(1), 3–22.

Higgins, G. E., Fell, B. D., & Wilson, A. L. (2007). Low self-control and social learning in understanding students' intentions to pirate movies in the United States. *Social Science Computer Review, 25*(3), 339–357.

Higgins, G. E., Wolfe, S. E., & Marcum, C. D. (2015). Music piracy and neutralization: A preliminary trajectory analysis from short-term longitudinal data. *International Journal of Cyber Criminology, 2*(2), 324–336.

Hinduja, S. (2008). Deindividuation and internet software piracy. *Cyberpsychology & Behavior, 11*(4), 391–398. https://doi.org/10.1089/cpb.2007.0048.

Hinduja, S., & Ingram, J. R. (2009). Social learning theory and music piracy: The differential role of online and offline peer influences. *Criminal Justice Studies, 22*(4), 405–420.

Hinduja, S., & Patchin, J. W. (2008). Cyberbullying: An exploratory analysis of factors related to offending and victimization. *Deviant Behavior, 29*, 129–156. https://doi.org/10.1080/01639620701457816.

Hirschi, T. (2004). Self-control and crime. In R. Baumeister & K. Vohs (Eds.), *Handbook of self-regulation: Research, theory, and applications* (pp. 537–552). New York: Guilford Press.

Hollinger, R. C., & Lanza-Kaduce, L. (1988). The process of criminalization: The case of computer crime laws. *Criminology, 26*(1), 101–126. https://doi.org/10.1111/j.1745-9124.1988.tb00834.x.

Holt, T. J. (2007). Subcultural evolution? Examining the influence of on- and off-line experiences on deviant subcultures. *Deviant Behavior, 28*, 171–198.

Holt, T. J., & Bossler, A. M. (2008). Examining the applicability of lifestyle-routine activities theory of cybercrime victimization. *Deviant Behavior, 30*(1), 1–25. https://doi.org/10.1080/01639620701876577.

Holt, T. J., & Bossler, A. M. (2013). Examining the relationship between routine activities and malware infection indicators. *Journal of Contemporary Criminal Justice, 29*(4), 420–436. https://doi.org/10.1177/1043986213507401.

Holt, T. J., & Bossler, A. M. (2014). An assessment of the current state of cybercrime scholarship. *Deviant Behavior, 35*(1), 20–40. https://doi.org/10.1080/01639625.2013.822209.

Holt, T. J., & Copes, H. (2010). Transferring subcultural knowledge on-line: Practices and beliefs of digital pirates. *Deviant Behavior, 31*(7), 625–654. https://doi.org/10.1080/01639620903231548.

Holt, T. J., & Turner, M. G. (2012). Examining risks and protective factors of on line identity theft. *Deviant Behavior, 33*, 308–323.

Holt, T. J., Bossler, A. M., & May, D. C. (2012). Low self-control, deviant peer associations, and juvenile cyberdeviance. *American Journal of Criminal Justice, 37*(3), 378–395. https://doi.org/10.1007/s12103-011-9117-3.

Holt, T. J., Freilich, J. D., & Chermak, S. M. (2017). Exploring the subculture of ideologically motivated cyber-attackers. *Journal of Contemporary Criminal Justice, 33*(3), 212–233. https://doi.org/10.1177/1043986217699100.

Irdeto (2017). *Infographic: When it comes to piracy – The world needs a tutor.* Downloaded on March 3, 2018 from: https://irdeto.com/index.html.

Jang, H., Song, J., & Kim, R. (2014). Does the offline bully-victimization influence cyberbullying behavior among youths? Application of general strain theory. *Concepts in Human Behavior, 31*, 85–93. https://doi.org/10.1016/j.chb.2013.10.007.

Kigerl, A. C. (2009). CAN SPAM act: An empirical analysis. *International Journal of Cyber Criminology, 3*(2), 566–589.

Kigerl, A. C. (2015). Evaluation of the CAN SPAM act: Testing deterrence and other influences of e-mail spammer legal compliance over time. *Social Science Computer Review, 33*(4), 440–458. https://doi.org/10.1177/0894439314553913.

Kowalski, R. M., Giumetti, G. W., Schroeder, A. N., & Lattanner, M. R. (2014). Bullying in the digital age: A critical review and meta-analysis of cyberbullying research among youth. *Psychological Bulletin, 140*(4), 1073–1137.

Kranenbarg, M. W., Holt, T. J., & van Gelder, J. (2017). Offending and victimization in the digital age: Comparing correlates of cybercrime and traditional offending-only, victimization-only, and the victimization-offending overlap. *Deviant Behavior*, 1–16. https://doi.org/10.1080/01639625.2017.1411030.

Leukfeldt, E. R., & Yar, M. (2016). Applying routine activities theory to cybercrime: A theoretical and empirical analysis. *Deviant Behavior, 37*(3), 263–280. https://doi.org/10.1080/01639625.2015.1012409.

Li, C. K. W., Holt, T. J., Bossler, A. M., & May, D. C. (2016). Examining the mediating effects of social learning on the low self-control- cyberbullying relationship in a youth sample. *Deviant Behavior, 37*(2), 126–138. https://doi.org/10.1080/01639625.2014.1004023.

Lowry, P. B., Zhang, J., Wang, C., & Siponen, M. (2016). Why do adults engage in cyberbullying on social media? An integration of online distribution and deindividuation effects with the social structure and social learning model. *Information Systems Research, 27*(4), 962–986.

Maimon, D., Alper, M., Sobesto, B., & Cukier, M. (2014). Restrictive deterrent effects of a warning banner in an attacked computer system. *Criminology, 52*(1), 33–59. https://doi.org/10.1111/1745-9125.12028.

Marcum, C. D., Higgins, G. E., Wolfe, S. E., & Ricketts, M. L. (2011). Examining the intersection of self-control, peer association and neutralization in explaining digital piracy. *Western Criminology Review, 12*(3), 60–74 Retrieved from https://www.researchgate.net/publication/228458057_Examining_the_Intersection_of_Self-control_Peer_Association_and_Neutralization_in_Explaining_Digital_Piracy.

Marcum, C. D., Higgins, G. E., Ricketts, M. L., & Wolfe, S. E. (2014). Hacking in high school: Cybercrime perpetration by juveniles. *Deviant Behavior, 35*(7), 581–591. https://doi.org/10.1080/01639625.2013.867721.

McQuade, S. C. (2006). *Understanding and managing cybercrime*. Upper Saddle River: Pearson Education.

Mitchell, O., & MacKenzie, D. L. (2006). The stability and resiliency of self-control in a sample of incarcerated offenders. *Crime & Delinquency, 52*(3), 432–449. https://doi.org/10.1177/0011128705280586.

Moon, B., McCluskey, J. D., & Perez McCluskey, C. (2010). A general theory of crime and computer crime: An empirical test. *Journal of Criminal Justice, 38*(4), 767–772. https://doi.org/10.1016/j.jcrimjus.2010.05.003.

Moore, R., & McMullan, E. C. (2009). Neutralizations and rationalizations of digital piracy: A qualitative analysis of university students. *International Journal of Cyber Criminology, 3*(1), 441–451 Retrieved from https://www.researchgate.net/publication/229020027_Neutralizations_and_rationalizations_of_digital_piracy_A_qualitative_analysis_of_university_students.

Morris, R. G. (2010). Computer hacking and the techniques of neutralization: An empirical assessment. In T. J. Holt & B. Schell (Eds.), *Corporate hacking and technology-driven crime: Social dynamics and implications* (pp. 1–17). New York: Information Science Reference.

Morris, R. G., & Higgins, G. E. (2009). Neutralizing potential and self-reported digital piracy: A multitheoretical exploration among college undergraduates. *Criminal Justice Review, 34*(2), 173–195. https://doi.org/10.1177/0734016808325034.

Morris, R. G., Johnson, M. C., & Higgins, G. E. (2009). The role of gender in predicting the willingness to engage in digital piracy among college students. *Criminal Justice Studies, 22*(4), 393–404. https://doi.org/10.1080/14786010903358117.

Mustaine, E. E., & Tewksbury, R. (1999). A routine activities theory explanation for women's stalking victimizations. *Violence Against Women, 5*(1), 43–62. https://doi.org/10.1177/10778019922181149.

Nagin, D. S. (2013). *Deterrence in the twenty-first century: A review of the evidence*. Carnegie Mellon University Research Showcase. Downloaded on April 4th, 2018 from: https://pdfs.semanticscholar.org/c788/48cc41cdc319033079c69c7cf1d3e80498b4.pdf.

Ngo, F. T., & Paternoster, R. (2011). Cybercrime victimization: An examination of individual and situational level factors. *International Journal of Cyber Criminology, 5*(1), 773–793.

O'Neill, M. E. (2000). Old crimes in new bottles: Sanctioning cybercrime. *George Mason Law Review, 9*, 237.

Patchin, J. W., & Hinduja, S. (2011). Traditional and nontraditional bullying among youth: A test of general strain theory. *Youth Society, 43*, 727–751.

Pratt, T. C., & Cullen, F. T. (2000). The empirical status of Gottfredson and Hirschi's general theory of crime: A meta-analysis. *Criminology, 38*(3), 931–964. https://doi.org/10.1111/j.1745-9125.2000.tb00911.x.

Pratt, T. C., Cullen, F. T., Blevins, K. R., Daigle, L. E., & Madensen, T. D. (2008). The empirical status of deterrence theory: A meta-analysis. In F. T. Cullen, J. P. Wright, & K. R. Blevins (Eds.), *Taking stock: The status of criminological theory* (pp. 367–396). New York: Taylor & Francis.

Pratt, T. C., Holtfreter, K., & Reisig, M. D. (2010). Routine online activity and internet fraud targeting: Extending the generality of routine activity theory. *Journal of Research in Crime and Delinquency, 47*(3), 267–296.

Raskauskas, J., & Stoltz, A. D. (2007). Involvement in traditional and electronic bullying among adolescents. *Developmental Psychology, 43*(3), 564–575. https://doi.org/10.1037/0012-1649.43.3.564.

Reyns, B. W. (2013). Online routines and identity theft victimization: Further expanding routine activities theory beyond direct-contact offenses. *Journal of Research in Crime and Delinquency, 50*(2), 216–238. https://doi.org/10.1177/0022427811425539.

Roberts, J. V., & Stalans, L. J. (1997). *Public opinion, crime, and criminal justice*. Boulder: Westview Press.

Seto, M. C. (2013). *Internet sex offenders*. Washington, DC: American Psychological Association.

Skinner, B. F. (1938). *The behavior of organisms: An experimental analysis*. Oxford: Appleton-Century.

Skinner, W. F., & Fream, A. M. (1997). A social learning theory analysis of computer crime among college students. *Journal of Research in Crime and Delinquency, 34*(4), 495–518.

Stalans, L. J., & Finn, M. A. (2016a). Defining and predicting pimps' coerciveness toward sex workers: Socialization processes. *Journal of Interpersonal Violence*, 1–24. https://doi.org/10.1177/0886260516675919.

Stalans, L. J., & Finn, M. A. (2016b). Consulting legal experts in the real and virtual world: Pimps' and johns' cultural schemas about strategies to avoid arrest and conviction. *Deviant Behavior, 37*(6), 644–664. https://doi.org/10.1080/01639625.2015.1060810.

Stalans, L. J., & Finn, M. A. (2016c). Introduction to special issue: How the internet facilitates deviance. *Victims and Offenders, 11*(4), 578–599.

Stewart, E. A., & Simons, R. L. (2010). Race, code of the street, and violent delinquency: A multilevel investigation of neighborhood street culture and individual norms of violence. *Criminology, 48*(2), 569–605.

Sykes, G. M., & Matza, D. (1957). Techniques of neutralization: A theory of delinquency. *American Sociological Review, 22*, 664–670.

Van Wilsem, J. (2013). Hacking and harassment- Do they have something in common? Comparing risk factors for online victimization. *Journal of Contemporary Criminal Justice, 29*(4), 437–453. https://doi.org/10.1177/1043986213507042.

Vazsonyi, A. T., Machackova, H., Sevcikova, A., Smahel, D., & Cerna, A. (2012). *The European Journal of Developmental Psychology, 9*(2), 210–227. https://doi.org/10.1080/17405629.2011.644919.

Vazsonyi, A. T., Mikuska, J., & Kelley, E. L. (2017). It's time: A meta-analysis on the self-control-deviance link. *Journal of Criminal Justice, 48*, 48–63. https://doi.org/10.1016/j.jcrimjus.2016.10.001.

Wall, D. S. (1998). Catching cybercriminals: Policing the Internet. *International Review of Law, Computers & Technology, 12*(2), 201–218.

Wilson, T., Maimon, D., Sobesto, B., & Cukier, M. (2015). The effect of a surveillance banner in an attacked computer system: Additional evidence for the relevance of restrictive deterrence in cyberspace. *The Journal of Research in Crime and Delinquency, 52*(6), 829–855. https://doi.org/10.1177/0022427815587761.

Wong-Lo, M., & Bullock, L. M. (2014). Digital metamorphosis: Examination of the bystander culture in cyberbullying. *Aggression and Violent Behavior, 19*, 418–422.

Yar, M. (2005). The novelty of 'cybercrime': An assessment in light of routine activity theory. *European Journal of Criminology, 2*(4), 407–427.

Cyber Aggression and Cyberbullying: Widening the Net

John M. Hyland, Pauline K. Hyland, and Lucie Corcoran

1 Introduction

This chapter provides an overview of current theories and perspectives within the field of cyberbullying, with a discussion of viewpoints regarding conceptualisation and operationalisation of cyberbullying and its position within the framework of aggression and cyber aggression. Specifically, a review of current theories of aggression will be presented and discussed, locating cyberbullying within this literature as a subset of aggression. Issues with defining the construct will be discussed with an argument for cyberbullying to be placed within the architecture of cyber aggression, due to an arguable over-narrowing of the parameters of cyberbullying. Subtypes of cyber aggression are presented, which include cybertrolling and cyberstalking, among others, and these subtypes are discussed in terms of definition, characteristics, and current debates within these fields. This chapter examines the need to broaden the scope of research with regard to Cyberbullying, including a need to adopt evidence-based approaches to intervention and prevention, and integrate more recent online models within associated fields such as mental health. Currently there is debate regarding legislation and, in particular, about setting the digital age of consent for Irish children; a pertinent concern when considering the implications for online presence and exposure to risks such as cyber aggression. This chapter highlights the breadth of prevention/intervention efforts relating to cyber aggression and emphasises the need for a multi-faceted response to this issue. Firstly, it is important to understanding the theoretical context of aggressive behaviour, which is the focus of the next section.

J. M. Hyland (✉) · P. K. Hyland · L. Corcoran
Department of Psychology, Dublin Business School, Dublin, Ireland
e-mail: john.hyland@dbs.ie

2 Theoretical Understanding of Aggression

Human aggression has been defined by Anderson and Bushman as "... any behaviour directed toward another individual that is carried out with the proximate (immediate) intent to cause harm" (Anderson and Bushman 2002, p. 28). This definition places importance on the deliberate intention to inflict harm on others, emphasising that accidental harm does not constitute aggressive behaviour as there is an absence of intent. Traditional and cyberbullying depicts this aggressive behaviour and as discussed later, carry many of its characteristics. Some of the key theories and theorists to consider when understanding aggressive behaviour include Freud (1920), the Frustration-Aggression Hypothesis (Dollard et al. 1939), Lorenz (1974), the Excitation Transfer Theory (Zillmann 1979, 1983), the Cognitive Neoassociation Theory (Berkowitz 1989, 1990, 1993), the Social Learning Theory (Bandura 1978, 1997), the Script Theory (Huesmann 1986, 1998), the Social Interaction Theory (Tedeschi and Felson 1994), and the General Aggression Model (Anderson and Bushman 2002; Anderson 1997).

From the psychoanalytic perspective, Freud (1920) argued that aggression was innate, that all humans are prewired for violence and that internal forces were causal factors of aggressive behaviour. Furthermore, these instincts, that of death (thanatos) and life (eros), conflict internally with one another, developing a destructive energy in the individual, which can only be reduced when this conflict is deflected onto other people through an aggressive act. Freud termed this 'catharsis', and as such, aggression towards others is a means to rebalance the individual by releasing the built-up energy. In response to Freud's account of aggression, drive theorists proposed a counterargument to that of innate aggression, where aggression was an external drive, as a consequence to circumstances outside of the individual, such as frustration, which incites a motivation to cause harm to others (Berkowitz 1989; Feshbach 1984). However, drive theorists argue that this drive for aggressive behaviour is not continuously present and increases in energy. Rather, only when an individual is prevented from satisfying a need, will the drive be activated. Specifically, the Frustration-Aggression Hypothesis (Dollard et al. 1939), posits that aggression is the product of external forces that create frustration within the individual. Aggression is born out of a drive to cease feelings of frustration where external factors have interfered in the individual's goal-directed behaviour. As such, it is this feeling of frustration that activates this drive, and in turn leads to aggressive behaviour. With regard to peer-directed aggression (bullying), this would suggest that the behaviour is the result of frustration brought on by a response to others.

However, drive theorists later expanded this stance of the Frustration-Aggression Hypothesis (Dollard et al. 1939) acknowledging that aggression was not limited to just frustration as the causal factor, as individuals engage in aggressive behaviours for reasons other than frustration alone. Furthermore, Krahé (2001) stated that not all frustration leads to aggression and can result in other emotional responses. Berkowitz (1989, 1990, 1993) proposed the Cognitive Neoassociation Theory to account for the flaws in the Frustration-Aggression Hypothesis (Dollard et al.

1939), where anger was the mediator between frustration and aggression, and the trigger for an aggressive act. Only when negative affect was evoked, would frustration result in aggression, however, it may be preceded by provocation or loud noises. These negative experiences and behaviours evoke responses associated with the fight or flight response, such as thoughts, memories, motor reactions and physiological responses. Consequently, a negative association may develop between the stimuli present during a negative event along with emotional and cognitive responses (Collins and Loftus 1975). Furthermore, when concepts of similar meaning are experienced this may evoke the associations and feelings and elicit similar emotional and cognitive responses. This highlights the complex context of many aggressive acts, and that the complex nature of aggression should be considered when attempting to counter school-based and cyber-based bullying.

From an ethological perspective, Lorenz (1974) proposed a more genetic model of aggression based on a fighting instinct, arguing that aggression is an unavoidable characteristic of human behaviour as it has been passed innately through generations of lineage, where the strongest males mated and passed on their genetic characteristics to their offspring. How this aggression manifests, is on the basis of individual and environmental factors such as the amount of aggression accumulated, and the extent to which the external stimuli can evoke an aggressive response. This theory is further held by Krahé (2001) in the field of sociobiology, where Darwin's (1859) 'Origins of Species' forms the basis of understanding social behaviour. From this perspective, aggression is adaptive, as its function is for defence purposes against attackers and rivals (Archer 1995; Buss and Shackelford 1997; Daly and Wilson 1994). Consequently, the propensities for aggression are passed on in line with the phylogeny of the species (Krahé 2001). The view of aggression in humans has been criticised as being too deterministic as it assumes that an individual will grow up to be violent if they have inherited the aggressive gene. It is at this point that behavioural genetics deviates, arguing that although an individual may be predisposed to aggression in their genetic make-up, it is the environmental factors that determine whether or not the aggression is occasioned and reinforced (Daly and Wilson 1994; Bleidorn et al. 2009; Hopwood et al. 2011; Johnson et al. 2005). This perspective carries important implications for countering school bullying and cyberbullying, as it would suggest that in many cases aggressive tendencies can be effectively reduced through intervention.

Zillman (1979, 1983) sought to understand aggression with the Excitation Transfer Theory. As physiological arousal dissipates over time, remnants in the individual from one emotionally evoking situation may be transferred to another. As such, if only a short time has passed between the two arousing events, arousal from the first event may be incorrectly assigned to the second. If anger is evoked, then the transferred arousal from the first event would lead to increased anger and a greater aggressive response misattributed to the second event, with the individual becoming angrier than what would be expected for that situation. Again, this has relevance to countering peer-directed aggression with children and young people, as there is recognition that bullying may be an indirect response to unrelated events.

Specifically, in the context of intervention and prevention, it places importance on emotion regulation in individuals when dealing with the behaviour.

In contrast to the evolutionary approach to understanding aggression, with the Social Learning Theory, Bandura (1978, 1997) adopts a stance that aggression is based on observational learning or direct experience. It is learned and imitated from social models and from observing social behaviour. It is through these models and past experiences that individuals learn aggressive behaviour, what constitutes retaliation or vengeance, and where and when aggression is permitted (Bandura 1986). This was demonstrated in Bandura's 'Bobo Doll' experiment, where children watched an individual as a 'model' act aggressively to the doll. These children later imitated this aggressive behaviour towards the doll without being reinforced to do so. When observing behaviour, the individual evaluates their competency to mimic the behaviour and makes assumptions about what is acceptable behaviour when provoked. Therefore, they develop an understanding about the observed behaviour which also allows for the behaviour to become generalised over a range of contexts. Similarly, Script Theory (Huesmann 1986, 1998), argues that scripts are learned based on observations or direct experience. Scripts such as aggressive scripts can be learned by children based on observations of violence portrayed in the mass media or based on people consider to be models of behaviour. These scripts provide guidance on how to behave in certain situations and what roles to assume in those situations. Once learned they are stored in semantic memory with causal links, goals and action plans. These can be retrieved and consulted on, to decide the role to assume the associated behaviour and outcomes to that script in that given scenario. In the context of conflict, when individuals increasingly consult these scripts and act aggressively to deal with conflict, the association to the script becomes stronger. As such the aggressive scripts become more acceptable and easier to access, therefore becoming generalised to more situations. However according the Social Information Processing (SIP) theory (Dodge 1980) some individuals have a 'hostile attribution bias,' where they tend to interpret ambiguous behaviour as having hostile motivations. This 'hostile attribution bias' activates the aggressive script, increasing the chance of selecting aggression as the reaction. Tedeschi and Felson (1994) place importance on social influence to explain aggression with the Social Interaction Theory, where motivation for aggression is based on higher level goals. Through coercion and intimidation, the victim's behaviour is changed for the individual's benefit, whether to gain something valuable, seek retribution for perceived wrong doings to gain a desired social identity. Again, there are implications here for modelling of bullying behaviour, the normalisation of aggression, and the impact this may have on developing young minds.

Following on from the Social Learning Theory (Bandura 1978, 1997), the General Aggression Model (Anderson and Bushman 2002; Bandura 1997) integrates many of these existing theories to develop a biosocial cognitive model to explain aggression. When an individual responds to overt aggression, the response is the result of a chain of events which is dictated by their characteristics. In its basic form, a reaction to a scenario is based on inputs (personal and situational factors), that influence routes in the individual (the internal states of affect, cognition, and arousal), and result in an outcome that is based on appraisal and decision making that is either thoughtful action or impulsive action (aggression). The personal factors may predispose some individuals to aggress. These factors include whether the individual is male or female, their traits, values, long-term goals and scripts, along with their attitudes and beliefs about violence. Feelings of frustration, drug use, incentives, provocation from others, exposure to aggressive cues along with anything that incites pain and discomfort in the individual, are all situational factors which can contribute to aggressive behaviour. It is the influence of these factors on an individual's internal state (arousal, affect and cognition), that produces overt aggression. Depending on the outcome of the appraisal of the situation, this can dictate whether an individual can control their anger or be impulsive and aggress. The actions in this process provide feedback to the individual for the current context but can also influence the development of the individual's personality. This process allows the individual to learn knowledge structures (scripts and schemas) that can influence behaviour. Repeatedly viewing violence in the media, along with other factors such as poor parenting, can result in aggressive personalities in adulthood (Huesmann and Miller 1994; Patterson et al. 1992) where aggressive-related knowledge structures have been developed, automatized and reinforced. These theories build a foundation to understand not only aggressive but also traditional and cyber bullying behaviour, creating an argument for both personal and environmental factors as predictors for involvement in such behaviour. Considering especially the importance placed on the environment in the development of the individual, involvement by both the school and the home are integral in dealing with prevention and intervention in aggressive and bullying (traditional and cyber) behaviour and the lasting impacts into adulthood.

In a recent publication examining aggressive behaviour in a cyberbullying context (and a broader context such as genocide), Minton claims one can "... predict that if (i) we do not share physical proximity with another person; and/or (ii) we socially distance ourselves from another person, such that we have no feelings of empathy with that other person, then we will be able to disregard our own agency in terms of our subsequent responsibility for our negative behaviour towards that other person" (Minton 2016, p. 110). Minton (2016), building on ideas proposed by thinkers like Lorenz and Milgram, highlights the importance of 'distance' from another person when carrying out aggressive behaviours, that is, we may be more inclined to carry out aggressive actions against others when we are distant from them. Distance can be physical, social, or moral. For instance, unlike other species, our weaponry is technologically advanced and so, we are not always required to get up close to our enemies in order to do them harm. Therefore, we can maintain physical distance from others when carrying out an attack (e.g., using a gun or firing a missile) and this reduces the threat of harm to ourselves and so inhibition of aggressive behaviour is somewhat mitigated. But Minton argues that it not just physical distance or proximity that can influence our behaviour, but also social distance. Social distance refers to seeing others as different or unequal, and in some cases reducing others to sub-human status. So as physical and social distance grow, our inhibitions and sense of responsibility may diminish. This makes the consequences of our aggressive behaviour more tolerable for ourselves. Minton's (2016) argument has clear implications for cyber-based aggression as the cyber world allows us to preserve physical distance from other users via technology, and furthermore, may create conditions in which we can portray/regard others as socially distant from ourselves. Indeed, Minton (2016) highlights the evidence that aggressive behaviour in a cyber context is correlated with moral disengagement and moral justification. Also considering the work of Latané and Darley, Minton suggests that another important influential factor is the bystander effect and the role of de-individuation in reducing one's sense of personal responsibility. With this in mind, it is conceivable that in the context of the world wide web one could more easily become part of a 'mob'. He ultimately argues that the role of physical proximity and social distance in relation to cyberbullying requires further investigation and that it would be beneficial to place greater emphasis on qualitative data when attempting to understand young people's involvement and experience in cyberbullying. One important building block for better understanding cyberbullying involvement is an appropriate definition of the phenomenon.

3 Cyberbullying: Definition, Conceptual, and Operational Issues

When addressing bullying in the online setting, 'cyberbullying' is a term that refers to a range of similar concepts such as internet harassment, online aggression, online bullying and electronic aggression (Dooley et al. 2009; Kowalski et al. 2008; Smith 2009; Tokunaga 2010). It is through the lens of traditional bullying that cyberbullying is understood, but with a unique venue (Dooley et al. 2009) and through electronic means (Sticca and Perren 2013). However, in doing so, it encounters similar problems to traditional bullying, as across languages and countries, the term cyber infers different meanings. For example, in Spain 'ciber' refers to computer networks (RAE 2018), where as in Germany 'cyber' refers to an online environment that is an extension of reality (Nocentini et al. 2010). This results in researchers developing several definitions for cyberbullying, and as such no uniform definition in the literature exists. For instance, Ybarra and Mitchell (2004) view it as an online, intentional and overt act of aggression towards another, whereas others define it as using the internet or other digital methods to insult or threaten others (Juvonen and Gross 2008). Smith et al. (2008) applied the features of Olweus's definition of traditional bullying to define cyberbullying, those of an intentional aggressive act to cause harm, power imbalance between the bully and victim, and repeated vicimisation (Grigg 2010). However, for cyberbullying it occurs with the use of technological devices (Dooley et al. 2009; Smith et al. 2008; Slonje and Smith 2008). Specifically, cyberbullying is "... an aggressive, intentional act carried out by a group or individual, using electronic forms of contact, repeatedly and over time against a victim who cannot easily defend him or herself" (Smith et al. 2008, p. 376).

Most of these features across both forms of bullying can be easily identified, however the feature of imbalance of power that is seen in traditional is somewhat different in the cyber setting. Imbalance of power can be viewed from the victim's perspective as being powerless in a given situation (Dooley et al. 2009). Moreover, powerlessness can be due to knowing the perpetrator in real life when their characteristics carry a threat to the victim (Slonje and Smith 2008). Furthermore, the aggressor can be perceived as a digital expert against whom the victim cannot defend him/herself (Vandebosch and Van Cleemput 2008). Similarly, the repeated nature of cyberbullying can have a unique presentation. For instance, a *single* act of posting/sending malicious content can lead to repeated victimization when the content is further disseminated by others, adding to the feeling of an imbalance of power for the victim (Menesini and Nocentini 2009). In addition, victims and bullies can re-read, re-view and re-experience an event (Law et al. 2012), also making it repeated in nature. Sometimes, there is no escape for the victim of cyberbullying as it can occur at any time (Walther 2007), and since it is through any electronic means it can occur anywhere, even in the privacy of the victim's home (Slonje and Smith 2008) allowing for no respite from the victimisation.

Although similar in terms of key features, traditional and cyberbullying do deviate in some respects. Due to the nature of online communication, the potential number of witnesses is larger with cyberbullying (Kowalski et al. 2008), the perpetrator has greater anonymity, there is less feedback between those involved in the behaviour, with fewer time and space limits (Slonje and Smith 2008), and reduced supervision (Patchin and Hinduja 2006). This can create a greater level of disinhibition and deindividualisation (Agatston et al. 2012; Davis and Nixon 2012; Patchin and Hinduja 2011; von Marées and Petermann 2012) as the perpetrator cannot see the consequences of their actions on the victim (Smith 2012). Without the face-to-face interaction seen in real life, cyberbullying can allow for emotional detachment and any empathy that would have otherwise been evoked in real life (Cassidy et al. 2013).

However, it must be noted that not all researchers view cyberbullying as a separate from traditional bullying. Olweus (2012) argues that it should only be understood in the context of traditional bullying and it is simply an extension of this behaviour to the cyber setting. He also argues that there is not an ever-increasing number of new victims and bullies, rather it can be the same individuals involved in traditional bullying, with some new involvement. Considering this, Olweus (2012) advises that school policies should centre on traditional bullying but also adapt the policies to have system-level strategies to deal with cyberbullying behaviour.

This overlapping nature of face-to-face bullying and cyberbullying has also been discussed by Patchin and Hinduja (Walther 2007), who indicated that the behaviour is 'moving beyond the schoolyard' and that individuals were victims of both online and offline bullying. This was echoed by Ybarra et al. (2007), with 36% of children experiencing both forms of the behaviour at the same time, and by Juvonen and Gross (RAE 2018), where 85% of cyber victims also experienced traditional school bullying. This has been further evidenced in the literature with correlations between the two forms of bullying (RAE 2018; Smith et al. 2008; Slonje and Smith 2008; Didden et al. 2009; Katzer et al. 2009). In terms of involvement, the bully both online and offline can be the same individual(s) or different (Ybarra et al. 2007). When the perpetrator is the same individual in cyber and traditional bullying, they are maximising the potential harm to the victim by employing online and offline methods (Tokunaga 2010). This overlapping nature of cyber and traditional bullying, and associated definitional issues, have implications for measurement and analysis, as the measurement tools should be able to account for both forms of the behaviour separately to report accurately on incidence rates.

However, an argument has emerged in relation to the concept of cyberbullying, in that it may not adequately capture all of the behaviours associated with it. Grigg (2010) proposes that the term cyber aggression is a more inclusive term for the sort of aggressive behaviours occurring in the online setting. She defines cyber aggression as "... intentional harm delivered by the use of electronic means to a person or a group of people irrespective of their age, who perceive(s) such acts as offensive, derogatory, harmful or unwanted" (p. 152). This definition accounts for such behaviours as flaming, stalking, trolling and other aggressive behaviours that employ electronic devices, or the Internet. The recognition of peer-directed cyber aggression, as opposed to a perhaps overly narrow and restrictive concept of cyberbullying, has also been advocated in a review by Corcoran, Mc Guckin and Prentice (Corcoran et al. 2015).

4 Cyber Aggression

Conceptually, 'aggression', of which 'cyber aggression' is a subset, involves the intention of causing harm to a targeted individual, as opposed to accidental or unintentional harm (Bushman and Anderson 2001; Geen 2001). Several forms exist, and there is variation in terms of motivation and provocation, including hostile, proactive, direct and indirect aggression. The following section will explore some forms of aggression 'online', which will subsequently be referred to as forms of cyber aggression. These forms will be more aligned with hostile aggression, though perpetrators may consider some of these as proactive (e.g., Political flaming). These behaviours can occur in both direct and indirect forms, for example online harassment may involve direct, continued, victimising of another individual through various mediums, whereas, exclusion may involve indirect aggression through ostracising an individual from a chatroom. Cyber aggression, defined earlier, involves intentional harm to a group or groups of individuals through electronic means. In terms of specific behaviours which are underpinned by such intentional harm, these include well-known acts such as bullying, stalking, and trolling, and employ tools for online engagement, such as smartphones, and personal laptops. The following sections provide an overview of some of these subcategories, including cyberbullying, cyberstalking, and cybertrolling.

4.1 Cyberbullying

Cyberbullying has received much focus in research over the last number of years, due in no small part to a number of well-known and tragic cases of suicide as a result of online victimisation. Therefore, understanding and educating people on cyberbullying, and developing effective interventions, has become a priority in many countries. Cyberbullying is considered by some researchers to be an extension of traditional bullying and has adopted a number of definitional characteristics from its more extensively researched cousin. These include factors such as an imbalance of power between the bully and the victim, and repeated, intentional victimisation of an individual or individuals. One issue which has emerged from considering cyberbullying within the general framework of traditional bullying, is the confusion over traditional features of the definition, such as repeated instances of victimisation. An important feature of cyberbullying concerns the repeated, sometimes viral nature of sharing potentially harmful material related to a victim. On many occasions, the sharers of this material are not explicitly connected to the original poster of the material, nor to the victim. This creates an issue with determining whether such targeted behaviour is an act of bullying, as the origin of the harmful material may only have been posted once, but through sharing, the victim is repeatedly abused. Examples such as this creates a difficulty with operationally defining cyberbullying in the same way as traditional bullying, and this has also been considered in previous research (e.g., Nocentini et al. 2010; Vandebosch and Van Cleemput 2008; Menesini et al. 2013). More recently, researchers such as Corcoran et al. (2015) have considered the fit of cyberbullying within the general framework of cyber aggression. General bullying behaviour, including cyberbullying, has highlighted the role of the bystander, something that Hyland et al. (2016) argue is an important factor in the context of cyber aggression. Another important consideration is the origin of the target individual or individuals, rather than the bully or bullies, something which has been stressed in recent literature (e.g., Langos 2012; Pyżalski 2012). Specifically, was the victim a member of the close peer group, or an individual not known to the bully personally such as a celebrity or an anonymous victim.

To date, a number of key associated correlates of involvement in cyberbullying have been identified, for both bullies and victims. These include poor school performance in victims (Patchin and Hinduja 2006), suicidal ideation in bullies and victims (Schenk and Fremouw 2012; Hinduja and Patchin 2010) and depression in bullies (Kokkinos et al. 2014). In terms of predictors of cyberbullying behaviour, cyberbullies tend to exhibit high rates of stress, depression, anxiety, and social difficulty compared with individuals not involved in such behaviour (Campbell et al. 2013). Cyberbullies also tend to demonstrate lower psychosocial adjustment, (Sourander et al. 2010), and increased difficulty at school (Wei and Chen 2009).

Willard (2007) operationalised cyberbullying in terms of seven behavioural categories, including, harassment, denigration, masquerading, outing/trickery, exclusion, flaming and cyberstalking. Harassment involves the repeated sending of messages to a particular individual or group. Specifically, Langos (2012) asserts that such behaviour can occur in various forms, such as SMS messaging, emails, websites, chatrooms, and instant messaging. Denigration relates to the posting of harmful or untrue statements about other people, whereas masquerading involves pretending to be the target individual in order to send offensive or provocative messages, which appear to come from that individual, and which are designed to bring negative attention to a victim or put them in the line of danger (Willard 2007). Outing/Trickery has some overlap with masquerading, but typically involves the sharing of personal information which has been shared with that person in confidence, again motivated to bring negative attention to the victim in question. Online exclusion is similar to traditional forms of exclusion, in that it involves denying an individual access to, or involvement in, a particular event. Traditionally, this may involve excluding individuals from social events such as games or meetups, whereas online it involves ostracizing an individual or a group from online spaces such as chatrooms or social networks (e.g., WhatsApp, Viber, etc.).

Finally, 'flaming', can be understood as hostile verbal behaviour, including insulting and ridiculing behaviour, towards an individual or group, within the context of computer-mediated communication (Hutchens et al. 2015). Many cases of flaming emerge as a result of a provocative post or comment on social media, which is sometimes referred to as 'flame bait' (Moor et al. 2010), and is designed to draw an individual into responding. Flaming can be observed on a number of online platforms such as YouTube (See Lingam and Aripin 2016), and Facebook (See Halpern and Gibbs 2013, for a comparison of both YouTube and Facebook), and across a number of specific contexts, such as Politics (Halpern and Gibbs 2013) and Gaming (Elliott 2012).

4.2 Cyberstalking

According to Foellmi et al. (2012) a consensus concerning a definition of cyberstalking has not been reached, but does seem to involve the wilful, malicious, repeated following or harassing of another person. Intent is another important component of stalking, and the behaviour should be interpreted as a credible threat to another individual, both of which are also important components of traditional and cyber forms of bullying. Moreover, debate has continued over whether cyberstalking is a new phenomenon or an extension of traditional stalking (Foellmi et al. 2012). This mirrors the debate researchers such as Corcoran et al. (2015) have contributed to regarding cyberbullying as an extension of traditional bullying. There is some variation in terms of defining cyberstalking, apart from considering it a subcategory of cyberbullying (e.g., Willard 2007). Some researchers have offered collective definitions of cyberstalking and cyberbullying (e.g., Short et al. 2016), as causing distress to someone, through electronic forms. Other researchers, while considering both phenomena related, have distinguished between cyberbullying and cyberstalking (e.g., Chandrashekhar et al. 2016). Chandrashekhar et al. (2016) assert that when cyberbullying includes secretly observing, following and targeting a specific person's online activities, it can be considered cyberstalking. Cyberstalking is not specific to particular populations but has been explored to a great extent in adolescents and young adults, individuals who have been exposed to the cyber age for much if not all of their lives. However, Chandrashekhar et al. (2016) comment that a number of other populations are at risk of such victimisation, including the disabled, the elderly, people who have been through recent breakups, and employers.

In terms of prevalence, Cavezza and McEwan (2014) reviewed rates in the student population across a number of studies and found variation between 1% and 41%. A large-scale analysis of incidence rates among 6379 individuals across a German social network (Dreßing et al. 2014), revealed that over 40% of individuals were harassed online at least once in their lifetime. However, when two other definitional factors were taken into account (continued harassment for more than 2 weeks and whether the incident provoked fear) this dropped to 6.3%. Moreover, it was reported that nearly 70% of cyberstalkers were male, almost 35% of incidences involved cyberstalking by an ex-partner, and females were significantly more likely to be victims of cyberstalking compared to males. Other studies have reported contrasting evidence on sex differences, such as Berry and Bainbridge (2017), who found no significant differences between males and females with regard to being victims of cyberstalking. With regard to predictors of cyberstalking, Ménard and Pincus (2012) found that childhood sexual maltreatment predicted both stalking and cyberstalking behaviour in both males and females. Interestingly, narcissistic vulnerability and interaction with sexual maltreatment predicted cyberstalking among males, with insecure attachment and alcohol expectancies predicting cyberstalking in females. Marcum et al. (2014) report that, among minors, lower levels of self-control predicted greater engagement in cyberstalking behaviour. Social involvement with deviant peers is also associated with such behaviour.

4.3 Trolling

A well-known piece of advice Internet users commonly come across when frequenting Twitter or YouTube posts is 'Don't feed the trolls'. Online 'Trolls' are Internet users who, according to Buckels et al. "...behave in a deceptive, or destructive manner in a social setting on the Internet with no apparent instrumental purpose" (Buckels et al. 2014, p. 97). According to Herring et al. (2002), specific trolling messages or posts can be categorised into one of three categories: (i) messages which seem to come from a place of sincerity; (ii) messages which are designed to predict outwardly negative reactions; and (iii) messages which are designed to waste time by provoking a futile argument. Buckels et al. (2014) amusingly compare trolls to well-known figures such as 'The Joker' in the Batman comics, who wreaks havoc over Gotham City, presumably just for amusement or to simply create anarchy.

Similar to other more contemporary online behaviours, the breadth of literature understanding the key characteristics of trolling is limited (Zezulka and Seigfried-Spellar 2016).

However, there are several key distinctions between cyberbullying and cybertrolling, which also relate to the considerations mentioned earlier with regard to the position of cyberbullying as directly targeting members of a peer group or individuals not directly peer to the bully. One such distinction, as Zezulka and Seigfried-Spellar (2016) note, is that trolling typically involves intentional harassment of individuals not knowing their victims, unlike cyberbullying, which does in a large part focus on specific known members of a peer group. This is one reason why trolling tends to occur in popular social media platforms such as Twitter, where opportunities for involvement in a wide variety of conversations and topics with unknown people and celebrities is available. Recent research by Buckels et al. (2014) explored specific personality correlates of online trolls, where they found that traits such as sadism, psychopathy, and Machiavellianism were positively correlated with self-reported enjoyment of trolling. Traits such as narcissism, which did positively correlate with enjoyment in debating topics of personal interest, was not correlated with trolling. More recently, Lopes and Yu (2017) extended the findings of Buckels et al. (2014) with regard to Psychopathy, which was found to significantly predict trolling. Also, and in line with previous research, narcissism did not predict trolling.

4.4 Cyberbullying, an Issue of Clarity

Much research has explored incidence, predictors and correlates of cyberbullying. However, and as evident from earlier coverage, there is much variation with regard to the terms used to describe cyberbullying and in particular cyberbullying behaviours. Aboujaoude et al. (2015) provide an overview of issues associated with terminology, where terms such as cyber harassment and cyberstalking, have been used interchangeably with cyberbullying (See Aboujaoude et al. 2015, for an illustration of other terms). This is in contrast to the classification of cyberbullying by Willard (2007) where harassment and stalking are specific sub-categories of cyberbullying, rather than interchangeable terms. Therefore, while cyberbullying is the most commonly used general term for this phenomenon, research is not referring to the behaviour with a common term, which may cause confusion when developing evidence-based interventions to tackle such problems in schools, workplaces, and other relevant contexts. Consideration of cyberbullying, cyberstalking and cyber trolling, as categories of cyber aggression, and subsequent alignment with hostile, direct, and indirect forms of online aggression, may offer an opportunity to provide clarity to this classification of behaviour.

5 Implications for Casting the Net Wide in Terms of Prevention/Intervention Efforts

When considering how best to prevent cyber aggression/cyberbullying from happening or to intervene once it has taken place, it is apparent that there are many approaches involving legal-, policy-, programme-, and education-based efforts. When attempting to evaluate the effectiveness of such efforts, the scientific research community have quite clear guidelines for assessment of quality. Mc Guckin and Corcoran (2016) set out an extensive list of criteria for evaluation of programmes which aim to counter cyberbullying. The core principles outlined include: the need for theory-driven and evidence-based intervention; advanced research methods; the importance of targeted intervention with clear parameters (e.g., behaviours to be tackled); outcome behaviours that are measurable; and sensitivity to the developmental stage of participants.

For researchers and practitioners attempting to implement prevention/intervention policy and programmes, there is a well-worn path in terms of countering school bullying. By the time the Internet became widely accessible in the Western world, there was already a wealth of knowledge regarding successful intervention in the form of school-based programmes to counter traditional bullying. Perhaps the most well supported (certainly a widely accepted) component of school-based programmes has been the Whole School Approach (Rigby et al. 2004). According to Smith, Schneider, Smith and Ananiadou, "The whole-school approach is predicated on the assumption that bullying is a systemic problem, and, by implication, an intervention must be directed at the entire school context rather than just at individual bullies and victims" (Smith et al. 2004p. 548). This approach recognises that there is an important social context to bullying and aggression that goes beyond those directly involved in the behaviour. The same context can be recognised in the cyber world with involvement of other Internet users in roles such as witness, voyeur, commentator, supporter of the victim, involvement in the mob etc. The Olweus Bullying Prevention Program (OBPP: Olweus 1993) targets bullying at school-level, classroom-level, individual-level, and community-level and is the first Whole School Approach model to be implemented and assessed on a large scale (Smith et al. 2004). There are also different perspectives and approaches which underpin different anti-bullying programmes. Some programmes focus on aspects of interpersonal contact such as enhancing social skills (e.g., S.S.GRIN: DeRosier and Marcus 2005), whilst others target bystander behaviour (e.g., Kiva: Kärnä et al. 2011), seeking to empower the witnesses of aggression. In fact, there are many approaches which have been implemented and evaluated, giving us a good body of evidence from which we can make informed choices about anti-bullying approaches. So, why not just select programmes such as these to address cyberbullying/cyber aggression in schools? This seems like the easy option when we recognise the common defining characteristics and overlap of involvement in cyberbullying and traditional bullying. However, cyberbullying, as stated earlier in this chapter, presents new and somewhat unique challenges to children and adolescents. Furthermore, as discussed, the concept of cyberbullying and its core characteristics may require further consideration. This means that we cannot take shortcuts and we have a duty to thoroughly consider how we can best counter cyber aggression among children and young people.

The good news is that there are ongoing efforts to develop novel approaches to countering cyberbullying and cyber aggression. Some of these attempts focus on training parents and practitioners to safely navigate the Internet and to prevent and address cyberbullying/cyber aggression when it occurs (e.g., the EU funded initiatives such as the CyberTraining programme [Project No.142237-LLP-1-2008-1-DE-LEONARDO-LMP] http://cybertraining-project.org] and the Cyber-Training-4-Parents programme [http://cybertraining4parents.org] Project number: 510162-LLP-1-2010-1-DE-GRUNDTVIG-GMP]). One approach to countering cyberbullying and cyber aggression could involve the gamification of interventions. The Friendly ATTAC programme (DeSmet et al. 2017) was developed for implementation with adolescents for the purpose of increasing positive bystander behaviour and reducing negative bystander behaviour in relation to cyberbullying. The programme design was based on behavioural prediction and change theory and evidence relating to bystander responses in cyberbullying situations. This was regarded as an important alternative to previous efforts based on knowledge from traditional bullying literature and previous neglect of behaviour change theory. The programme was developed in accordance with the Intervention Mapping Protocol (DeSmet et al. 2017) which sets guidelines for development of behaviour change programmes, including theory-based intervention and evaluation of programmes. In order to implement the programme, the researchers used a serious game intervention which is a type of organised play that is delivered via computer technology for the purposes of entertainment, instruction provision, training, or attitude change. This is an approach which has already been used in a number of anti-bullying programmes, such as the Kiva programme (Kärnä et al. 2011) which was developed as an addition to a Whole School Approach to bullying. The intervention was delivered via a game which allowed participants to navigate a cyberbullying problem (ugly person page) and was found ultimately to have "... significant small, positive effects on behavioural determinants and on quality of life, but not in significant effects on bystander behaviour or (cyber-)bullying vicitmization or perpetration" (DeSmet et al. 2017, p. 341). Behavioural determinants included variables such as self-efficacy and moral disengagement. Overall, the authors conclude that, although further development is required, the Friendly ATTAC game was successful in some respects such as enhancing positive bystander self-efficacy, prosocial skills, intention to respond positively as a bystander, and quality of life.

Menesini and colleagues (see 2016; Palladino et al. 2016) have implemented and evaluated a school-wide programme called 'Noncadiamointrappola!' ("*Let's not fall into the trap*") with Italian teenagers which aims to combat traditional bullying and cyberbullying and endorse positive engagement with technology. They suggest that it is sensible to develop anti-bullying programmes to also address cyberbullying as we have evidence of overlap between the two behaviours. However, they acknowledge that there are unique features of cyberspace which require specific considerations. The programme includes online support, encourages positive behaviours online, and uses peers as educators. The programme is evidence informed and includes the student voice in the design (an important factor highlighted by Välimäki et al. 2012). The authors have adapted the programme since an earlier implementation in 2009/2010 and found it to be more effective following adaptations to aspects such as the emphasis on bystander and victim roles, and peer-led activities delivered face-to-face. Menesini et al. (2016) reported a decrease in bullying, victimisation, and cyber victimisation in the experimental group compared to the control group. The experimental group also exhibited greater tendency towards more adaptive and less maladaptive coping responses. Similar to the work of DeSmet et al. (2017), they examined additional variables and found that variables like empathy and anti-bullying attitudes are important in predicting bystander responses. One important aspect of this study was that it provided support for peer-led intervention; an approach that has had mixed support with regard to effectiveness.

Gunther et al. (2016) emphasise the paucity of evidence-based interventions. They also recognise the reluctance of children and young people to report experiences of victimization to parents and practitioners, as well as tendency to seek help anonymously and via the Internet. On the basis of these tendencies and taking the framework of e-mental health initiatives (Internet-based interventions), they implemented a programme which attempts to reach young people online. Such an approach allows for reduced stigma and the possibility to seek help regardless of time or location. The appropriateness of such a programme for CBT treatment of anxiety is highlighted by Gunther et al. (2016). They recommend exploring the inclusion of cyberbullying content in a mental health intervention or in a programme for cyberbullied young people who also experience mental health difficulties. They also suggest that there is potential in blended care (combination of online and face to face delivery). These three intervention approaches do not point us in the direction of a "best" intervention approach. Rather they highlight the diversity of intervention approaches. What we know from application of theory and research is that we must be sensitive to uniqueness of human beings in terms of situational and personal factors. Therefore, a one size fits all approach simply will not do, and therefore the variety of approaches to prevention/intervention is to be welcomed. Although researchers and educators tend to focus on education-based and psycho-education-based interventions, there are also sometimes calls for a punitive response to cyber aggression. But how should we begin to police today's 'wild west'? The Internet is often characterised as a land of high opportunity, high risk, and high anonymity. These features make it more difficult to regulate, moderate, and police.

There is legislation specific to cyberbullying in some jurisdictions (e.g., see Seth's Law, 2011: http://e-lobbyist.com/gaits/text/354065 and Brodie's Law: http://www.justice.vic.gov.au/home/safer+communities/crime+prevention/bullying+-+brodies+law), and this raises questions as to whether the appropriate response to cyber aggression is education, criminalisation, or both. Szoka and Thierer (2009) argue that education is preferable to criminalisation. One reason they propose is that a cyberbullying law may lead to differing repercussions for traditional bullying and cyberbullying (e.g., counselling as a response to traditional bullying, and imprisonment as a response to cyberbullying). They recommend awareness raising and training as an alternative and highlight the possibility that prosecution of young people can lead to stigma in later life. Their argument for properly considering the consequences before implementing such legislation should also move us to consider the possible consequences of settling on an inadequate definition or concept. We must have a comprehensive understanding of cyberbullying as a behaviour if we are to consider legislating to deter cyberbullying. Levick and Moon (2010) also highlight the potential for black and white laws to be interpreted in a manner that was not anticipated. They use the example of young people who sext their peers being prosecuted under existing child pornography laws. Moreover, in some instances there is existing law which can serve to protect against cyberbullying.

In an Irish context (see the Education [Welfare] Act 2000), schools have a legal obligation to include bullying in their code of behaviour. Guidelines for schools in relation to countering bullying have been recently updated (Department of Education and Skills 2013) to include specific types of bullying, including cyberbullying, homophobic bullying, and race-based bullying. However, there are also other non-cyberbullying-specific Irish laws which are relevant to cyber aggression, such as legislation relating to misuse of the telephone (Post Office [Amendment] Act 1951) the violation of which can result in prosecution. Furthermore, the age of digital consent is currently under review in an Irish context. Setting the age of consent at 13 years would restrict websites from using the personal data of younger children.

In a consultation paper on the digital age of consent, the Department of Justice and Equality (2016) in Ireland stated that children of insufficient maturity and understanding can be more susceptible to online risks such as grooming and cyberbullying and they emphasise the need to safeguard children. Furthermore, they highlight the roles of parents/guardians in this context.

The Psychological Society of Ireland (PSI: Psychological Society of Ireland 2018) has contributed to the work of the Oireachtas Joint Committee on Children and Youth Affairs with regard to the matter of digital age of consent in Ireland. The PSI states that there is not sufficient evidence to conclude that there is a direct negative causal relationship between social media activity and young people's mental health and furthermore, that there is potential for benefits to be reaped from online communications. Moreover, the importance of not being overly reliant on anecdotal evidence is emphasised. Highlighting the complexity of human psychology, the PSI refer to the various determinants of how and why one experiences distress, including psychological, social, behavioural, and individual factors. Offering support for the digital age of consent to be set at 13 years, the PSI states that "Rather than blanket restriction and regulation of technology, guided and scaffolded exposure to technology is recommended if young people are to develop into experienced, skilled and safe users of technology." (Psychological Society of Ireland 2018, p.129). Again, it seems that an educational response has an important place in terms of safeguarding children and preparing them for responsible behaviour as digital citizens.

6 Conclusions

It is evident from the review of aggression theory that the causes of aggression are many and varied. This is important to consider when attempting to understand cyber-based aggression; that is, there is not necessarily one particular causal factor. However, Minton raises some important contextual aspects of cyberspace – primarily the physical remoteness that new information and communication technologies allow. The context of cyberspace has also had important implications when attempting to define cyberbullying. Indeed, given the unique features of the cyber context, the term cyber aggression may be an appropriate widening of the net with regard to conceptual parameters. However, as stated above, the cyber context is not an isolated sphere in the sense that there is overlap in experiences and social networks in the physical world and the Internet. However, there is an argument for recognising cyber aggression without the constraints of traditional bullying behaviours, given the unique context of cyberspace. This chapter highlights the variety of peer-directed aggressive behaviours under examination, including cyberbullying, cyberstalking, and cybertrolling; forms of behaviour which could be considered sub-types of cyber aggression. Furthermore, interventions which focus on cyberbullying specifically are varied with focus on education, counselling, prevention, and in some cases, prosecution or legislative or policy reform. Whilst all of these approaches have an important role to play in safeguarding children and adolescents (and adults), there is a thread running through the literature which leads back to education as a central component in tackling aggression online. Ultimately this chapter leads to the conclusion that we must approach cyber aggression with a broad perspective theoretically, conceptually, and in terms of prevention and intervention.

References

Aboujaoude, E., Savage, M. W., Starcevic, V., et al. (2015). Cyberbullying: Review of an old problem gone viral. *Journal of Adolescent Health, 57*(1), 10–18. https://doi.org/10.1016/j.jadohealth.2015.04.011.

Agatston, P., Kowalski, R., & Limber, S. (2012). Youth views on cyberbullying. In J. W. Patchin & S. Hinduja (Eds.), *Cyberbullying prevention and response: Expert perspectives* (pp. 57–71). New York: Routledge.

Anderson, C. A. (1997). Effects of violent movies and trait hostility on hostile feelings and aggressive thoughts. *Aggressive Behavior, 23*(3), 161–178. https://doi.org/10.1002/(SICI)1098-2337(1997)23:33.0.CO;2-P.

Anderson, C. A., & Bushman, B. J. (2002). Human aggression. *Psychology, 53*(1), 27–51. https://doi.org/10.1146/annurev.psych.53.100901.135231.

Archer, J. (1995). What can ethology offer the psychological study of human aggression? *Aggressive Behavior, 21*(4), 243–255. https://doi.org/10.1002/1098-2337(1995)21:4<243::AID-AB2480210402>3.0.CO;2-6.

Bandura, A. (1978). Social learning theory of aggression. *The Journal of Communication, 28*(3), 12–29.

Bandura, A. (1986). *Social foundations of thought and action: A social cognitive theory.* New Jersey: Prentice-Hall.
Bandura, A. (1997). *Self-efficacy: The exercise of control.* New York: Freeman.
Berkowitz, L. (1989). Frustration-aggression hypothesis: Examination and reformulation. *Psychological Bulletin, 106*(1), 59–73. https://doi.org/10.1037/0033-2909.106.1.59.
Berkowitz, L. (1990). On the formation and regulation of anger and aggression: A cognitive-neoassociationistic analysis. *The American Psychologist, 45*(4), 494–503. https://doi.org/10.1037/0003-066X.45.4.494.
Berkowitz, L. (1993). Pain and aggression: Some findings and implications. *Motivation and Emotion, 17*(3), 277–293. https://doi.org/10.1007/BF00992223.
Berry, M. J., & Bainbridge, S. L. (2017). Manchester's cyberstalked 18–30s: Factors affecting cyberstalking. *Advances in Social Sciences Research Journal, 4*(18), 73–85. https://doi.org/10.14738/assrj.418.3680.
Bleidorn, W., Kandler, C., Riemann, R., et al. (2009). Patterns and sources of adult personality development: Growth curve analyses of the NEO PI-R scales in a longitudinal twin study. *Journal of Personality and Social Psychology, 97*(1), 142–155. https://doi.org/10.1037/a0015434.
Buckels, E. E., Trapnell, P. D., & Paulhus, D. L. (2014). Trolls just want to have fun. *Personality and Individual Differences, 67*, 97–102. https://doi.org/10.1016/j.paid.2014.01.016.
Bushman, B. J., & Anderson, C. A. (2001). Is it time to pull the plug on hostile versus instrumental aggression dichotomy? *Psychological Review, 108*(1), 273–279.
Buss, D. M., & Shackelford, T. K. (1997). Human aggression in evolutionary psychological perspective. *Clinical Psychology Review, 17*(6), 605–619. https://doi.org/10.1016/S0272-7358(97)00037-8.
Campbell, M. A., Slee, P. T., Spears, B., et al. (2013). Do cyberbullies suffer too? Cyberbullies' perceptions of the harm they cause to others and to their own mental health. *School Psychology International, 34*(6), 613–629. https://doi.org/10.1177/0143034313479698.
Cassidy, W., Faucher, C., & Jackson, M. (2013). Cyberbullying among youth: A comprehensive review of current international research and its implications and application to policy and practice. *School Psychology International, 34*(6), 575–612. https://doi.org/10.1177/0143034313479697.
Cavezza, C., & McEwan, T. E. (2014). Cyberstalking versus off-line stalking in a forensic sample. *Psychology, Crime & Law, 20*(10), 955–970. https://doi.org/10.1080/1068316X.2014.893334.
Chandrashekhar, A. M., Muktha, G. S., & Anjana, D. K. (2016). Cyberstalking and cyberbullying: Effects and prevention measures. *Imperial Journal of Interdisciplinary Research, 2*(3), 95–102.
Collins, A. M., & Loftus, E. F. (1975). A spreading-activation theory of semantic processing. *Psychological Review, 82*(6), 407–428. https://doi.org/10.1037/0033-295X.82.6.407.
Corcoran, L., Mc Guckin, C. M., & Prentice, G. (2015). Cyberbullying or cyber aggression? A review of existing definitions of cyber-based peer-to-peer aggression. *Societies, 5*(2), 245–255. https://doi.org/10.3390/soc5020245.
Daly, M., & Wilson, M. (1994). Evolutionary psychology of male violence. In J. Archer (Ed.), *Male violence* (pp. 253–288). London: Routledge.
Darwin, C. (1859). *On the origin of species.* London: Murray.
Davis, S., & Nixon, C. (2012). Empowering bystanders. In J. W. Patchin & S. Hinduja (Eds.), *Cyberbullying prevention and response: Expert perspectives* (pp. 93–109). New York: Routledge.
Department of Education and Skills. (2013). Anti-bullying procedures for primary and post-primary schools. Retrieved from https://www.education.ie/en/Publications/Policy-Reports/Anti-Bullying-Procedures-for-Primary-and-Post-Primary-Schools.pdf.
Department of Justice and Equality. (2016). Data protection safeguards for children ('digital age of consent'). Consultation paper. Retrieved from http://www.justice.ie/en/JELR/Consultation_paper_Digital_Age_of_Consent.pdf/Files/Consultation_paper_Digital_Age_osf_Consent.pdf.

DeRosier, M. E., & Marcus, S. R. (2005). Building friendships and combating bullying: Effectiveness of S.S.GRIN at one-year follow-up. *Journal of Clinical Child Adolescent, 34*(1), 140–150. https://doi.org/10.1207/s15374424jccp3401_13.

DeSmet, A., Bastiaensens, S., Van Cleemput, K., et al. (2017). The efficacy of the friendly attac serious digital game to promote prosocial bystander behavior in cyberbullying among young adolescents: A cluster-randomized controlled trial. *Computers in Human Behavior, 78*, 336–347. https://doi.org/10.1016/j.chb.2017.10.011.

Didden, R., Scholte, R. H., Korzilius, H., et al. (2009). Cyberbullying among students with intellectual and developmental disability in special education settings. *Developmental Neurorehabilitation, 12*(3), 146–151. https://doi.org/10.1080/17518420902971356.

Dodge, K. A. (1980). Social cognition and children's aggressive behavior. *Child Development, 51*, 620–635.

Dollard, J., Doob, L. W., Miller, N. E., et al. (1939). *Frustration and aggression*. New Haven: Yale University Press.

Dooley, J. J., Pyżalski, J., & Cross, D. (2009). Cyberbullying versus face-to-face bullying: A theoretical and conceptual review. *The Journal of Psychology, 217*(4), 182–188. https://doi.org/10.1027/0044-3409.217.4.182.

Dreßing, H., Bailer, J., Anders, A., et al. (2014). Cyberstalking in a large sample of social network users: Prevalence, characteristics, and impact upon victims. *Cyberpsychology, Behavior and Social Networking, 17*(2), 61–67. https://doi.org/10.1089/cyber.2012.0231.

Elliott, T. P. (2012) Flaming and gaming– Computer-mediated-communication and toxic disinhibition. Dissertation, University of Twente.

Feshbach, S. (1984). The catharsis hypothesis, aggressive drive, and the reduction of aggression. *Aggressive Behavior, 10*(2), 91–101. https://doi.org/10.1002/1098-2337(1984)10:2<91.

Foellmi, M., Cahall, J., & Rosenfeld, B. (2012). Stalking: What we know and what we need to know. In B. Winder & P. Banyard (Eds.), *A psychologist's casebook of crime: From arson to voyerism* (pp. 209–224). London: Palgrave Macmillan.

Freud, S. (1920). *Beyond the pleasure principle*. New York: Bantam Books.

Geen, R. G. (2001). *Human aggression* (2nd ed.). Oxford: Taylor & Francis.

Grigg, D. W. (2010). Cyber-aggression: Definition and concept of cyberbullying. *Australian Journal of Guidance and Counselling, 20*(2), 143–156. https://doi.org/10.1375/ajgc.20.2.143.

Gunther, N., Dehue, F., & Thewissen, V. (2016). Cyberbullying and mental health: Internet-based interventions for children and young people. In C. Mc Guckin & L. Corcoran (Eds.), *Bullying and cyberbullying: prevalence* (pp. 189–200). New Jersey: Psychological Impacts and Intervention Strategies. Nova Publishers.

Halpern, D., & Gibbs, J. (2013). Social media as a catalyst for online deliberation? Exploring the affordances of facebook and youtube for political expression. *Computers in Human Behavior, 29*(3), 1159–1168. https://doi.org/10.1016/j.chb.2012.10.008.

Herring, S., Job-Sluder, K., Scheckler, R., et al. (2002). Searching for safety online: Managing "trolling" in a feminist forum. *The Information Society, 18*(5), 371–384. https://doi.org/10.1080/01972240290108186.

Hinduja, S., & Patchin, J. W. (2010). Bullying, cyberbullying, and suicide. *Archives of Suicide Research, 14*(3), 206–221. https://doi.org/10.1080/13811118.2010.494133.

Hopwood, C. J., Donnellan, M. B., Blonigen, D. M., et al. (2011). Genetic and environmental influences on personality trait stability and growth during the transition to adulthood: A three-wave longitudinal study. *Journal of Personality and Social Psychology, 100*(3), 545–556. https://doi.org/10.1037/a0022409.

Huesmann, L. R. (1986). Psychological processes promoting the relation between exposure to media violence and aggressive behavior by the viewer. *Journal of Social Issues, 42*(3), 125–139. https://doi.org/10.1111/j.1540-4560.1986.tb00246.x.

Huesmann, L. R. (1998). The role of social information processing and cognitive schema in the acquisition and maintenance of habitual aggressive behavior. In R. Geen & E. Donnerstein (Eds.), *Human aggression: Theories, research and implications for policy* (pp. 73–109). New York: Academic.

Huesmann, L. R., & Miller, L. S. (1994). Long-term effects of repeated exposure to media violence in childhood. In L. R. Huesmann (Ed.), *Aggressive behavior* (pp. 153–186). New York: Springer Science and Business Media.

Hutchens, M. J., Cicchirillo, V. J., & Hmielowski, J. D. (2015). How could you think that?!?!: Understanding intentions to engage in political flaming. *New Media & Society, 17*(8), 1201–1219. https://doi.org/10.1177/1461444814522947.

Hyland, P. K., Hyland, J. M., & Lewis, C. A. (2016). Conceptual and definitional issues regarding Cyberbullying: A case for using the term cyber aggression? In C. Mc Guckin & L. Corcoran (Eds.), *Bullying and cyberbullying: prevalence, psychological impacts and intervention strategies* (pp. 29–49). New Jersey: Nova Publishers.

Johnson, W., McGue, M., & Krueger, R. F. (2005). Personality stability in late adulthood: A behavioral genetic analysis. *Journal of Personality, 73*(2), 523–552.

Juvonen, J., & Gross, E. F. (2008). Extending the school grounds? Bullying experiences in cyberspace. *Journal of School Health, 78*(9), 496–505. https://doi.org/10.1111/j.1746-1561.2008.00335.x.

Kärnä, A., Voeton, M., Little, T. D., et al. (2011). A large-scale evaluation of the KiVa antibullying program: Grades 4–6. *Child Development, 82*(1), 311–330. https://doi.org/10.1111/j.1467-8624.2010.01557.x.

Katzer, C., Fetchenhauer, D., & Belschak, F. (2009). Cyberbullying: Who are the victims? A comparison of victimization in internet chatrooms and victimization in school. *Journal of Media Psychology-German, 21*(1), 25–36. https://doi.org/10.1027/1864-1105.21.1.25.

Kokkinos, C. M., Antoniadou, N., & Markos, A. (2014). Cyber-bullying: An investigation of the psychological profile of university student participants. *Journal of Applied Developmental Psychology, 35*(3), 204–214. https://doi.org/10.1016/j.appdev.2014.04.001.

Kowalski, R. M., Limber, S., & Agatston, P. W. (2008). *Cyberbullying: Bullying in the digital age*. Malden: Blackwell Publishers.

Krahé, B. (2001). *The social psychology of aggression*. East Sussex: Psychology Press.

Langos, C. (2012). Cyberbullying: The challenge to define. *Cyberpsychology, Behavior and Social Networking, 15*(6), 285–289. https://doi.org/10.1089/cyber.2011.0588.

Law, D. M., Shapka, J. D., Hymel, S., et al. (2012). The changing face of bullying: An empirical comparison between traditional and internet bullying and victimization. *Computers in Human Behavior, 28*(1), 226–232. https://doi.org/10.1016/j.chb.2011.09.004.

Levick, M., & Moon, K. (2010). Prosecuting sexting as child pornography: A critique. *Valparaiso University Law Review, 44*(4), 1035–1054.

Lingam, R. A., & Aripin, N. (2016). "Nobody Cares, Lah!" The phenomenon of flaming on youtube in Malaysia. *Journal of Business Society Review Emergency Economics, 2*(1), 71–78. https://doi.org/10.26710/jbsee.v2i1.20.

Lopes, B., & Yu, H. (2017). Who do you troll and why: An investigation into the relationship between the dark triad personalities and online trolling behaviours towards popular and less popular Facebook profiles. *Computers in Human Behavior, 77*, 69–76. https://doi.org/10.1016/j.chb.2017.08.036.

Lorenz, L. (1974). *Civilised man's eight deadly sins. Harcourt, Brace*. New York: Jovanovich.

Marcum, C. D., Higgins, G. E., & Ricketts, M. L. (2014). Juveniles and cyber stalking in the United States: An analysis of theoretical predictors of patterns of online perpetration. *International Journal of Cyber Criminology, 8*(1), 47–56.

Mc Guckin, C., & Corcoran, L. (2016). Intervention and prevention programmes on cyberbullying: A review. In R. Navarro, Y. Santiago, & E. Larrañaga (Eds.), *Cyberbullying across the globe: Gender, family and mental health* (pp. 221–238). London: Springer International Publishing. https://doi.org/10.1007/978-3-319-25552-1.

Ménard, K. S., & Pincus, A. L. (2012). Predicting overt and cyber stalking perpetration by male and female college students. *Journal of Interpersonal Violence, 27*(11), 2183–2207. https://doi.org/10.1177/0886260511432144.

Menesini, E., & Nocentini, A. (2009). Cyberbullying definition and measurement: Some critical considerations. *The Journal of Psychology, 217*(4), 230–232. https://doi.org/10.1027/0044-3409.217.4.230.

Menesini, E., Nocentini, A., Palladino, B. E., et al. (2013). Definitions of cyberbullying. In P. K. Smith & G. Steffgan (Eds.), *Cyberbullying through the new media: Findings from an international network* (pp. 23–36). Oxfordshire: Psychology Press.

Menesini, E., Palladino, B. E., & Nocentini, A. (2016). Noncadiamointrappola! Online and school based program to prevent cyberbullying among adolescents. In T. Völlink, F. Dehue, & C. Mc Guckin (Eds.), *Cyberbullying and youth: From theory to interventions. Current issues in social psychology series* (pp. 156–175). London: Psychology Press/Taylor & Francis.

Minton, S. J. (2016). Physical proximity, social distance, and cyberbullying research. In C. Mc Guckin & L. Corcoran (Eds.), *Bullying and cyberbullying: Prevalence, psychological impacts and intervention strategies* (pp. 105–118). New York: Nova Publishers.

Moor, P. J., Heuvelman, A., & Verleur, R. (2010). Flaming on youtube. *Computers in Human Behavior, 26*(6), 1536–1546. https://doi.org/10.1016/j.chb.2010.05.023.

Nocentini, A., Calmaestra, J., Schultze-Krumbholz, A., et al. (2010). Cyberbullying: Labels, behaviours and definition in three European countries. *Australian Journal of Guidance and Counselling, 20*(2), 129–142. https://doi.org/10.1375/ajgc.20.2.129.

Olweus, D. (1993). *Bullying at school: What we know and what we can do*. Oxford: Blackwell.

Olweus, D. (2012). Cyberbullying: An overrated phenomenon? *The European Journal of Developmental Psychology, 9*(5), 520–538. https://doi.org/10.1080/17405629.2012.682358.

Palladino, B. E., Nocentini, A., & Menesini, E. (2016). Evidence-based intervention against bullying and cyberbullying: Evaluation of the NoTrap! program in two independent trials. *Aggressive Behavior, 42*(2), 194–206. https://doi.org/10.1002/ab.21636.

Patchin, J. W., & Hinduja, S. (2006). Bullies move beyond the schoolyard: A preliminary look at cyberbullying. *Youth Violence and Juvenile Justice, 4*(2), 148–169. https://doi.org/10.1177/1541204006286288.

Patchin, J. W., & Hinduja, S. (2011). Traditional and nontraditional bullying among youth: A test of general strain theory. *Youth Society, 43*(2), 727–751. https://doi.org/10.1177/0044118X10366951.

Patterson, G. R., Reid, J. B., & Dishion, T. J. (1992). *Antisocial boys*. Oregon: Castalia.

Psychological Society of Ireland. (2018). Psychological society of Ireland submission to the oireachtas joint committee on children and youth affairs. *The Irish Journal of Psychology, 44*(6), 128–129.

Pyżalski, J. (2012). From cyberbullying to electronic aggression: Typology of the phenomenon. *Emotional Behavioural Difficulties, 17*(3–4), 305–317. https://doi.org/10.1080/13632752.2012.704319.

RAE. (2018). *Diccionario de la lengua Española* [Dictionary of the Spanish language]. http://dle.rae.es/?id=98ULSyc. Accessed 16 Apr 2018.

Rigby, K., Smith, P. K., & Pepler, D. (2004). Working to prevent school bullying: Key issues. In P. K. Smith, D. Pepler, & K. Rigby (Eds.), *Bullying in schools: How successful can interventions be?* (pp. 1–12). Cambridge: Cambridge University Press.

Schenk, A. M., & Fremouw, W. J. (2012). Prevalence, psychological impact, and coping of cyberbully victims among college students. *Journal of School Violence, 11*(1), 21–37. https://doi.org/10.1080/15388220.2011.630310.

Short, E., Barnes, A. B. J., Zhraa, M. C., et al. (2016). Cyberharassment and cyberbullying; Individual and institutional perspectives. *Annual Review of Cybertherapy Telemedicine, 14*, 115–122.

Slonje, R., & Smith, P. K. (2008). Cyberbullying: Another main type of bullying? *Scandinavian Journal of Psychology, 49*(2), 147–154. https://doi.org/10.1111/j.1467-9450.2007.00611.x.

Smith, P. K. (2009). Cyberbullying: Abusive relationships in cyberspace. *The Journal of Psychology, 21*(4), 180–181. https://doi.org/10.1027/0044-3409.217.4.180.

Smith, P. K. (2012). Cyberbullying and cyber aggression. In S. R. Jimerson, A. B. Nickerson, M. J. Mayer, et al. (Eds.), *Handbook of school violence and school safety: International research and practice* (2nd ed., pp. 93–103). New York: Routledge.

Smith, J. D., Schneider, B. H., Smith, P. K., et al. (2004). The effectiveness of whole-school antibullying programs: A synthesis of evaluation research. *School Psychology Review, 33*(4), 547–560.

Smith, P. K., Mahdavi, J., Carvalho, M., et al. (2008). Cyberbullying: Its nature and impact in secondary school pupils. *Journal of Child Psychology and Psychiatry, 49*(4), 376–385. https://doi.org/10.1111/j.1469-7610.2007.01846.x.

Sourander, A., Klomek, A. B., Ikonen, M., et al. (2010). Psychosocial risk factors associated with cyberbullying among adolescents: A population-based study. *Archives of General Psychiatry, 67*(7), 720–728. https://doi.org/10.1001/archgenpsychiatry.2010.79.

Sticca, F., & Perren, S. (2013). Is cyberbullying worse than traditional bullying? Examining the differential roles of medium, publicity, and anonymity for the perceived severity of bullying. *Journal of Youth and Adolescence, 42*(5), 739–750. https://doi.org/10.1007/s10964-012-9867-3.

Szoka, B., & Thierer, A. (2009). Cyberbullying legislation: Why education is preferable to regulation. *Progress Freedom Foundation, 16*(12), 1–26. https://doi.org/10.2139/ssrn.1422577.

Tedeschi, J. T., & Felson, R. B. (1994). *Violence, aggression, and coercive actions*. Washington, DC: American Psychological Association.

Tokunaga, R. S. (2010). Following you home from school: A critical review and synthesis of research on cyberbullying victimization. *Computers in Human Behavior, 26*(3), 277–287. https://doi.org/10.1016/j.chb.2009.11.014.

Välimäki, M.A. Almeida, D., Cross, M., et al (2012). Guidelines for preventing cyber-bullying in the school environment: A review and recommendations: COST Action IS0801: Cyberbullying: Coping with negative and enhancing positive uses of new technologies, in relationships in educational settings. https://sites.google.com/site/costis0801/guideline.

Vandebosch, H., & Van Cleemput, K. (2008). Defining cyberbullying: A qualitative research into the perceptions of youngsters. *Cyberpsychology & Behavior, 11*(4), 499–503. https://doi.org/10.1089/cpb.2007.0042.

von Marées, N., & Petermann, F. (2012). Cyberbullying: An increasing challenge for schools. *School Psychology International, 33*(5), 467–476. https://doi.org/10.1177/0143034312445241.

Walther, J. B. (2007). Selective self-presentation in computer-mediated communication: Hyperpersonal dimensions of technology, language, and cognition. *Computers in Human Behavior, 23*(5), 2538–2557. https://doi.org/10.1016/j.chb.2006.05.002.

Wei, H. S., & Chen, J. K. (2009). Social withdrawal, peer rejection and victimization: An examination of path models. *Journal of School Violence, 8*(1), 18–28. https://doi.org/10.1080/15388220802067755.

Willard, N. E. (2007). *Cyberbullying and cyberthreats: Responding to the challenge of online social aggression, threats, and distress*. Illinois: Research Press.

Ybarra, M. L., & Mitchell, K. J. (2004). Online aggressor/targets, aggressors, and targets: A comparison of associated youth characteristics. *Journal of Child Psychology and Psychiatry, 45*(7), 1308–1316. https://doi.org/10.1111/j.1469-7610.2004.00328.x.

Ybarra, M. L., Diener-West, M., & Leaf, P. J. (2007). Examining the overlap in internet harassment and school bullying: Implications for school intervention. *Journal of Adolescent Health, 41*(6), S42–S50. https://doi.org/10.1016/j.jadohealth.2007.09.004.

Zezulka, L. A., & Seigfried-Spellar, K. C. (2016). Differentiating cyberbullies and internet trolls by personality characteristics and self-esteem. *Journal of Digital Forensics, Security, and Law, 11*(3), 5. https://doi.org/10.15394/jdfsl.2016.1415.

Zillmann, D. (1979). *Hostility and aggression*. New Jersey: Erlbaum.

Zillmann, D. (1983). Arousal and aggression. In R. Geen & E. Donnerstein (Eds.), *Aggression: Theoretical and empirical reviews* (pp. 75–101). New York: Academic.

Part II
Cyber-Threat Landscape

Policies, Innovative Self-Adaptive Techniques and Understanding Psychology of Cybersecurity to Counter Adversarial Attacks in Network and Cyber Environments

Reza Montasari, Amin Hosseinian-Far, and Richard Hill

1 Introduction

In today's cyber security environment, there is a growing number of threats resulting from old and new sources. The speed, diversity and frequency of such attacks are generating cyber security challenges that have never been witnessed before (Godin 2017). Moreover, the essence and purpose of the attacks are evolving in that they are becoming more politically and economically motivated (Godin 2017; Jahankhani et al. 2014). Several critical infrastructures such as industrial control systems are attractive targets for cyber-attacks (Knowles et al. 2015). Therefore, identifying, assessing and protecting assets and resources from harm are of utmost importance (Haley 2008). With the increasing number of new types of attack techniques such as zero-day exploit attack and advanced persistent threats, network security is encountering severe "easy-to-attack and hard-to-defend" challenges (HackerWarehouse 2017; Jajodia et al. 2011). Adversaries have time benefit to scan and acquire information on targeted systems before carrying out attacks. The longer an attacker is within a system, the more difficult it is for the cyber-defenders to contain and expel them from their cyber domain. The more time the attacker has, the safer environments they can create and hide within them. They can install

---

R. Montasari (✉)
Department of Computer Science, The University of Huddersfield, Huddersfield, UK
e-mail: R.Montasari@hud.ac.uk

A. Hosseinian-Far
Department of Business Systems & Operations, University of Northampton, Northampton, UK
e-mail: Amin.Hosseinian-Far@northampton.ac.uk

R. Hill
Head of Department of Computer Science, University of Huddersfield, Huddersfield, UK
e-mail: R.Hill@hud.ac.uk

modified backdoors to dominate and threaten network systems after vulnerabilities have been discovered through the benefits of asymmetric information.

As enterprises of different sizes encounter rapidly growing frequency and sophistication of cyber-attacks, such threats have had detrimental effects on network security, compliance, performance, and availability. Moreover, many of such threats have eventuated in the theft or exposure of sensitive data. A cyber-attack can have devastating effect on an enterprise's viability, and the results of such attacks can have lasting impact on its brand with negative long-standing effects on customer trust and loyalty. Many victim organisations have also experienced collateral damages including: fines, lawsuits, credit problems and reduced stock prices. The public revelation resulting from a breach goes beyond the IT realm, affecting every aspect of business within the organisation. Advanced Persistent Threats (APTs), sophisticated malware and targeted attacks are some of the new, constantly evolving threats that enterprises face when searching for cracks in enterprise IT systems. Various enterprise technologies – such as smart mobile devices, web applications, portable storage, virtualization, cloud-based technologies – present cybercriminals with convenient support network of attack vehicle.

At the same time, many systems are developed with set limits and presumptions without the capability to adapt when assets change suddenly, new threats emerge or unfound vulnerabilities are exposed (Salehie et al. 2012). The features offered by the existing defence methods are not capable of determining all kinds of network attacks to protect systems proactively (Lei et al. 2017). Current defence methods such as firewalls and intrusion detection systems are always behind adversaries' sophisticated exploitation of systems' susceptibility. The existing cyber defences are mainly static and are administered by slow processes such as testing, security patch deployment and also human-in-the-loop monitoring. Consequently, attackers can methodically explore target networks, premeditate their attacks, and continue for a long time inside compromised networks and hosts with an assurance that those networks will change slowly. This is due to the fact that hosts, networks and services that are mainly developed for the purposes of availability and uniformity do not reconfigure, adapt or regenerate apart from ways to support maintenance and uptime requirements.

Many systems are developed with set limits and presumptions without the capability to adapt when assets change suddenly, new threats emerge or unfound vulnerabilities are exposed. Thus, in order to address such changes, systems must be developed such that they are capable of enabling various security countermeasures dynamically (Salehie et al. 2012). Moreover, to tackle cyber-security threats more effectively, enterprises will also need to have more robust cyber security policies and systems that will enable the reinforcing of the defence and make the cyber-defenders more effective when responding to attacks. In addition, enterprises will need to have a new, more adaptive, integrated approach based on the foundations of prediction, prevention, detection and response so as to address the limitations of traditional enterprise IT systems security. Such robust policies and systems must be developed and updated to facilitate various security countermeasures dynamically.

This paper surveys the latest research on the foundation of Adaptive Enterprise Security (AEC). To this end, it discusses potential security policies and strategies that are easy to develop, are established, and have a major effect on an enterprise's security practices. These policies and strategies can then efficiently be applied to an enterprise's cyber policies for the purposes of enhancing security and defence. The study also discusses various adaptive security measures that enterprises can adopt to continue with securing their network and cyber environments. To this end, the paper continues to survey and analyse the effectiveness of some of the latest adaptation techniques deployed to secure these network and cyber environments.

The remainder of this paper is structured as follow: The next section, Sect. 2, discusses potential security policies and strategies that have a major impact on an enterprise's security practices. Section 3 discusses various adaptive security measures, while Sect. 4 surveys and analyses some of the latest adaptation techniques employed to secure network and cyber environments. The final section, Sect. 5, presents the conclusions. Two main contributions of this paper are the scope of the discussion – no surveys of similar scope currently exist – and the provision of a research agenda focused on security matrix for adaptive network and cyber security.

2 Security Policies and Strategies

In situations where an enterprise needs to develop a more effective cyber defence stance, there will be a priority of work that must be undertaken to ensure achievement. The first phase for an enterprise is to establish a robust governance that employees will adhere to and trust. In order to accomplish this, the main leadership within an enterprise must engage in the cyber defence governance panel. The high-ranking officials' agreeing and signing off on decisions will highlight to the employees the significance of the cyber defence to the enterprise (Godin 2017). Such approach will also enable the employees to remember that the cyber threat is always present and that the safeguarding measures are supported by the high-level leadership.

The second phase in developing a robust governance model must include a vigorous training. There already exist many Good Practice Guides providing the details on how to create a new or improve existing cyber awareness and skills for enterprise systems (Peltier 2016; PA Consulting Group (PACG) 2015a,b; Stouffer et al. 2015; Bada et al. 2014; ENISA 2009; Symantec Inc and Landitd Ltd 2009). These documents often place a high emphasise on the frequency and consistency of the training. Such guides enable employees to perform in accordance with established security policy and to report incident with confidence that they are doing the right thing at the right time. Moreover, creating a robust governance requires the development of some kinds of a recognition system whereby employees are rewarded for the fact that they have acted responsibly to stop incidents or attacks or any other exploits that enhances the defence of the enterprise (Godin 2017).

The second priority of actions for the enterprise must be the collection, processing, and distribution of actionable intelligence to the company's cyber defence team. Assessing the laborious task of selecting and establishing relationships at the early stage will be valuable to the enterprise in the long run. There will be various sources of information and partners that an enterprise should search for. External sources consist of agencies such as the UK National Cyber Security Centre, which is the governmental agency that helps networks of national significance and all sectors of industry against sophisticated attacks (NCSC 2017), the wider public sector and academia. Some of the services offered by the NCSC include helping the enterprises:

- determine the extent of the incident,
- work to ensure the immediate impact is managed,
- provide recommendations to remediate the compromise and increase security across the network,
- produce an incident report to describe the scope of the problem, the technical impact, mitigation activities and an assessment of business impact, and
- give an Impact Assessment – where the incident affects partners or customers.

Enterprises should also have a policy of creating a relationship with law enforcement agencies that are responsible for cybercrime. In some cases, if possible, the enterprise's cyber governance panel must also provide a seat for a law enforcement liaison to participate. Such liaison will assist with providing a consistent direction from the enterprise's direction and will facilitate and accelerate communications in case of attacks (Godin 2017). Then, there need to be (within the Service Level Agreement (SLA) between the company and their ISP agreements on communication lines) information allocation, and accountabilities during the periods of disaster. For instance, this should cover the actions to be taken to ensure business continuity and disaster recovery. Nowadays, enterprises are increasingly adopting cloud services that necessitate some kinds of SLA with the cloud service provider.

In the final phase, there must be a robust policy to create information dissemination amongst the enterprise and other companies within the same industry or companies that deploy identical IT equipment. In such relational situations, enterprises, however, will need to balance out issues such as complying with Intellectual Property and at the same time, also maintaining competitive advantage even if they are disseminating information of cyber-attacks. Sharing information on cyber-attacks is economically valuable to both parties. An example of such cooperation between different enterprises is of that between the auto and financial industries by developing joint Cyber security centers (Godin 2017).

The most effective way to distribute information while protecting Intellectual Property is to adopt a standard to exchange information such as STIX and TAXII (US-CERT 2017). The most excellent source of information is often within the enterprise, themselves. This consists of the IT infrastructure and employees. The acquisition and examination of logs and their behaviour are vital in establishing an effective cyber security stance and a swift and robust cyber defence response. Collecting intelligence from employees is also essential; for instance, employees must be asked to report malevolent emails or social engineering attempts. Also, providing employees with instructions on reporting mechanism must become part of the security awareness training within the enterprise. Such training must cover (1) what to report, (2) when to report, and (3) whom to report to Godin (2017). Publically providing employees with awards in situations where they have operated according to the training enhances such actions. This is highly likely to lead to more participation of the employees, in turn resulting in the formation of impetus and enthusiasm for cyber security amongst the employees.

The final decision associated with intelligence gathering is the execution of a system that will collect, organise, combine, and conduct an initial examination of all the acquired information. Such a system can function as an intelligence. Nevertheless, the technology will be extraneous as long as there is a systematic approach with a feedback mechanism to the formation of intelligence that will enable the cyber-security team to detect and prevent an intrusion, find the intruders, and react to safeguard the system in a speedier manner and with more precision. The next policy decision must be about the size of the cyber-security team that an enterprise requires to safeguard a robust defence that is capable of both preventing and reacting effectively. Godin (2017) suggests the ratio of 20–25 per 1000 employees and IT equipment combined (Godin 2017). For instance, an enterprise with 10,000 employees and 15,000 pieces of IT equipment (25,000 combined) will require a cyber-security team of 500–625. This team will consist of system administrators, service desk personnel, technicians, and Cyber security experts.

However, it should be noted that an enterprise cannot be expected to terminate all nefarious activities in their network or cyber space. However, there exist steps that an enterprise can undertake to assure that they are not part of the problem. When attempting to use the principles of neutralisation, the major effort must be placed on splitting the connections amongst the attackers' systems. To this end, two measures will need to be adopted. The first measure is to ensure that there does not exist a link between the attacker's systems by acting as a node or a transit point. The policies discussed previously will enable enterprises to ensure that their network and cyber domains do not become a refuge for cyber-criminals (Apostolaki et al. 2017; Godin 2017). The second measure is to carry out a supply chain analysis to ensure its integrity. Such measures will enable the enterprises to avoid providing refuge or resources to cyber-criminals (Nagurney et al. 2017; Markmann et al. 2013).

3 Adaptive Security Measures

Security requirements is about extracting, representing and examining security goals and their relationships with other security elements such as critical assets, threats, attacks, risk, and countermeasures (Nhlabatsi et al. 2012; Salehie et al. 2012; Moffett and Nuseibeh 2003). However, such elements can dynamically change as the functioning environment or the requirements change. Unfortunately, current security requirements engineering techniques are not capable of identifying and dealing with runtime changes that particularly affect security (Chen et al. 2014). Therefore, adaptive security is needed to address such runtime changes. The main goal of an adaptive security is to identify and analyse different kinds of changes at runtime that might have a negative impact on system security and activate countermeasures offering an acceptable level of protection (Pasquale et al. 2014; Nhlabatsi et al. 2012). For instance, integrating a valuable asset into the system might require a higher level of protection which in turn demands stronger countermeasures. Security objectives might change, new threats and attacks might arise, new system vulnerabilities might be found, and current countermeasures might become ineffective. In such situations, adaptive security must be capable of addressing the impacts of such changes, which might undermine the system and harm its resources (Salehie et al. 2012).

When designing and implementing adaptive security systems, three main models must be considered. These include Asset Model, Objective Model, and Threat Model (Aagedal et al. 2002). The Asset Model signifies assets and their relationships (Moffett and Nuseibeh 2003). In the context of a network, assets signify individual nodes on the network, such as servers, routers, and laptops. Asset ranges signify a group of network nodes addressable as adjacent block of IP addresses. Zones signify allocations of the network itself and are also defined by an adjacent block of addresses. The attacks that target a network might damage or impair the connected assets as well.

On the other hand, the Objective Model signifies the main goals which a system must attain and disintegrates them into functional and non-functional requirements. Such a model consists of security objectives including Confidentiality, Integrity, Availability and Accountability (CIAA) (Salehie et al. 2012). Security objectives consist of a hierarchical structure and can be disintegrated into operational countermeasures which include various operations to alleviate security risks (Moffett and Nuseibeh 2003; Stoneburner et al. 2002). Some security objectives cannot be satisfied without sacrificing other non-functional requirements such as performance and usability (Salehie et al. 2012). The countermeasures used to impose the satisfaction of security objectives cannot be chosen without taking into account their side effects. For instance, if a system deploys a higher level of encryption algorithm, such countermeasure might create deterioration in system performance or usability.

Similarly, a threat model consists of threat agents, threat goals, and attacks. Threat agents can be natural (e.g., flood), human (e.g., hacker), or environmental (e.g., power failure) (Salehie et al. 2012; Stoneburner et al. 2002; Hosseinpournajarkolaei 2014). Assets are associated with the threat objectives that they inspire, whereas threat objectives are connected with the attacks that are carried out for their attainment. Threat objectives signify motivations of threat agents to attack a system (Salehie et al. 2012). Attacks are activities whereby threat goals can be attained and as a result assets would be harmed (Nhlabatsi et al. 2012; Lamsweerde 2004; Stoneburner et al. 2002). Thus, threats can be modelled as "operationalizations of threat goals" (Salehie et al. 2012).

Often, it is difficult to ascertain the security of the design process of network systems resulting in security weaknesses. In addition, the static implementation of the current network information systems presents the attackers with adequate time to scan and identify systems' vulnerability. Thus, it will be increasingly challenging for the traditional static defence systems effectively to withstand unknown system hardware and software weaknesses, and to avert possible backdoor attacks and the growing sophisticated and intelligent network intrusion penetrations. Therefore, such a situation aggravates the asymmetry between the offense and the defence in the network. A new technology titled Adaptive Cyber Defence (ACD) challenges attackers with changing attack surfaces and system configurations, compelling attackers constantly to re-evaluate and revise their cyber activities. Despite the usefulness of technologies such as Moving Target Defence, Dynamic Diversity, and Bio-Inspired Defence (discussed later in the paper), all these technologies presume static and aleatory but non-adversarial environments (Cybenko et al. 2014). Cybenko et al. argue that in order to reach full potential, scientific foundations need to be developed in order for system resiliency and robustness in adversarial environments to be rigorously defined, quantified, measured and extended in a laborious and reliable manner (Cybenko et al. 2014).

Therefore, by countering an attack in a timely fashion, an adaptive security aims to reduce the effect and extent of potential threats. This consists of the possibility of responding to "zero-day" attacks, in which a threat is so new that there does not yet exist a patch or other countermeasure. Despite the fact that adaptive security measures are evolving, an adaptive method can be developed by utilising technologies available today. This remainder of the section presents concepts related to adaptive security and the manner in which the method enhances system survivability. It discusses adaptive security and the reason why the method is beneficial, reviews its features and principles, and also discusses a design approach. To this end, this section addresses the following topics:

1. Objectives and Components of Adaptive Security,
2. Complex Adaptive Systems in Security Design,
3. Structural Approach Based on Adaptive Security, and
4. Design Approach to an Adaptive Security Model.

3.1 Objectives and Components of Adaptive Security

In the context of IT infrastructure and cyber-security, an Adaptive Security approach aims to contain active threats and also counterpoise potential attack vectors. Similar to other security architectures, Adaptive Security Model aims:

- to decrease threat intensification and limiting the potential dissemination of failures,
- to make the target of an attack smaller,
- to reduce the rate of attacks,
- to respond to an attack quickly,
- to stop attacks that attempt to restrict resources, and
- to address attacks aimed at compromising data or system integrity.

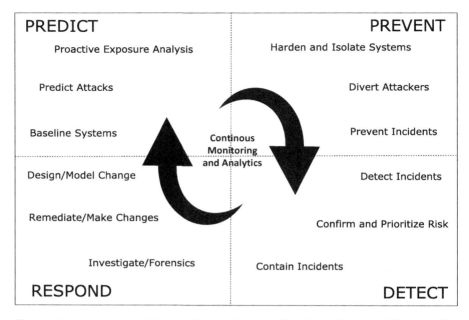

Fig. 1 Adaptive security architecture. (Adapted from MacDonald and Firstbrook 2014 as cited by Vectra 2016)

Furthermore, in addition to supporting SLAs, the main aim of an adaptive security approach is to maintain system and data integrity, facilitate reliability and assurance. Similar to all other types of security approaches, the adaptive security ultimately is aimed at ensuring that data and processing resources are trustworthy, reliable, available, and functioning within satisfactory boundaries. Also, one of the main principles of the adaptive security is survivability which is the ability of a system to accomplish its mission in a timely way when attacks, failures and accidents take place (MacDonald and Firstbrook 2014). In order to ensure the survival of a system, it is imperative first to distinguish system elements (i.e. things that must survive) against elements that are considered sacrificial. For the purposes of this paper, a system is deemed to have survived if it endures to accomplish its business goals within planned Service Level Agreements (SLAs) (Weise 2008). An Adaptive Security Architecture encompasses four crucially important capabilities as depicted in Fig. 1 (Vectra 2016; MacDonald and Firstbrook 2014).

Prediction Those enterprises that have access to the latest threat intelligence and trends are better equipped to predict and avoid attacks. Training employees to distinguish tactics deployed in attacks boosts prognostic analysis in addition to the capability to learn from past mistakes by forensically examining breaches (Jahankhani and Hosseinian-Far 2014, 2017). Moreover, penetration testing can also assist with revealing the weak spots in enterprises' IT systems security.

Prevention The main goal of prevention will be to diminish attack surface regardless of the attack being traditional, signature-based anti-malware, device controls or patching application vulnerabilities. Tightening systems and deploying as many as hurdles in the way of attackers as possible are two main aspects of an all-embracing approach that includes restricting the capability of attacks to propagate and decrease their impact.

Detection Advanced attacks can remain undetected for many months and even years. According to a research conducted by Kaspersky Lab (2016), some attacks can remain undetected up to 200 days (Kaspersky 2016). Technologies for incident detection underlined by the best threat analysis enhances incident detection. The most effective detection strategy is often developed based on the capability to figure out behaviours and sequences of events that indicate a breach has occurred.

Response Efficient enterprise security should include the capability to respond to and reduce the effects of a breach. This can include: (1) "if/then" policy for procedures that can be automated such as patching, and (2) post-breach examination or the utilization of incident-response expert teams to halt, reduce and investigate attacks, breaches and other security incidents. In order to be effective, these capabilities must work together as a multi-tiered system. Some of the main attributes of an all-inclusive, adaptive enterprise security architecture are intelligence-driven, threat focused, integrated, holistic and strategy-driven.

3.2 Complex Adaptive Systems in Security Design

Complexity is the major barrier in designing secure IT architectures and effectively fighting security threats. Complex systems are not understood by anyone. Therefore, if no one can comprehend more than a portion of a complex system, then no one can foresee all the ways that a system can be penetrated by an attacker (Weise 2008; Elkhodary and Whittle 2007; Geer et al. 2003). Averting insecure operating modes in complex systems is challenging and unlikely to do without incurring a significant cost. This denotes that the defenders or enterprises have to counter all possible attacks; the attacker only needs to identify one insecure means of attack. A potential solution to the increased complexity of IT security infrastructure is a Complex Adaptive System, which is an active network of various distributed and decentralized agents that continuously interrelate with and learn from one another. A security architecture that impersonates a Complex Adaptive System can be efficient in that it can adapt and respond continuously to emerging and changing security threats.

3.3 Structural Approach Based on Adaptive Security

In order to detect threats effectively, IT systems will need to understand a baseline of what is deemed normal behaviour and what is not considered normal. The notion of self and non-self are central in IT systems. Functional systems effectively distinguish between what is native to the system and what is not native. What is not native is considered as a threat and eradicated. An IT system is automatically capable of safeguarding itself by accurately detecting and dealing with threats and suspicious activity and differentiating these from legitimate components, protocols and operational processes within IT infrastructure. An IT infrastructure intended for survivability must present the following characteristics (Weise 2008; Janssen and Kuk 2006):

- The flexibility to respond to new and diverse threats,
- The capability of being self-detecting, self-governing, self-recovering and self-protecting,
- A basis on a formalised security model with enforcement mechanisms that enforce security policy compliance,
- The ability to identify unauthorised resource modification such as data, files, file systems, operating systems and configurations, and also launch remedial actions such as (a) quarantining resources for the purposes of digital forensic investigations so that the system can learn from the attack, and (b) providing other resources to substitute for compromised systems in order to facilitate service continuity, and
- Applying remedial actions as required.

Adaptive security takes advantage of architectural and operational principles from different disciplines. The following principles are applicable to information systems. These principles are identified (Jones 2015; Weise 2008; Wilkinson 2006; Janssen and Kuk 2006; De Castro and Timmis 2002) as valuable features that are valuable in IT systems to decrease exposure to threats, contain the degree of threats and fight them in a timely manner.

Pattern Recognition IT systems need to be able to address sophisticated pattern matching techniques in order to detect regular and irregular behaviour in code, command/response dialogues, communication protocols, etc.

Uniqueness IT systems need to be able to address sophisticated pattern matching techniques in order to detect regular and irregular behaviour in code, command/response dialogues, communication protocols, etc. Uniqueness discourages the existence of monocultures that can be vulnerable to a common computer virus. It also equips diverse IT systems with the essential robustness to survive targeted threats.

Self-Identity IT systems isolate and eliminate what does not belong according to baseline manifests and security policy. This includes support for intra/inter-systems communication and sharing information on threats, countermeasures, security policies, and trust relationships between systems and IT infrastructure.

Diversity In IT systems, diversity displays itself through various control mechanisms such as compartmentalization through operating system virtualization of Trusted Platform Module (TPM)-based hardware trust anchors (Weise 2008).

3.4 Design Approach to an Adaptive Security Model

An automatic system that integrates an insusceptible response ability could be a reasonable design approach in developing a secure Adaptive Security Model. One way of utilising adaptive principles is through Defence-in-Depth security architecture that implements various strategies. Diversity can be accomplished by applying mechanisms such as clustering, redundant hardware or numerous kinds of firewall appliances from multiple vendors. In this method, if one components fails to respond to a certain threat, it is probable that other components do not capitulate. In this way, the survivability of the system is preserved. Similarly, the property of elasticity can be maintained via virtualization techniques. Employing virtualization technologies, infrastructure systems can categorise various system services in secure execution containers. These containers can be deployed to separate service instances. This denotes that if a threat alters a service in one container, it will not affect the implementation of running services in other containers. This will ensure that services within IT infrastructure can continue.

At the same time, response mechanisms could quarantine the affected container and contain the attack's impact. The main difference in an adaptive security architecture from the existing state-of-the-art practices is that adaptive security approaches are implemented not only to defend against known threats but also to predict unknown threats (Jones 2015). The following outlines one possible way of implementing an adaptive security architecture in both cyberspace and network security environments. This method should be incorporated into a larger context of the complete security architecture. Moreover, it must take place within the framework of other security features such as application, system, network design, and quality assurance and configuration validation to assure that all components and design elements adhere to the overall security policy (Weise 2008).

The followings provide an outline of the steps required to design and implement an adaptive security model (Jones 2015; Weise 2008; Janssen and Kuk 2006; De Castro and Timmis 2002):

- Delineate threats and its features that are necessary to avoid or destroy. A threat feature is likely to comprise the entire threat structure. It could also be a specific activity displayed by an entity or process.
- Ascertain satisfactory behaviour, trusted components and activities that must be differentiated from a threat. This step is crucial to stop Denial-of-Service (DoS) attacks.

- Characterise triggers that could scan for suspicious activities and to launch threat detection sensors that will warn the larger IT infrastructure of possible threats and prepare threat response mechanisms.
- Carry out redundancy for main functions.
- Describe threat response mechanisms that do not culminate in terminating the host.
- Outline a recovery process through which systems are able to reconfigure and restart themselves adaptively. This process must also consist of a learning and knowledge dissemination mechanism in order for infrastructure to learn how to evade analogous threats in the future.
- Outline feedback abilities that will enable the threat response mechanism to authenticate threats in order for them to respond only to valid and realistic threats. Such feedback mechanisms assist with ensuring that the triggers and threat response mechanisms recognise the security setting within which they function. This will facilitate the preferred adaptive behaviour.

Not every infrastructure should have every threat features delineated. The purpose should be to develop a varied set of systems, each of which can have different threat response abilities. By filling the fundamental building blocks of threats and threat responses, individual systems will be capable of adapting to threats and respond to these threats accordingly. Once the response is successful, the individual system can then disseminate that knowledge with other reliable systems that have not undergone the original threat. It is expected that sacrificial components are implemented into the complete IT infrastructure. Thus, a threshold of acceptable harms should be established and monitored.

4 Adaptation Security Techniques

As stated previously, current cyber defences are mainly static providing adversaries with opportunities to probe the targeted networks with the assurance that those networks will change slowly, if at all. Often, adversaries are not concerned with time to develop reliable exploits and premeditate their attacks since their targets are static and almost undistinguishable (Cybenko et al. 2014; Wang and Wu 2016). In order to address such situations, researchers in the domain of security have started to explore different approaches that make networked information less analogous and less predictable (Anderson and McGrew 2017; Tague 2017; Wang and Wu 2016; DeBruhl and Tague 2014; Cybenko et al. 2014; Jajodia et al. 2012, 2011). The main reason for Adaptation Techniques (AT) is to design systems with similar functionalities but randomised manifestations. Adaptation methods are normally deployed to deal with various phases of potential attacks (Cybenko et al. 2014). In contrast, various defence undertakings could have various Confidentiality, Integrity, Availability and Accountability (CIAA) requirements (Salehie et al. 2012). For instance, if a cyber-attack on Availability were assessed to be present or imminent, adaptation mentors for preserving availability would be given priority over methods for improving confidentiality or integrity. Analogous functionality enables authorised usage of networks and services in predictable, formal ways at the same time that randomised manifestations make it cumbersome for adversaries to develop exploits remotely. Preferably, each exploit would need the same amount of the effort by the adversary. The remainder of this section aims to survey and analyse some of the latest Adaptation Techniques proposed by the research community. This examination has been restricted to only six techniques due to the space constraints.

Instances of the Adaptation Techniques (AT) include the following notions to the degree that they implicate system adaptation for security and resiliency purposes (Anderson and McGrew 2017; Tague 2017; DeBruhl and Tague 2014; Cybenko et al. 2014; Jajodia et al. 2012, 2011):

- Randomized Network Addressing and Layout,
- Network Moving Target Defence (MTD),
- Inference-Based Adaptation,
- ACD Framework Based on Adversarial Reasoning,
- OS Fingerprinting Multi-Session Model Based on TCP/IP, HTTP and TLS,
- Address Space Layout Randomization,

4.1 Randomized Network Addressing and Layout

Randomized instruction set and memory layout restrict the degree to which a single buffer overflow based penetration could be utilised to breach a collection of hosts. This, however, at the same time, makes it more challenging for cyber-defenders (e.g. systems administrators or software developers) to debug and update hosts due to the fact that all the binaries are different. Additionally, randomised instruction set and memory layout techniques will not present the adversaries with a difficult challenge to determine a network's layout and its available services. Analogous examination can be carried out for each of the above techniques. For instance, randomising network addresses will present attackers with more challenges to conduct reconnaissance on a target network remotely. However, it does not create any difficulty for the adversary to penetrate a particular host after it has been identified and reachable. Another example can relate to that of a mission such as the generation of a daily Air Tasking Order (ATO) (Cybenko et al. 2014), which could prioritize confidentiality and integrity to safeguard details of future sorties over availability in order for the network layout and addressing to be used to perplex potential adversary at the expense of network performance.

4.2 Network Moving Target Defence

Network Moving Target Defence (NMTD) is employed to enhance the efficiency of defensive mode and facilitate a dynamic, non-deterministic and non-sustained runtime environment (Lei et al. 2017; Sun and Jajodia 2014; Jajodia et al. 2012, 2011). The NMTD is an innovative Adaptation Technique that changes the adversarial patterns amongst attack and defence with an end-point information hopping. It disrupts the dependency of the attack chain on the consistency of the network operating environment by multi-level dynamical changes (Lei et al. 2017). One of the significant elements of the NMTD is the Endpoint Hopping Techniques, which have received extensive attention (Lei et al. 2017; Xu and Chapin 2009). Although such techniques are useful, they do not enable the full potentials of NMTD hopping resulting in limiting their use in simple network threat such as APT and zero-day attacks.

There exist two main issues with the existing end-point hopping research. The first significant problem is that the advantages from hopping defence is reduced because of the insufficient dynamic of network hopping triggered by self-learning inadequacy in reconnaissance attack strategy culminating in the blindness of hopping mechanism selection. The second main problem is that because of the restricted network resources and high overhead, the availability of hopping mechanism is low. Thus, to address such issues, Network Moving Target Defence based on Self-Adaptive End-Point Hopping Technique (SEHT) has been proposed (Lei et al. 2017). The SEHT was developed to address the lack of hopping mechanisms capable of self-adaptive to scanning attacks, and also to describe the restraints of hopping formally which increases the availability of hopping mechanisms in order to ensure the low hopping overhead. The SEHT is claimed to be capable of counterweighing the defensive value of end-point information hopping and service quality of network system, based on adversary strategy awareness.

Through their theoretical and experimental results reported in their research paper, it appears that Lei et al. (2017) have addressed the blindness issue of hopping mechanism associated with defence by applying hopping triggering based on adversary strategy awareness (Lei et al. 2017). The aim of this solution is to direct the choice of hopping mode by discriminating the scanning strategy, which improves targeted defence. Lei at al. (2017) also employ "satisfiability modulo" theories to describe hopping constraints formally in order to ensure low hopping overhead (Lei et al. 2017).

4.3 Inference-Based Adaptation

Inference-Based Adaptation techniques focus on tackling stronger attacks in wireless communications where observant attackers can attain significant gains by incorporating knowledge of the network under attack. In these situations, cybercriminals are capable of adapting parameters and behaviours to offset system dynamics, hinder detection, and save valuable resources. Thus, robust wireless communication protocols that can survive such adaptive attacks require new techniques for near-real-time defensive adaptation, allowing the defenders similarly to change their parameters in response to perceived attack impacts. One of such latest new techniques is Inference-Based Adaptation Techniques for Next Generation Jamming and Anti-Jamming Capabilities (DeBruhl and Tague 2014; Tague 2017).

4.4 ACD Framework Based on Adversarial Reasoning

ACD framework that deploys adversarial reasoning is aimed at dealing with several limitations of traditional game-theoretic analysis such as empirically defining the game and the players. The framework utilises control-theoretic analysis to bootstrap game analysis and to quantify the robustness of candidate actions (Cybenko et al. 2014). This framework comprises of four parts, each of which has a different purpose. The aim of Part 1 is to design and implement a subsystem which takes two inputs including streaming observations of the networked system and also external intelligence about possible adversaries. The purpose of Part 2 is to employ empirical methods to activate a game model from which it acquires "strategically optimised defence actions". The goal of Part 3 is to focus on identifying and adding innovative adaptation mechanisms into the defence strategy space. Part 4 aims to conduct trade off analysis which will consider not only functionality, performance, usability and exploitation but also robustness, stability, observability and resilience.

4.5 OS Fingerprinting Multi-session Model Based on TCP/IP, HTTP and TLS

Enterprise networks encounter various menace activities such as attacks from external devices (Cheswick et al. 2003), contaminated internal devices (Virvilis and Gritzalis 2013) and unauthorized devices (HackerWarehouse 2017; Wei et al. 2007). One important traditional method of defence is Passive Operating System Fingerprinting (POSF), which detects the operating system of a host merely through the observation of network traffic. POSF discloses vital information such as intelligence to the defenders of heterogeneous private networks. Meanwhile, cyber-criminals can employ fingerprinting to explore networks. Therefore, cyber-defenders require obfuscation techniques to thwart these attacks. POS Fingerprinting techniques emerged almost two decades ago in order to deal with remote devices sending network attack traffic (Spitzner 2008). As a result it was quickly adopted by the open source community (Zalewski 2014). Subsequently, research community built upon Passive OS Fingerprinting further. For instance, Lippmann et al. (2003) as cited by

Anderson and McGrew (2017) presented the notion of Near-Match Fingerprints, employed machine learning classifiers to produce them, and ascertained the OS groups that were distinguishable through fingerprinting (Lippmann et al. 2003; Anderson and McGrew 2017). Tyagi et al. (2015) deployed passive OS Fingerprinting of TCP/IP to identify unauthorized operating systems on private internal networks (Tyagi et al. 2015).

The data structures originally employed in fingerprinting originated from TCP/IP headers. However, the latest research has applied characteristics from HTTP headers (Mowery et al. 2011; Zalewski 2014) and unencrypted fields from the TLS/SSL handshake (Durumeric et al. 2017; Husák et al. 2015). These characteristics can be examined independently when only a single session's data is available, which is not unusual in some scenarios. Despite the fact that it is valuable for cyber-defender (e.g. network administrators) to apply Passive Fingerprinting to detect operating systems on their networks, cyber-criminals have also adopted these techniques to seek for possible victims (Anderson and McGrew 2017). Due to the fears resulting from malevolent use of detection, cyber-defenders have attempted to identify new methods to apply obfuscation to overcome the technique. Although these techniques have been useful in that are capable of obscuring individual session or raw data structures that a cyber-defender controls; nevertheless, they are less ineffective in the multi-session model. This is because it is unusual for a cyber-defender to be capable of rewriting all conceivable network protocols which are being transmitted from different devices.

An analogous adaptive technique is Active OS Fingerprinting, in which one or more packets are transmitted to a device so as to activate a visible response (Anderson and McGrew 2017). Passive and Active Fingerprinting was formalised by Shu and Lee, who also devised the Parameterized Extended Finite State Machine (PEFSM) to model behaviour when numerous messages were transmitted and received (Shu and Lee 2006). Likewise, Greenwald and Thomas investigated Active Fingerprinting and demonstrated that information gain can be employed to reduce the number of probes that were required (Greenwald and Thomas 2007). Kohno et al. employed passive observations of the TCP Timestamp option to fingerprint individual devices according to their clock skew (Kohno et al. 2005). Similarly Formby et al. presented Cross-Layer Response Times to fingerprint devices passively on enterprise networks (Formby et al. 2016).

Although all the aforementioned techniques associated with Operating System Fingerprinting are beneficial, they are not adaptive and can be disruptive to a network workflows. However, a new technique, entitled "OS Fingerprinting Multi-Session Model Based on TCP/IP, HTTP and TLS" developed by Anderson and McGrew (2017), appear to have addressed the shortcoming of the previous techniques. The OS Fingerprinting Multi-Session Model Based on TCP/IP, HTTP and TLS is a strictly "passive" technique which is both adaptive and much less disruptive to networks and applications. Moreover, this technique is easier to be assimilated into network monitoring workflows and facilitates backward-looking discovery. These techniques employ data features from TLS in addition to TCP/IP and HTTP protocols in a multi-session model, which is pertinent whenever several sessions can be observed within a time window.

By employing TCP/IP, HTTP, and TLS features combined within the multi-session model, accurate fingerprinting is possible, even to the extent of minor version detection. A machine learning classifier is capable of addressing the multitude of data features efficiently providing more accuracy than single session fingerprints. The incorporation of TLS fingerprints for operating system identification is predominantly vital since the TLS-encrypted HTTPS protocol substituted for HTTP, and the traditional User-Agent strings will no longer be visible. The multi-session model enables cyber-defenders easily to include additional, explicit fingerprinting data types, which are important characteristics of an adaptive fingerprinting scheme. The multi-session model based on TLS, HTTP, and TCP/IP can detect vulnerable operating systems with higher accuracy, and that fingerprinting can be both adaptive and robust even when confronted with levels of data feature obfuscation that could be observed on an enterprise network.

4.6 Address Space Layout Randomization

Address Space Layout Randomization (ASLR) is often carried out offline at application code compile time in order for a decision to utilise ASLR to be open-loop in the control sense. The ASLR techniques stop attackers from locating target functions by randomizing the process layout. Previous ASLR techniques protected only against single-target brute force attacks, which worked by locating a single, supreme system library function such as execve(). However, such techniques were not adequate to guard against chained return-into-lib(c) attacks that invoke a series of system library functions. Thus, the research community built upon this technique to address its shortcomings. For instance, Xu and Chapin proposed the Island Code Transformation (ICT) that addresses chained return-into-lib(c) attacks (Xu and Chapin 2009). A code island is a chunk of code that is isolated in the address space from other code blocks. This code not only randomises the base pointers used in memory mapping but also maximizes the entropy in function layout. There are various other types of Adaptation Techniques, the descriptions of which are outside the scope of this paper due to the space constraint. These include, for instance:

- Bio-Inspired Defences.
- Randomized Instruction Set and Memory Layout,
- Randomized Compiling,
- Just-in-Time Compiling and Decryption,
- Dynamic Virtualization,
- Workload and Service Migration, and
- System Regeneration.

4.7 Discussion on the Existing Adaptation Techniques

From the survey and the analysis of the above discussed Adaptation Techniques (ATs), it can be deduced that there exist various potential trade-offs when considering fundamental assignment, the perceived attack type as well as the system adaptation methods present by means of AT methods. It can also be deduced that although there are various ATs, the settings in which they are valuable to the defenders can differ significantly. Often, the major focus of research on ATs has been on engineering particular new techniques in contrast with comprehending their overall functionality and costs, when they can be most beneficial, what their potential inter-relationship can be. Despite the fact that each AT is likely to have some design accuracy, the discipline is still based on ad hoc approaches in relation to comprehending the entirety of ATs and their augmented use.

5 Human Factors and Psychology of Attack

One of the main factors that should be considered when developing a security policy within a firm is the contemplation of human factors and the psychology of cyber-attacks. The existing research on the social and psychological factors of cyber-attacks are conducted mostly by computer security and forensics specialists rather than by the social scientists (McAlaney et al. 2016). Some of the reasons and motivations behind cyber-attacks are commonly listed in numerous sources as financial motivations, enjoyment and personal fun, political reasons also known as 'Hacktivism' (Ludlow 2013), disruption, etc. Nevertheless, considering the behaviour and psychology of cybercriminals would not be sufficient. Gaining an understanding of the behavioural requirements of the victims is also necessary.

Nudge theory first introduced by James Wilk (1999) discusses how small suggestions could influence the decision making of an individual or a group in favour of the proposer's intentions. Such theoretical underpinning could be potentially used by the adversary to gain positive compliance of a victim e.g. in a social engineering scenario.

Another psychological and behavioural notion that could assist cybercriminologists is the COM-B System (Michie et al. 2011). Within this framework, motivation is influenced by capability and opportunity. Opportunity, capability and motivation are all influenced by behaviour. Understanding the causality within of elements within the cybercrimiology context will pave the way to develop a holistic cyber security policy which would ultimately assist businesses to be able to defend against potential cybercrimes (Fig. 2).

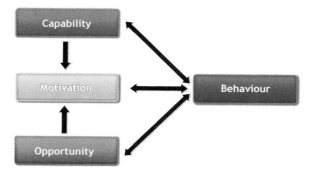

Fig. 2 COM-B behavioral model (Michie et al. 2011)

6 Conclusions

Cyber-attackers are constantly devising new and sophisticated attacks, while traditional cyber-security approaches can only deal with known attacks and might prevent those attacks only temporarily and partially. Thus, new scientific foundation and the corresponding technologies are required in order to deal effectively with adaptive and dynamic cyber operations given that adversaries are increasingly becoming sophisticated. The efficiency of any cyber-defence system adaptation technology are unlikely to be quantified in a laborious way without such a scientific foundation.

Furthermore, there can be a significant improvement in security and a more effective cyber defence by employing established security policies and strategies such as those discussed in this paper. The use of such a solution provides an opportunity for the cyber defenders to have a new set of tools for both network and cyber environments that are established to be beneficial to enterprises. The policies and strategies discussed in this paper will enable enterprises to have a more robust security posture. Implementing these steps will ensure that the principles of carrying out operation are valued. This can be materialized, firstly, by ensuring that the enterprises possess a robust governance model that will encourage participation and compliance from both employees and managers. The cooperation between the members of the leadership team in relation to a common approach and set of objectives will help to consolidate the role and standing of cyber governance panel boards. Secondly, the distribution of intelligence both internally and externally will enable boosting the enterprise's network and cyber security stance and reducing the response time of the cyber defenders. Thirdly, having an appropriate ratio of cyber defenders to the employees as suggested by Godin (2017) is essential to provide and uphold a sense of security and to benefit from the collected intelligence. Lastly, by adopting complexity theory's principles of systems analysis, the cyber defenders will be able to focus on protecting the points between systems that are vital to survival with higher effectiveness and with less trial and error.

Finally, to mitigate the limitations associated with traditional cyber-defence systems, it is imperative to design and implement new adaptive network and cyber security systems to combat attacks in these domains more effectively, such as those described in this paper. Such adaptive security systems based on intelligent Adaptive Techniques (such as those described in this paper) can also help to fuse information from various sources more effectively and also to profile cyber attackers more efficiently.

References

Aagedal, J. O., Den Braber, F., Dimitrakos, T., Gran, B. A., Raptis, D., & Stolen, K. (2002). Model-based risk assessment to improve enterprise security. In *The 6th International Conference on Enterprise Distributed Object Computing* (pp. 51–62).

Anderson, B., & McGrew, D. (2017). *OS fingerprinting: New techniques and a study of information gain and obfuscation.* Cisco Systems, Inc. arXiv preprint arXiv: 1706.08003.

Apostolaki, M., Zohar, A., & Vanbever, L. (2017). Hijacking bitcoin: Routing attacks on cryptocurrencies. In *IEEE Symposium on Security and Privacy (SP)* (pp. 375–392).

Bada, M., Creese, S., Goldsmith, M., Mitchell, C., & Phillips, E. (2014). *Computer security incident response teams (CSIRTs) an overview*. Global Cyber Security Capacity Centre (pp.1–23).

Chen, B., Peng, X., Yu, Y., Nuseibeh, B., & Zhao, W. (2014). Self-adaptation through incremental generative model transformations at runtime. In *The 36th International Conference on Software Engineering* (pp. 676–687).

Cheswick, W. R., Bellovin, S. M., & Rubin, A. D. (2003). *Firewalls and Internet security: Repelling the Wily Hacker* (2nd ed.). London: Addison-Wesley Longman Publishing.

Cybenko, G., Jajodia, S., Wellman, M. P., & Liu, P. (2014). Adversarial and uncertain reasoning for adaptive cyber defense: Building the scientific foundation. In *International Conference on Information Systems Security* (pp. 1–8). Cham: Springer.

DeBruhl, B., & Tague, P. (2014). Keeping up with the jammers: Observe-and-adapt algorithms for studying mutually adaptive opponents. *Pervasive and Mobile Computing, 12*, 244–257.

De Castro, L. N., & Timmis, J. (2002). *Artificial immune systems: A new computational intelligence approach*. London: Springer Science & Business Media.

Durumeric, Z., Ma, Z., Springall, D., Barnes, R., Sullivan, N., Bursztein, E., Bailey, M., Halderman, J. A., & Paxson, V. (2017). The security impact of HTTPS interception. In *Symposium (NDSS'17) on Network and Distributed Systems* (pp.1–14).

Elkhodary, A., & Whittle, J. (2007). A survey of approaches to adaptive application security. In *International Workshop on Software Engineering for Adaptive and Self-Managing Systems* (p. 16).

ENISA, Symantec Inc., Landitd Ltd. (2009). *Good practice guide network security information exchanges* (Special Publication (ENISA) – Rev. 1).

Formby, D., Srinivasan, P., Leonard, A., Rogers, J., & Beyah, R. A. (2016). Who's in control of your control system? Device fingerprinting for cyber-physical systems (NDSS). InternetSociety.org.

Geer, D., Bace, R., Gutmann, P., Metzger, P., Pfleeger, C., Querterman, J., & Scheier, B. (2003). *CyberInsecurity: The cost of monopoly-how the dominance of microsoft's products poses a risk to security*. Washington, DC: Computer and Communications Industry Association.

Godin, A. (2017). *Using COIN doctrine to improve cyber security policies*. Available at: https://www.sans.org/reading-room/whitepapers/policyissues/coin-doctrine-improve-cyber-security-policies-37557. Accessed August 26, 2017.

Greenwald, L. G., & Thomas, T. J. (2007). Toward undetected operating system fingerprinting. In *USENIX Workshop on Offensive Technologies (WOOT)* (pp. 1–10)

HackerWarehouse. (2017). *MiniPwner penetration testing toolbox*. Available at: http://hackerwarehouse.com/product/minipwner/. Accessed 28th Aug 2017.

Haley, C., Laney, R., Moffett, J., & Nuseibeh, B. (2008). Security requirements engineering: A framework for representation and analysis. *IEEE Transactions on Software Engineering, 34*(1), 133–153.

Hosseinpournajarkolaei, A., Jahankhani, H., & Hosseinian-Far, A. (2014). Vulnerability considerations for power line communication's supervisory control and data acquisition. *International Journal of Electronic Security and Digital Forensics, Inderscience, 6*(2), 104–114.

Husák, M., Cermák, M., Jirsík, T., & Celeda, P. (2015). Network-based HTTPS client identification using SSL/TLS fingerprinting. In *2015 10th International Conference on Availability, Reliability and Security (ARES)* (pp. 389-396).

Jahankhani, H., & Hosseinian-Far, A. (2017). Challenges of cloud forensics. In V. Chang et al. (Eds.), *Enterprise security* (pp. 1–18). Cham: Springer.

Jahankhani, H., & Hosseinian-Far, A. (2014). Digital forensics education, training, and awareness. In *Cyber crime and cyber terrorism investigator's handbook* (Vol. 1, pp. 91–100). Waltham: Elsevier.

Jahankhani, H., Al-Nemrat, A., & Hosseinian-Far, A. (2014). Cyber crime classification and characteristics. In *Cyber crime and cyber terrorism investigator's handbook* (Vol. 1, pp.149–164). Massachusetts: Elsevier.

Jajodia, S., Ghosh, A. K., Swarup, V., Wang, C., & Wang, X. S. (2011). *Moving target defense: Creating asymmetric uncertainty for cyber threats* (Vol. 54). New York: Springer Science & Business Media.

Jajodia, S., Ghosh, A. K., Subrahmanian, V. S., Swarup, V., Wang, C., & Wang, X. S. (2012). *Moving target defense II: Application of game theory and adversarial modeling* (Vol. 100). New York: Springer Science & Business Media.

Janssen, M., & Kuk, G. (2006). A complex adaptive system perspective of enterprise architecture in electronic government. In *The 39th Annual Hawaii International Conference on System Sciences* (Vol. 4, pp. 71b–71b).

Jones, M. T. (2015). *Artificial intelligence: A systems approach*. Massachusetts: Jones & Bartlett Learning, Sudbury, MA.

Kaspersky Lab. (2016). *Kaspersky security solutions for enterprise: Securing the enterprise*. Available at: http://media.kaspersky.com/pdf/b2b/. Accessed August 15, 2017.

Knowles, W., Prince, D., Hutchison, D., Disso, J. F. P., & Jones, K. (2015). A survey of cyber security management in industrial control systems. *International Journal of Critical Infrastructure Protection, 9*, 52–80.

Kohno, T., Broido, A., & Claffy, K. C. (2005). Remote physical device fingerprinting. *IEEE Transactions on Dependable and Secure Computing, 2*(2), 93–108.

Lamsweerde, A. V. (2004). Elaborating security requirements by construction of intentional anti-models. In *26th International Conference on Software Engineering* (pp. 148–157).

Lei, C., Zhang, H. Q., Ma, D. H., & Yang, Y. J. (2017). Network moving target defense technique based on self-adaptive end-point hopping. *Arabian Journal for Science and Engineering, 42*, 1–14.

Lippmann, R., Fried, D., Piwowarski, K., & Streilein, W. (2003). Passive operating system identification from TCP/IP packet headers. In *IEEE Workshop on Data Mining for Computer Security* (pp. 40–49).

Ludlow, P. (2013). *What is a 'Hacktivist'? NYTimes*. Available at: https://opinionator.blogs.nytimes.com/2013/01/13/what-is-a-hacktivist/.

MacDonald, N., & Firstbrook, P. (2014). *Designing an adaptive security architecture for protection from advanced attacks*. Available at: https://www.gartner.com/doc/2665515/designing-adaptive-security-architecture-protection. Accessed August 14, 2017.

Markmann, C., Darkow, I. L., & von der Gracht, H. (2013). A Delphi-based risk analysis? Identifying and assessing future challenges for supply chain security in a multi-stakeholder environment. *Technological Forecasting and Social Change, 80*(9), 1815–1833.

McAlaney, J., Thackray, H., & Taylor, A. (2016). The social psychology of cybersecurity. *The British Psychological Society, 29*, 686–689.

Michie, S., van Stralen, M. M., & West, R. (2011). The behaviour change wheel: A new method for characterising and designing behaviour change interventions. *Implementation Science, 6*(42) (pp. 2–11).

Moffett, J., & Nuseibeh, A. (2003). A framework for security requirements engineering. Report-University of York, Department of Computer Science YCS (pp. 1–30).

Mowery, K., Bogenreif, D., Yilek, S., & Shacham, H. (2011). Fingerprinting information in javascript implementations. In *Proceedings of W2SP* (pp.180–193).

Nagurney, A., Daniele, P., & Shukla, S. (2017). A supply chain network game theory model of cybersecurity investments with nonlinear budget constraints. *Annals of Operations Research, 248*(1–2), 405–427. IGI Global.

NCSC. (2017). *The National Cyber Security Centre: A part of GCHQ*. Available at: https://www.ncsc.gov.uk/. Accessed August 28, 2017.

Nhlabatsi, A., Nuseibeh, B., & Yu, Y. (2012). Security requirements engineering for evolving software systems: A survey. In K. M. Khan (Ed.), *Security-aware systems applications and software development methods* (pp. 108–128). Hershey: IGI Global.

PA Consulting Group (PACG). (2015a). *Security for industrial control systems – Improve awareness and skills: A good practice guide* (PACG Special Publication). PA Consulting Group, 123 Buckingham Palace Road, London SW1W 9SR, United Kingdom.

PA Consulting Group (PACG). (2015b). *Security for industrial control systems: Improve awareness and skills – A good practice guide* (Special Publication (CPNI), Rev. 1). PA Consulting Group, 123 Buckingham Palace Road, London SW1W 9SR, United Kingdom.

Pasquale, L., Ghezzi, C., Menghi, C., Tsigkanos, C., & Nuseibeh, B. (2014). Topology aware adaptive security. In *The 9th International Symposium on Software Engineering for Adaptive and Self-Managing Systems* (pp. 43–48).

Peltier, T. (2016). *Information security policies, procedures, and standards: Guidelines for effective information security management*. CRC Press. Taylor and Francis Group: New York.

Salehie, M., Pasquale, L., Omoronyia, I., Ali, R., & Nuseibeh, B. (2012). Requirements-driven adaptive security: Protecting variable assets at runtime. In *20th IEEE International Conference on Requirements Engineering* (pp.111–120).

Shu, G., & Lee, D. (2006). Network protocol system fingerprinting – A formal approach. In *25th IEEE International Conference on Computer Communications* (pp. 1–12).

Spitzner, I. (2008). *Know your enemy: Passive fingerprinting*. Available at: https://www.honeynet.org/papers/finger. Accessed August 23, 2017.

Stoneburner, G., Goguen, A., & Feringa, A. (2002). *Risk management guide for information technology systems and underlying technical models for information technology security*. Pennsylvania: Diane Publishing Company.

Stouffer, K., Pillitteri, V., Lightman, S., Abrams, M., & Hahn, A. (2015). *Guide to industrial control systems (ICS) security* (Special Publication (NIST SP)-800-82 Rev 2). Gaithersburg, MD.

Sun, K., & Jajodia, S. (2014). Protecting enterprise networks through attack surface expansion. In *ACM Workshop on Cyber Security Analytics, Intelligence and Automation, 2014* (pp. 29–32).

Symantec Inc and Landitd Ltd. (2009). *Good practice guide network security information exchanges*. ENISA.

Tague, P. (2017). *Inference-based adaptation techniques for next generation jamming and anti-jamming capabilities*. Available at: https://www.cylab.cmu.edu/research/projects/2013/inference-based-adaptation-jamming.html. Accessed August 27, 2017.

Tyagi, R., Paul, T., Manoj, B. S., & Thanudas, B. (2015). Packet inspection for unauthorized OS detection in enterprises. *IEEE Security & Privacy, 13*(4), 60–65.

US-CERT. (2017). *Information sharing specifications for cybersecurity*. Available at: https://www.us-cert.gov/Information-Sharing-Specifications-Cybersecurity?. Accessed August 24, 2017.

Vectra. (2016). *How vectra enables the implementation of an adaptive security architecture*. Available at: https://info.vectranetworks.com/hubfs/how-vectra-enables-the-implementation-of-an-adaptive-security-architecture.pdf?t=1487862985000. Accessed August 28, 2017.

Virvilis, N., & Gritzalis, D. (2013). The big four-what we did wrong in advanced persistent threat detection. In *8th International Conference on Availability, Reliability and Security (ARES)* (pp. 248–254).

Wang, L., & Wu, D. (2016). Moving target defense against network reconnaissance with software defined networking. In *International Conference on Information Security* (pp. 203–217).

Wei, W., Suh, K., Wang, B., Gu, Y., Kurose, J., & Towsley, D. (2007). Passive online rogue access point detection using sequential hypothesis testing with TCP ACK-pairs. In *7th ACM SIGCOMM Conference on Internet Measurement* (pp. 365–378).

Weise, J. (2008). *Designing an adaptive security architecture* (pp.1–18). Sun Global Systems Engineering Security Office.

Wilk, J. (1999). Mind, nature and emerging science of change: An introduction to metamorphology. In G. C. Cornelis (Ed.), *Metadebates on science* (Vol. 24, pp. 71–87). Dordrecht: Springer Netherlands.

Wilkinson, M. (2006). Designing an 'adaptive' enterprise architecture. *BT Technology Journal, 24*(4), 81–92.

Xu, H., & Chapin, S. J. (2009). Address-space layout randomization using code islands. *Journal of Computer Security, 17*(3), 331–362.

Zalewski, M. (2014). p0f – Passive OS fingerprinting tool. Available at: http://lcamtuf.coredump.cx/p0f3/. Accessed August 16, 2017.

The Dark Web

Peter Lars Dordal

The **dark web** consists of those websites that cannot be accessed except through special **anonymizing** software. The most popular anonymizing system is Tor, originally an acronym for The Onion Router, but there are others, such as Freenet and I2P (below). While there are legitimate uses of dark websites (the New York Times has one, to allow sources to communicate confidentially), the dark web is perhaps best known for attracting criminal enterprises engaged in the sale of contraband. Products such as stolen credit-card data and child pornography are easily delivered via the Internet, but the dark web has also attracted merchants selling illegal drugs, armaments and other physical items.

Tor dark-web addresses end in the suffix ".onion", e.g. nytimes3xbfgragh.onion or facebookcorewwwi.onion. The challenge of anonymizing software is to figure out how to deliver traffic to such public addresses without allowing anyone to trace the traffic. Tor was designed to achieve anonymity for users and servers even from government-level attempts at unmasking. In the past two decades governments have gotten much better at monitoring the Internet; see the attacks outlined in "Traffic Correlation" below. However, most if not all hidden-site discoveries to date have relied on operational errors rather than any fundamental weaknesses in the Tor protocol.

All Internet traffic is transmitted via chunks of data called **packets** that are delivered to an attached **IP address**. Given a server IP address, the approximate location of the server is easy to discover using standard networking software; the exact location is straightforward for authorities to obtain. An immediate consequence is that a public dark-web address can never be associated with the site's IP address.

P. L. Dordal (✉)
Department of Computer Science, Loyola University Chicago, Chicago, IL, USA
e-mail: pld@cs.luc.edu

1 Virtual Private Networks

As a first step in describing anonymization, we consider VPNs. These provide limited anonymity for users, but none for servers. Suppose user Alice wishes to access server Bob, but not let Bob know it was her. To achieve this, she contracts with VPN provider Victoria. Alice prepares a packet addressed to Bob, and attaches to it an additional header sending the packet to Victoria. Victoria removes this additional header, rewrites the sender address to refer to Victoria rather than Alice, and sends it on to Bob. As far as Bob can tell, the packet came from Victoria; there is no evidence of Alice's participation.

Bob then sends the reply back to Victoria, which recognizes from context (specifically, from the "port numbers" associated with Alice's Internet TCP connection) that it must be forwarded on to Alice. Victoria does so, and Alice receives Bob's reply.

Nothing in this scheme provides any anonymity for Bob, whose real IP address must be available to Alice at the start. However, nothing in the packets seen by Bob identifies Alice. Alice is thus able to browse Bob's website anonymously.

Alice's identity can be easily unmasked by the authorities, however. Victoria's IP address was seen by Bob, so Victoria can be identified. If the authorities now show up at Victoria's with a subpoena, there is a good chance they will find log records showing that customer Alice, identified by her IP address if nothing else, had Victoria send packets on to Bob. If Victoria has kept no records, then Alice's original interaction with Bob is untraceable. However, the authorities can likely compel Victoria to record future connections to Bob; if Alice tries again, she is revealed.

Alice's identity can also be discovered by monitoring traffic at Victoria (perhaps from the vantage point of Victoria's Internet Service Provider), and looking for correlations between arriving and departing packets. If every packet arriving from Alice is followed by a packet sent on to Bob, and vice-versa, then Alice is unmasked. This approach has the advantage that Victoria need not be involved or even notified.

In the current-day Internet, some VPNs pride themselves on the anonymity they provide for their customers. Some advertise not keeping any logs, or at least not logging per-connection information. Some accept anonymous payment in cryptocurrencies such as Bitcoin. Some locate servers in jurisdictions outside the United States and Europe.

2 The Tor Approach

The strategy used by Tor can be loosely described as a three-stage VPN, with the added element of encryption to prevent the VPN stages from learning more than the minimum necessary about one another. The VPN stages are known as **Tor nodes**, described below. The basic three-stage approach conceals only the user, not the server, but a variation allows anonymity for both endpoints.

The ideas behind Tor, and in particular the concept of "onion routing", were developed by Paul Syverson, David Goldschlag and Michael Reed at the US Naval Research Laboratory; (Syverson et al. 1997) is their survey paper describing their work. Their ideas were strongly influenced by Chaum (1981). A stated early goal was the support of anonymous web surfing and anonymous emailing. In 1997, criminal misuse of anonymity was not widespread. Anonymous email services such as anon.penet.fi existed through much of the early development of Tor; the developers were aware of these services and their limitations.

The first "production" version of Tor was released as an open-source project in 2003 (with an alpha version the year before). In 2004, Roger Dingledine, Nick Mathewson and Paul Syverson published a description of the "second generation" Tor mechanism (Dingledine et al. 2004). This remains generally current, though there have been technical updates.

In 2006 a nonprofit organization The Tor Project was formed; it continues to manage development of the Tor software. The primary funder of The Tor Project has been the US government (Levine 2014), with the stated goal of supporting democracy activists in authoritarian countries.

2.1 Tor Circuits and Nodes

The basic building block of Tor is the bidirectional **Tor circuit**, built around a chain of **Tor nodes**, usually of length three. One end of the circuit connects to Alice, and the other end connects to a public website. The Tor circuit will, like a VPN, prevent the website from identifying Alice.

As such a circuit must have a public IP address as its remote endpoint, it cannot by itself provide anonymity for servers. We will return to this point below, but the short answer is that to access anonymous servers, both the client and the server create a Tor circuit, and these meet somewhere in the middle.

The lifetime of a Tor circuit is on the order of 10 min. That is enough for one complete web connection and its immediate followups. A single Tor circuit may also be used to contact multiple websites. After 10 min or so (the exact time is chosen randomly), the client creates a new Tor circuit, though if a circuit is in continuous use as part of a large file transfer then it stays in place for as long as necessary.

A Tor client user can browse the web with little fear that the sites contacted will be able to determine the user's IP address and thus the user's identity (though see below at "Potential Attacks"). A Tor client user can browse sensitive information, or can upload leaked files to the press, or can send and receive email (through an ordinary free email account or through a special Tor-only email account), all with negligible risk of identification by any but the most committed adversaries. In these and other cases, we assume for the moment that the server end of the Tor connection is a public Internet website.

Tor nodes are usually run by volunteers who are concerned about Internet privacy. Tor nodes do not operate in secrecy; the list of all Tor nodes is of necessity public, as users must have this list to create their Tor circuits. In 2018, there were a little over 6000 Tor nodes. The limiting factor of a typical Tor node is how much bandwidth the administrator is willing to devote to Tor traffic.

If Alice wishes to connect using Tor to a public IP address, say owned by Bob, then Alice's first task is to pick three (for a Tor circuit of length three) Tor nodes, which we will call Tammy, Terrell and Tim. Alice picks these nodes by downloading the list of all Tor nodes and then choosing the three at random. (Alice may choose her nodes so they meet additional bandwidth and stability requirements, though that slightly reduces the randomness.) The first node on the list here, Tammy, is Alice's "guard" node; to reduce the effectiveness of some correlation attacks (below), Alice may wish to use the same small set of guard nodes for several weeks. The last node on the list, Tim, must be a Tor node that has agreed to serve as a Tor **exit node**, below.

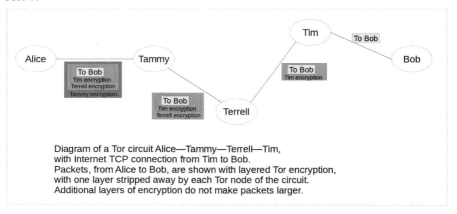

Diagram of a Tor circuit Alice—Tammy—Terrell—Tim, with Internet TCP connection from Tim to Bob.
Packets, from Alice to Bob, are shown with layered Tor encryption, with one layer stripped away by each Tor node of the circuit.
Additional layers of encryption do not make packets larger.

Once the circuit is built, Alice can send a packet to Bob by way of, in succession, Tammy, Terrell and Tim. Furthermore, none of the Tor nodes is aware of the IP address of any of the other non-adjacent nodes; that is, Tammy knows the IP address of Alice and of Terrell, Terrell knows the IP addresses of Tammy and Tim, and Tim knows the IP addresses only of Terrell and Bob.

As with a VPN, the authorities can likely identify Tim as having exchanged packets with Bob. However, the rules for Tor nodes prevent Tim from keeping any logs of its packet exchanges, and so the connection cannot be traced back to Terrell unless Tim has been subpoenaed or compromised. Even if this is the case, Terrell and Tammy would also have to have been compromised in order to trace the connection all the way back to Alice. This is unlikely, given Alice's random selection of Tammy, Terrell and Tim, though it does remain a theoretical risk. Another risk is a statistical attack, detailed below at "Traffic Correlation", though as of today that too is mostly theoretical.

Bob can see Alice's connection as coming from Tim, and so can determine that the connection in question probably is using Tor (only probably, because host Tim might be used for non-Tor purposes as well). Some public websites do place some restrictions on what can be done via Tor connections; for example, Wikipedia limits editing over Tor, except in special circumstances.

Tor packets (often called cells) all have a fixed size; smaller messages are extended by padding, and larger messages are split over two or more Tor packets. Fixed-size messages make it harder to deanonymize users based on packet-size traffic analysis.

The length of a Tor circuit, normally three, can in principle be changed. However, this is seldom a straightforward configuration option; it usually requires recompiling the software, Increasing the circuit length may not result in material increases in privacy, on the theory that if the first and last nodes of the Tor circuit are compromised then traffic-correlation attacks have a reasonable probability of success regardless of the number of intermediate nodes.

2.2 Sending Packets

To set up the Tor circuit, Alice first contacts Tammy, and negotiates an appropriate cryptographic session key (as opposed to Tammy's public key), using Diffie-Hellman-Merkle key exchange. Alice then tells Tammy that the next hop is Terrell, and Tammy forwards packets from Alice on to Terrell.

Alice now repeats the key negotiation with Terrell. At no point is Terrell aware that the start of the Tor circuit is Alice; the key negotiation between Terrell and Alice is conducted via Tammy as an intermediary. Once the Alice—Terrell key is negotiated, Alice tells Terrell, over an encrypted channel unreadable by Tammy, that Tim is the next hop.

The final step is for Alice to negotiate a session key with Tim. Again, Tim knows only that the communications are coming from Terrell; Tim has no idea of Alice's real identity. Similarly, Tammy knows nothing about Tim.

At the head of each Tor packet is a two-byte "circuit identifier", which is sent unencrypted. When Alice sends a packet to Tammy, she prefixes it with the circuit ID she used to set up the initial contact with Tammy. Tammy uses this circuit ID to look up the appropriate encryption key for this leg, and to look up the next hop in the circuit, Terrell. Tammy then sends the packet on to Terrell, updating the circuit ID to the value Tammy negotiated with Terrell. The circuit ID is examined and updated by each Tor node until the end of the circuit.

Alice is now ready to send a packet to Bob. She includes Bob's address (but not her own), and encrypts the packet with the key she shares with Tim. She then re-encrypts everything with the key she shares with Terrell. Finally, she encrypts a third time with the key she shares with Tammy. After this last encryption, she attaches the circuit ID agreed on with Tammy.

The packet is then sent to Tammy, who decrypts it with the key shared with Alice. Tammy sees, from the circuit ID, that the next hop is Terrell, so Tammy sends it on. Terrell receives the packet and decrypts it with the Terrell-Alice key; the packet is still encrypted with the Tim-Alice key. Terrell sees from the circuit ID that the next hop is Tim, and sends it on. Tim removes the final layer of encryption, and sees that the final destination is Bob. Tim then sends this packet on to Bob.

Alice's packet to Tim is effectively wrapped in three layers of encryption. These layers are stripped away, one by one. This layering is notionally like the layering of an onion, which gives rise to the name Onion Routing. The layered encryption prevents any one of the Tor nodes from finding out more than they need to know of the others. For example, even if Tammy and Tim are both compromised, they cannot together trace the connection definitively back to Alice, because the Tim—Bob packets do not match the Alice—Tammy packets due to the differing layers of encryption. (However, an attacker now does have a good chance of unmasking Alice probabilistically due to traffic correlations between transmissions from Tammy to Terrell and those from Terrell to Tim; see below at "Traffic Correlation".) In general, Alice's anonymity is has at least some protection as long as at least one of the Tor nodes on her circuit is not compromised.

Because each Tor packet appears completely different on each of the Alice—Tammy, Tammy—Terrell and Terrell—Tim links, due to the different number of layers of encryption, an outsider has no hope of correlating packets based on content, and thus tracing the connection back to Alice. There *is* a potential risk that such a correlation can be achieved by looking closely at traffic patterns and volumes; see below. In addition to the Tor encryption, Tor traffic on each circuit link is often also protected by Transport Layer Security (TLS) encryption, the standard Internet-connection encryption mechanism.

Actual connections over Tor must be made using the TCP protocol, used for the vast majority of Internet traffic. When Alice wishes to open a new TCP connection to Bob, she sends to Tim—via the circuit—a special Tor setup packet known as Relay Begin. This packet contains Bob's site address and the desired TCP port number (e.g. port 80 for web traffic). It is Tim that opens the TCP connection. Afterwards, Alice sends data to Bob (and vice-versa) through Relay Data packets. Tim converts Relay Data packets from Alice to the Tim—Bob TCP connection, and vice-versa.

Tim has access to the plaintext of the Relay Begin packet. Tim may have access to the plaintext of later packets, but it is common that Alice will negotiate an encrypted connection with Bob (using TLS), so that not even Tim can read the actual contents of further Alice—Bob exchanges.

When Bob wants to reply to Alice's request, Bob sends his data to Tim. Tim encrypts it with the key Tim shares with Alice, and sends it on to Terrell. Terrell adds another layer to the onion, re-encrypting the packet with the key Terrell shares with Alice, and sends it on to Tammy. Tammy adds a third layer, and sends the packet on to Alice, who is able to decrypt everything in the order Tammy, Terrell, Tim.

2.3 Tor Exit Nodes

The last node in the circuit, Tim, is in effect the public face of the circuit. If Alice is participating in the online sharing of a copyrighted work, or if Alice is accessing illegal content, then it is Tim that appears to be the guilty party. Managers of Tor exit nodes are routinely served copyright-related subpoenas, and are occasionally raided by the authorities. This is why there are, as of 2018, about 6000 Tor nodes but only about 800 exit nodes.

The Tor Project maintains a standard exit-node notice, which many exit nodes post online (The Tor Project, Exit Notice). It contains an explanation of Tor, and various disclaimers:

> This router maintains no logs of any of the Tor traffic...
> Attempts to seize this router will accomplish nothing...
> If you are a representative of a company who feels that this router is being used to violate the DMCA, please be aware that this machine does not host or contain any illegal content.

A Tor exit node can have a complex *exit policy*, designed to limit the sites accessible via the node. An overall bandwidth limitation is nearly universal; an exit node can also enforce a lower per-connection bandwidth limitation. Exit policies may also block certain protocols and certain IP address ranges. While most nodes allow web access, via TCP port 80, it is not uncommon to block the sending of bulk email by denying Tor connections to TCP port 25. Exit-node policies are available online, allowing a Tor user to choose an exit node compatible with his or her objectives.

It is the bandwidth limitations put in place by Tor-node administrators—of both exit nodes and internal nodes—that are the primary cause of Tor's slowness. The triple-forwarding of each packet contributes a fixed delay, of perhaps a couple hundred milliseconds, but delays from this source do not accumulate proportionally as download sizes and web-page sizes increase. However, if a Tor node caps a connection's bandwidth at 100 KB/s, then a 2 MB page will take 20 s to load. The bandwidth-limitation issue has the greatest impact on those sharing large files, such as full-length movies.

2.4 Anonymous Servers

As described so far, Tor only supports anonymous clients. It is also possible, and central to the idea of the Dark Web, to support anonymous *servers* (The Tor Project). The basic strategy is that the server and the client each create Tor circuits to an agreed-upon *introduction point*, and communicate via that point. Although the introduction point may be public, the server cannot be traced back from it any more than the client can.

Anonymous servers were originally called *hidden services*, though the less-perjorative term *onion services* has recently become popular.

Suppose Bob wishes to create an onion service. The first step is to create a public/private keypair, using RSA. The .onion service name is then based on an 80-bit secure hash of the public key. When users contact the .onion site, they will receive the public key, and can verify that it matches the site name.

For a first try, Bob might generate q4wgkcm22kdafxgb.onion as his public key. This is not very memorable, and so most sites try to generate large numbers of keys until they get one that begins with something human-readable. Normally, to get, say, four specific characters takes on order $32^4 \simeq 1{,}000{,}000$ tries. The .onion name for the New York Times, nytimes3xbfgragh.onion, likely took something like 30 billion tries. Facebook's onion address is facebookcorewwwi.onion; the first eight characters are intentional and the second eight just happened to end up as something that makes superficial sense in English. Facebook has claimed this was due to luck, but, given the computational resources available to them, their idea of luck may differ from that of normal-sized entities. (The New York Times has an onion address to support the leaking of documents; their onion server would remain hidden even if the authorities immediately demanded access to all their non-hidden servers. It is not clear why Facebook needs an .onion address, as users must login and so are not anonymous.)

After Bob generates his keys, and configures his web server, it is time for him to get his site out there. To do this he picks a set of Tor nodes—not necessarily exit nodes—that are known as *introduction points*. Bob builds Tor circuits to each of them. Bob then uploads to a public distribution service (built into Tor) the list of these introduction points, together with Bob's public key. Bob signs this list with his private key.

Alice obtains this information, and sets up her circuit Tammy to Terrell to Tim as before. This time Tim is asked to serve as a *rendezvous point*, but Tim need not be an exit node as Alice's traffic leaving Tim will not in any way be publicly visible.

Alice then sends Tim a secret password, and, via a second Tor circuit, contacts one of Bob's introduction points and sends the rendezvous point and the secret password. Both are encrypted with Bob's public key.

Bob now creates a circuit to the rendezvous point, Tim. In setting up this circuit, Tim is the ultimate destination. We will suppose Bob's circuit is Ty to Tula to Toni. After Bob verifies that Tim knows the secret password, and thus is legitimately the far end of Alice's part of the circuit, Alice and Bob can begin communicating via the combined circuit Tammy—Terrell—Tim—Toni—Tula—Ty.

Bob and Alice do not use one of Bob's introduction points in their combined circuit because the introduction points are publicly known. Avoiding them is more cryptographically sound, and also means that Bob's introduction points cannot be accused of carrying Bob's content. Tor exit nodes, by comparison, are accused of this sort of thing regularly.

2.5 Anonymity and Browsers

All Tor provides is connection anonymity. It is possible, however, that Tor user Alice may be identified by her browser. First, the browser may send cookies to Alice's computer. Second, browser *fingerprinting* techniques, perhaps based on the unique set of fonts and plugins Alice has installed, may allow the browser to be identified uniquely to Bob (see below at Application Attacks). If Alice uses the same browser to browse non-anonymously, and if Bob shares the fingerprint information with other sites Alice has visited, then Alice's real IP address may be revealed.

To avoid this, Tor users typically use a browser that has been specially configured to resist common and not-so-common identification attacks. Fingerprinting is likely to be blocked, and all browser cookies should be deleted at the end of the session. So-called "private browsing" is often made the default. Secondarily, using a different browser with Tor than for "public" browsing makes it very hard to connect the Tor use to the public use.

The browser most commonly bundled with the Tor package is based on the Firefox Gecko browser engine, which supports a broad range of strong privacy features (many of which are not enabled by default in the standard version of Firefox). On Apple systems the Gecko engine is not available, and so either an alternative browser is used, or the user runs Tor in a virtual machine.

Tor itself simply creates network connections; ultimately, the end-user can use whatever applications he or she trusts. Users can, for example, with a little technical knowledge use Tor with any browser. This way the Tor system does not require anyone to trust the application software distributed with it.

2.6 Legitimate Uses of Tor

There is a long history of governments defending citizen surveillance with the argument "if you have nothing to hide then you have nothing to fear". Some government agencies have long been suspicious of any use of Tor. There are, however, many everyday situations in which users might be more comfortable using Tor than a conventional web browser. For some of these, a VPN might serve as well, but Tor is free while VPNs are not.

Legitimate uses of Tor start with ordinary browsing for information that may be quite personal or sensitive. Someone searching for information about "HIV" or "addiction" might be very averse to public discovery or tracking.

Victims of stalking can use Tor to avoid revealing their IP address, and thus their location.

Ordinary people browsing non-sensitive topics might also want to use Tor if they simply wish to avoid relentless tracking by advertisers and large Internet companies. Some of this can be achieved by using an ordinary browser in "private" or "incognito" mode, but not all; in particular, private-mode browsing still reveals the client's IP address.

Persons engaged in political activism that is legal but that nonetheless attracts untoward government scrutiny—members of an antiwar group, for example, or Occupy Wall Street—might use Tor to read online manifestos and to communicate with fellow activists.

Tor is the tool of choice for those leaking government information, including whistle-blowers. Those leaking non-governmental information would probably be safe simply with a temporary email account, but Tor is sometimes used along with that for additional security. While leaking government information is sometimes against the law, such leaking is frequently viewed as a public good. The SecureDrop system, designed to support anonymous communications to the press and supported by many newspapers, is based on Tor.

Law-enforcement officers often use Tor so that their Internet use does not appear to be coming from an IP-address block assigned to the police. Citizens sometimes use Tor to report tips to the police anonymously.

Tor provides a lifeline for pro-democracy activists living under authoritarian regimes; this is in fact the US government's usual official argument in favor of continued Tor funding. Activists can keep in touch with one another and can disseminate news and images without risk of arrest.

There are frequent claims that US agents, and foreigners recruited by them for spying, use Tor to communicate. It is difficult to evaluate the volume of such traffic. It seems likely, though, that if the US government does make significant use of Tor for this and related purposes, then it would be likely to encourage other, non-espionage uses in order to provide a significant volume of cover traffic. The benefits of such cover traffic remain even if those other uses are of questionable legality.

Tor is often used for copyright infringement; for example, to allow someone to access online content via a peer-to-peer service without revealing their IP address. (Configuring bittorrent to use Tor securely in this matter is quite difficult, and is not recommended; bittorrent clients are notorious for leaking real IP addresses. Tor's bandwidth limitations also make it problematic for large file downloads.) While much of this might not be considered a "legitimate use", it is usually not criminal, and in some cases may be defensible on Fair Use grounds.

There are also some legitimate uses of server-side anonymity. News organizations, for example, often use Tor onion servers for submissions from whistle-blowers and leakers so as to largely eliminate the risk of document seizure by the authorities.

As another example, consider a website supporting online discussion of sensitive topics, such as addiction or even ordinary medical issues. Such a site might allow users to log in; keeping the site anonymous will eliminate any risk that the authorities will demand information about site users. This may encourage users to participate, and to open up about their experiences. A site catering to stalking victims might maintain an onion server to enforce the privacy of its users.

The use of Sci-Hub to obtain scientific papers, while clearly copyright infringement, is sometimes justified on the grounds that such infringement has no negative effect on the incentives for content creation. Sci-Hub has lost most of its traditional domain names to governmental seizure, but its onion service remains available worldwide.

2.7 Anonymity and Crime

Tor's browser anonymity, without onion services, enables a variety of antisocial actions. Things that an isolated Tor user can achieve, without anonymous confederates, include the harassment of others with impunity, the posting of embarrassing revenge content, or the infringement of copyrights through peer-to-peer networks. With confederates, Tor users can exchange illegal content such as child pornography with one another.

Adding onion servers enables additional illegal actions; for example, user-to-server copyright infringement (though this has been widespread even with publicly identifiable servers) and large-scale free sharing of illegal content. The existence of onion servers also enables a wide range of "political" crimes; that is, offending various governments. Terrorists can use Tor for recruitment and training.

However, most traditional criminal activities usually involve the exchange of money, and for these Tor alone is of limited use. However, the development of Bitcoin in 2009, providing a mostly anonymous form of currency, has enabled the rise of Tor-based e-commerce sites that sell illegal products and services. The most common illegal item sold appears to be drugs, but weapons, stolen credit cards, hacking services and thugs for hire are also available.

At least one illegal Tor-based online marketplace, known as The Farmer's Marketplace, did attempt to use conventional payment systems such as Paypal. Despite attempts to obfuscate the delivery of funds, the US Drug Enforcement Agency was able to trace the flow of money and shut down the site.

2.8 Alternatives to Tor

The Invisible Internet Project, or I2P, is an alternative to Tor. It was founded in 2004. I2P is designed for hidden services only, not for anonymous browsing of public websites. I2P's circuits are all one-way; every endpoint creates at least one *in-tunnel* (or circuit) and one *out-tunnel*. If Alice and Bob wish to communicate, Alice's out-tunnel connects to Bob's in-tunnel and vice-versa. This makes traffic-correlation attacks much harder, as a typical observer will see information flowing in one direction only.

I2P also supports the connection of a single out-tunnel to the in-tunnels of multiple destinations. If Alice wants to communicate to Bob, Charlie and Debra, she can consolidate all her outbound traffic to any of the three into a single out-tunnel. The end point of that tunnel will forward the packets to their correct destination. This technique is called *garlic routing*, the idea being that Alice's bundled packets represent a head of garlic, broken into individual cloves at the exit of Alice's out-tunnel.

The Freenet system, first released in 2000, is another alternative to Tor. Like Tor and I2P, it relies on a cloud of Freenet nodes to handle routing. With Freenet, however, hidden data is also stored in that cloud. The data is distributed throughout the cloud; any one file may be split up over multiple nodes. Popular data is likely to be cached by multiple Freenet nodes.

By default, Freenet looks for hidden data anywhere in the Freenet network. Freenet also has a "darknet" mode, in which data is retrieved only from nodes on a manually generated list of trusted nodes.

3 Potential Attacks

The goal of DarkNet attacks is to breach one endpoint's anonymity, but not necessarily to be able to read the encrypted traffic. Even relatively weak evidence may be useful. For example, if, after collecting online evidence, the authorities believe there is a 10% chance Alice might be one of the persons connecting regularly to an online narcotics marketplace, they might then monitor what is being delivered to Alice's home, or carefully examine discarded wrappings. For onion services, the goal is to identify the physical location of the server involved, or the identity of one or more of its administrators. (In all the cases described below, the onion server was discovered first, which eventually led to the unmasking of the administrators.)

3.1 Traffic Correlation

The biggest deanonymization risk to most Tor users relies on traffic correlation; that is, by looking for transmission patterns at one point in the network that are repeated very soon after at another point, thus suggesting, over time, that the traffic is connected. Traffic correlation attacks tend to be easier when the two points in question are close to one another, but this is not essential if the attacker has sufficient resources. Some of these attacks require sufficiently high levels of resources and network access that they can only be executed by a government-level actor, but that may be small comfort. See (Syverson et al. 2000; Johnson et al. 2013).

Perhaps the simplest attack is discovery of the circuits passing through a single Tor node T. The attacker monitors all traffic entering and leaving T, recording the source IP address of each arriving packet and the destination address of each departing packet. If the attacker notices, over time, that whenever a packet arrives from A, another packet departs for B within 50 ms, and vice-versa, that is very suggestive evidence that there is a Tor circuit through A, T and B. It may help if there are patterns to the traffic; for example, perhaps A regularly ends 5 packets and receives back 8, followed 200 ms later by another 14. If this (5,8,14) pattern shows up for only one of the other addresses T communicates with, it is likely that this other address represents B.

If A is user Alice, then the attacker has identified the first and second Tor nodes of Alice's circuit. If A and B are other Tor nodes, the attacker has identified an entire three-node Tor path, but not the user endpoints.

The information garnered by this attack is comparable to what would be obtained if the attacker had completely compromised node T, or was actually running node T. However, in isolation, discovery of the circuit neighbors at a single Tor node does not deanonymize any user. The real risk, below, is if this attack is perpetrated simultaneously against other Tor nodes.

Correlation-based circuit-neighbor discovery is not a sure thing. The Tor node T likely has many simultaneous circuits; based on data from (The Tor Project, Metrics), an order-of-magnitude estimate is 100. As each circuit lasts only 10 min, there might not be time to deanonymize all the circuits before they expire.

The Internet Service Provider of node T is easily able to carry out this kind of correlation attack, possibly at the request of the ISP's government. If an ISP is induced to launch this surveillance attack against one Tor node in its domain, it is likely to attack all of them. The attacker may also run some Tor nodes directly, gaining the same circuit-neighbor information. If the first and last nodes of Alice's circuit from earlier, Tammy and Tim, are surveilled or compromised, then Alice is deanonymized. If Alice has been accessing an onion service, it might take attacks on four such nodes to reveal Alice's connection to that service.

It is also potentially possible, though harder, to launch a larger-scale traffic-correlation attack that monitors Alice's traffic to and from the Tor node she is connected to, and also monitors a large number of exit nodes for matching traffic. For the latter, cooperation of a number of ISPs would likely be required. Initially, Alice's contribution to the exit-node traffic will be lost in the noise. However, over time, some correlations may appear. Again, specific traffic patterns may help. Although this attack is less certain, and generally more expensive, a success means that Alice is completely deanonymized.

To make this job easier, the ISP of a Tor node might even apply "traffic shaping" to that node's outbound traffic, to create recognizable patterns. For example, traffic from Tammy to other Tor nodes might be saved up and sent in batches a few tens of milliseconds apart. If this burst signature is then seen at another Tor node, that is evidence of a Tor circuit to that second node through Tammy.

Exit nodes may be monitored by their ISPs. It is also, however, straightforward for a committed adversary to set up a large number of exit nodes. This may have, in fact, been done, by various governmental agencies.

This kind of larger-scale attack might not even need exit-node monitoring. Websites can be profiled by the number of packets they send and receive (Hintz 2013). Suppose a connection to a particular public site involves 1 packet sent to the site, 7 sent back, 3 sent to, and then 17 sent back. That (1,7,3,17) signature would likely still be apparent even if the site were accessed via Tor; the only question would be how many other sites have the same signature. Extending the length of the signature, or including timing information on the delays between packet exchanges, may make this kind of signature significantly more trustworthy. Building a database of signatures for a large number of websites is quite straightforward. While this attack is largely hypothetical today, work continues on making it effective.

Earlier, we claimed that if the first and last nodes of Alice's Tor circuit, Tammy and Tim, are compromised, then Alice can still not be definitively deanonymized, because the traffic Tim sends to Terrell cannot be matched with certainty to the traffic Terrell sends to Tammy due to the layer of encryption added by Terrell. However, if Tammy and Tim *are* compromised, traffic correlation is likely to unmask Alice with a high degree of probability. Investigators will look for bursts of packets sent by Tammy to some other node X, followed soon after by a very similar burst from X to Tim. At this point it is usually quite easy to infer that X is Terrell.

If a number of Tor nodes are, collectively, connecting to a host leased in a cloud datacenter that does not run any public services, that might lend support to the hypothesis that the host in question is hosting a Tor onion service. This situation can readily be monitored by the datacenter itself.

In theory, correlation attacks are straightforward to prevent, by having Tor nodes introduce random delays when forwarding packets, and by having the nodes also send considerable volumes of "fake" traffic. Neither of these approaches is practical, however; the first increases the delays experienced by Tor users to unacceptable levels, and the second uses up Tor-node bandwidth that is already in short supply.

In none of the specific examples described below in "Tor Identity Breaches" do correlation attacks appear to have played a role.

3.2 DNS Leaks

Suppose Alice wishes to connect to, say, hackforums.net. The site's name must be looked up, using the Domain Name System, to determine its IP address. The correct way to do this is for Alice to set up her Tammy—Terrell—Tim circuit, as before, and then have her exit node, Tim, do the DNS lookup. Alice sends to Tim a Relay Begin message containing the string form of the site name, "hackforums.net". If Alice's software is configured incorrectly, though, it is possible that Alice will send the DNS query directly to her local ISP, which will return the IP address ("B" 2014). Her local ISP will likely keep a log record of this request, and Alice's attempt to access the site is revealed.

The browser bundled with most Tor software distributions is correctly configured to do remote-end DNS lookups, but Tor is also used with other, non-web protocols, such as email clients, Usenet news readers and the secure shell (ssh) login client. Configuring these so that DNS lookup works safely with Tor can be complex.

Relatedly, if at the beginning of Alice's Tor session she uses a conventional browser to search for "nytimes onion address", an observer might suspect that Alice went to nytimes3xbfgragh.onion to leak something. This is especially true if it is already known that someone has leaked documents that were available only to Alice's work department of a dozen persons.

Attacks exist that monitor the DNS names looked up by a set of Tor exit nodes, through eavesdropping, but with this approach it is somewhat harder to tie a given request to Alice.

3.3 Application Attacks

If Alice is browsing the Internet using Tor, her browser is probably the one packaged with Tor: a derivative of Firefox. Web browsers in general are notorious for having vulnerabilities. If Bob is running an onion service, odds are it involves Apache, MySQL and PHP. All three of those introduce vulnerabilities of the sort that Tor provides no protection against.

On the client side, many browser plugins leak information. Tor browsing should not use insecure plugins. Sometimes, though, Tor users have been talked into installing deanonymization plugins using the ruse that the plugin is a "security enhancement".

There are a large number of techniques for fingerprinting browsers. Most of these techniques are heavily used in the normal-browsing world, by advertisers and their allies. A server may extract the lists of fonts and plugins; many browsers are uniquely determined by this. Another fingerprinting technique involves drawing an image on an offscreen "canvas" and checking for subtle rendering details. Yet another technique involves precise timing measurements of mouse movements, which serves to fingerprint the client human user, not the client system. None of these fingerprint techniques reveal the identity of the user by themselves, but if the same user generates the same fingerprint via public browsing, the jig is up. The Tor browser attempts to block most known fingerprinting techniques, though the user is often asked if the blocking should continue, and it is easy to click "no" by mistake.

The onion server, if compromised, may be able to serve Javascript to the Tor clients that extracts information about the clients, such as their public IP address. A Javascript program can be downloaded which instructs the user machine to contact a designated server via the machine's public IP address. The usual Tor browser configuration includes settings to block most Javascript, but these settings can be changed. Though it is used less commonly than in the past, the Adobe Flash plugin will also run Javascript.

Web servers, regardless of whether they are used with Tor, are subject to a wide range of attacks. SQL injection can lead to database compromise. If the database includes usernames, order history and shipping addresses, a great many client users are exposed.

A common approach to expose the server itself is to find a flaw or misconfiguration that exposes the public IP address of the server. This is likely what happened in the Silk Road, Playpen and Hansa Market cases below. Ironically, onion servers do not actually need public IP addresses; they can be behind a network-address-translation firewall, and be assigned only a private IP address that is useless for tracking.

3.4 Metadata

Suppose someone uses Tor to upload an image anonymously. Now suppose that the EXIF image metadata was left attached to that image, and that it contains the GPS coordinates of where the picture was taken, and the name of the owner of the camera. Anonymity is lost, through no fault of Tor.

4 Tor Identity Breaches

Eldo Kim was an undergraduate at Harvard University. During finals week in December 2013, someone used Tor to email a bomb threat to the Harvard authorities. At the time the email was sent, Harvard's network logs showed that Kim's laptop was the only campus device that had connected to any Tor node. That pretty much pinpointed Kim, who confessed when confronted by the authorities. Note that Kim would likely not have been unmasked had Tor been more popular on campus. (Brandom 2013)

4.1 The Silk Road

The Silk Road, silkroad6ownowfk.onion, was the first contraband marketplace to exist on the Dark Web. The site was launched in early 2011, the identity of the owner was eventually discovered to be Ross Ulbricht. Ulbricht ran the site using the alias Dread Pirate Roberts, or "dpr". The site primarily sold drugs, and also stolen credit-card numbers and online account credentials; sales of child pornography or of violent services were not allowed.

In its early months, the Silk Road faced a problem common with anonymous transactions: the seller may fail to deliver. To avoid this, the Silk Road instituted both a seller-review system and an escrow system. Under the escrow system, a purchaser who did not receive their ordered merchandise had some chance of obtaining a refund. Sellers on the Silk Road had to pay a fee to participate.

The Silk Road was known to the US Federal Bureau of Investigation and Drug Enforcement Administration early on, but they had no way to find out who ran it or where the server was. DEA agent Carl Force was, however, able to become an online confidante of Ulbricht (as dpr), under the alias "Nob". Force had no idea, of course, about Ulbricht's real identity, or where the server was located.

Eventually the DEA was able to identify Curtis Green as a Silk Road employee, or at least as a customer, perhaps through seized packages. In January 2013 the FBI raided Green's home. After Green contacted Ulbricht about the arrest, Ulbricht allegedly tried to hire his online confederate Nob—actually Force—$80,000 to murder Green (Greenberg 2015). The DEA organized fake photos of Green's death, and Ulbricht paid up. Ulbricht was never tried for this allegation. (After Ulbricht's trial, Force was convicted of stealing from the government some of the Bitcoins that were seized during the investigation.)

Over time, Ulbricht made a series of errors in *operational security*; these are issues that did not involve fundamental weaknesses in the Tor protocol. In March 2013, he posted a technical question about Tor on the StackOver.com site, specifically about how to connect to a Tor onion service using the cURL software package. He used an alias, "altoid", but gave his email address as rossulbricht@gmail.com (Ulbricht 2013). The email address, and the alias, were changed very soon after.

By June 2013, the FBI had figured out the location of the Silk Road Tor server, below; this is believed to be the result of a server configuration error (below). Ulbricht had been careful to pay for the server using Bitcoin and fake identification, so the discovery of the server did not lead quickly back to him.

Also in June 2013, Gary Alford, an Internal Revenue Service agent attached to the DEA, went searching for Internet posts touting the Silk Road when it was first getting started. He found one by a user with alias "altoid", and connected that to the StackExchange.com post above. As of this point, Ulbricht was on the list of suspects (Popper 2015).

In July 2013, a package of nine fake IDs was intercepted at the US border (Hern 2013). When investigators went to the address they had been shipped to, Ulbricht was there. Worse, a picture resembling Ulbricht was on the IDs. This strengthened the DEA's suspicions about Ulbricht; he was not arrested.

Another serious error was that on isolated occasions Ulbricht logged into the Silk Road server without using Tor. Some of these logins were from an Internet café a few blocks from Ulbricht's home. Once the FBI located the server, below, they were able to log these connections, and use them to determine Ulbricht's general location.

On October 1, 2013, Ulbricht was arrested in a public library in San Francisco. Two FBI agents created a distraction, while others grabbed Ulbricht's laptop, which was configured to encrypt itself had Ulbricht had enough time to close the lid. The Silk Road server was also seized and shut down at this time.

Ulbricht was convicted of narcotics trafficking and related offenses in February 2015. Numerous site dealers, and probably some site customers, were also eventually arrested and convicted.

From a technical perspective, the most interesting question is how the FBI was able to locate the Silk Road's server, which is what led eventually to Ulbricht. The FBI's official explanation, presented in an affidavit by Chris Tarbell (Tarbell 2014), was that the login page contained a misconfigured CAPTCHA software widget; these components ask the user to, for example, type in the letters appearing in a distorted image, and are intended to prevent automatic logins. Tarbell stated that he tried sending a variety of data combinations to the login page (a technique known as "fuzzing"), and at some point one of the replies contained an IP address that did not belong to a Tor node. When he attempted to connect directly to that IP address, a CAPTCHA identical to the Silk Road's came up.

While misconfigured systems can do odd things, CAPTCHA widgets do not normally return IP addresses at all. This has lead to suspicions that the FBI may not have been telling the full story (Krebs 2014). One possibility is that they were able to install some form of malware on the server; another possibility is that some input error (perhaps not on the login page at all) forced the site's PHP programming language to dump all its state as an error message, including the public IP address. It is also possible that Ulbricht made modifications to the CAPTCHA widget that did not quite work as planned.

At Ulbricht's trial, his legal team tried to argue, among other things, that he had simply set up a web server, and wasn't responsible for what was sold on it. However, the prosecution presented detailed message logs indicating that Ulbricht had a close hand in managing the site. He was convicted in February 2015, and sentenced by Judge Katherine Forrest to life in prison without parole. Ulbricht's legal team has claimed that the severity of the sentence was based in part on the uncharged allegation that Ulbricht had conspired to have Green murdered; Judge Forrest did mention the alleged murder-for-hire plot at the sentencing hearing (Judge Forrest 2015).

Ulbricht's team had also argued that the FBI raided the server, located in Iceland, without a warrant. Counterarguments include the fact that the server was not under US jurisdiction, that the raid was led by Icelandic authorities, that the server was leased (from a cloud provider) and operating contrary to the provider's terms of service, and that Ulbricht has never claimed he had an ownership interest in the server.

Running the Silk Road took considerable technical skill; Ulbricht would not have wanted for traditional, legal employment. It does not appear that Ulbricht created his site so that he would be able to obtain illegal drugs for himself more easily, though (Bearman and Hanuka 2015) reports that in his youth he was a moderately heavy user of cannabis. One motivating factor, surely, was the promise of considerable wealth; Ulbricht's total earnings amounted to tens of millions of dollars, at a minimum.

However, Ulbricht was also a committed libertarian. On his LinkedIn page (Ulbricht 2015) he wrote

> I want to use economic theory as a means to abolish the use of coercion and agression amongst mankind... The most widespread and systemic use of force is amongst institutions and governments, so this is my current point of effort... I am creating an economic simulation to give people a first-hand experience of what it would be like to live in a world without the systemic use of force.

Ulbricht's "economic simulation" was a marketplace in which buyers might purchase drugs free of governmental "coercion" and "systemic use of force".

4.2 Silk Road 2

A month after Silk Road's seizure, the site reopened as Silk Road 2.0. The reopened site claimed to be under control of former Silk Road administrators. Three Silk Road 2.0 administrators were arrested the following month, probably traced by their activity on the original Silk Road. In November 2014, as part of Operation Onymous, Blake Benthall was arrested as the alleged owner of Silk Road 2.0, and the site closed (Wikipedia, Operation Onymous; O'Neill 2014). Several other onion-service marketplaces were also closed, though not the two largest marketplaces, Agora and Evolution. The seizure of Silk Road 2.0 is believed to have been enabled by operational-security errors by Benthall, and perhaps also by over-reliance on the only partial anonymity provided by Bitcoin transactions.

In 2015, the Evolution marketplace was closed by its owners as part of an "exit scam": the owners walked away with an estimated $12 million in Bitcoin held in the site's buyer-escrow fund (Brandom 2015; Krebs 2015). The Agora marketplace closed voluntarily a few months later, with the owners citing increased security concerns about Tor itself.

4.3 Playpen

The onion service known as Playpen started in August 2014 as a marketplace for child pornography. An FBI investigation began soon after, and received a significant boost in December 2014 when a source reported to the FBI that under some conditions the site leaked its real IP address (Rumold 2016). The FBI was then able to track the site to a data center in Virginia, and, from there, was able to identify the site's owner, Steven Chase.

The site was seized on February 20, 2015, and Chase was arrested. However, the FBI kept the site running until March 4 in order to collect information about the users.

Playpen's customers would have had no reason to supply a shipping address (*cf* users of Hansa Market, below), so the FBI tried a different approach. They took advantage of a vulnerability in the version of the Firefox-based browser then bundled with Tor, and were able to obtain IP addresses of about 1300 users. No information about the details of the vulnerability have been released, but it seems likely that it involved execution of code remotely installed on users' computers by the Tor server.

The FBI obtained a search warrant in the Virginia district where the server was found, signed by Magistrate Judge Theresa Buchanan. The warrant allowed for the deployment of a "network investigative technique", or NIT, against any computer that logged into the Playpen server (Crocker 2016).

The warrant was controversial in that it did not specify the locations of the user computers to be searched, and most of them turned out to be outside the Virginia district in question. Rule 41 of the Federal Rules for Criminal Procedure specified at the time that warrants could be issued "to search for and seize a person or property located *within* the district."

Rule 41 was amended at the end of 2016 specifically to allow the use of tools like the one used by the FBI in this case; a warrant may now be issued for the search of computers in any location if "the district where the [computer] is located has been concealed through technological means" (Federal Rules of Criminal Procedure).

The courts have not yet fully resolved the Fourth Amendment issues at stake here, or whether the amended Rule 41 passes Constitutional muster.

In March 2017, in the Playpen case *United States v. Jay Michaud*, the government dropped the charges when the FBI was ordered to reveal the precise technical details of the "network investigative technique". However, the indictment was dismissed without prejudice, allowing the government to re-file the charges at a later date, presumably after the point when the browser vulnerability is patched and therefore of no further use (Newman 2017).

Stephen Chase, unlike Ross Ulbricht, has issued no philosophical manifesto justifying his site. The forfeiture order following his conviction (Judge Voorhees 2017) lists no cash or cryptocurrency assets, indicating that Chase earned little if any money from the site. Several other child-pornography websites have been run on a free-exchange basis, suggesting their founders shared the paraphilia of their customers.

4.4 AlphaBay and Hansa Market

AlphaBay and Hansa Market were two competing Tor-based online e-commerce sites. Both primarily sold illegal drugs.

On July 4, 2017, the an international law-enforcement operation led by the FBI seized AlphaBay's servers, operating in Canada and the Netherlands, and arrested the site's owner, Alexandre Cazes, in Thailand (Greenberg 2018). Exactly how the authorities found the servers has not been released, but operational-security errors appear to have played a large role. An email address used by AlphaBay, for example, had been used previously by Cazes for a legitimate business, and the pseudonym Cazes used on the site had also been used elsewhere by him previously.

Cazes was found dead in his cell a week after his arrest, apparently by suicide.

At the time of the shutdown, AlphaBay had 350,000 product listings, according to the FBI, versus about 14,000 for the Silk Road at the point it was shut down (FBI 2017).

AlphaBay customers went scrambling for new sources. Most ended up at Hansa Market, which had been the second-largest online drug marketplace before the AlphaBay closure. But on July 20, 2017, Hansa Market too shut down. Worse, for buyers and sellers, it turned out that Hansa Market had been operating under complete control by the Dutch police since June 20, as part of a Dutch-German-American operation. The Dutch team had also figured out how to identify large numbers of buyers and sellers.

According to Dutch police, sometime in late 2016 an independent computer-security firm first got wind of the possible location of a Hansa Market development server, used for testing new software before it was migrated to the production servers, and notified the Dutch authorities (Greenberg 2018). How this discovery was made has not been released, but the development server is believed to have accepted non-Tor connections and thus would have been vulnerable to IP-address scanning.

When the Dutch authorities began monitoring the server, in a Dutch data center, they soon discovered one of Hansa Market's production servers in the same data center, and other Hansa servers in Germany. Searching those servers revealed references to two administrators' real identities.

Shortly after their discovery, these Hansa Market servers went dark, as Hansa Market itself was migrated to different servers. However, in April 2017 the police were able to track a Bitcoin payment, via blockchain analysis, from the suspected administrators to a data center in Lithuania. That data center turned out to be hosting the new Hansa Market servers.

When the servers were taken over in June, the police configured them to save all messages sent through the site. The site continued to strip EXIF metadata from images uploaded by dealers, but now began logging it first; this data often included GPS coordinates. Sellers were sent cryptographic-key files in Excel format; when opened, these files contained a macro that contacted the authorities (Dutch National Police Corps 2017).

Hansa Market continued to operate normally, to all appearances, though the police did ban the online sale of the exceptionally dangerous drug fentanyl. That decision, however, was initially proposed by existing Hansa Market moderators (Krebs 2017; Popper 2017).

At one point after the AlphaBay seizure, Hansa Market was getting so many new registration requests that the police had to temporarily disallow new registrations.

At the time Hansa Market was finally shut down, the police had information on tens of thousands of customers, and hundreds of dealers. Information on customers came from the messages they sent dealers; for about 10,000 customers, one of their messages included a shipping address. It is not expected, however, that more than a fraction of those customers will actually face prosecution.

References

"B", David. (2014, January 29). *Common darknet weaknesses 3: DNS leaks and application level problems, privacy PC*. Available at http://privacy-pc.com/articles/common-darknet-weaknesses-3-dns-leaks-and-application-level-problems.html. Accessed Mar 2018.

Bearman, J., & Hanuka, T. (2015, April). The rise and fall of silk road. *Wired Magazine*. Available at https://www.wired.com/2015/04/silk-road-1/.

Brandom, R. (2013, December 18). FBI agents tracked Harvard bomb threats despite Tor. *The Verge*. Available at https://www.theverge.com/2013/12/18/5224130/fbi-agents-tracked-harvard-bomb-threats-across-tor. Accessed Feb 2018.

Brandom, R. (2015, January 21). *Feds found Silk Road 2 servers after a six-month attack on Tor. The Verge.* Available at https://www.theverge.com/2015/1/21/7867471/fbi-found-silk-road-2-tor-anonymity-hack. Accessed Feb 2018.

Chaum, D. (1981, February). Untraceable electronic mail, return addresses, and digital pseudonyms. *Communications of the ACM, 24*(2), 84–89.

Crocker, A. (2016, September 28). *Why the warrant to hack in the playpen case was an unconstitutional general warrant.* Electronic Frontier Foundation. Available at https://www.eff.org/deeplinks/2016/09/why-warrant-hack-playpen-case-was-unconstitutional-general-warrant. Accessed Feb 2018.

Dingledine, R., Nick Mathewson, N., & Syverson, P. (2004, August). Tor: The second-generation onion router. In *Proceedings of the 13th USENIX security symposium*, San Diego, California.

Dutch National Police Corps. (2017, July 20). *Underground Hansa Market taken over and shut down.* Available at https://www.politie.nl/en/news/2017/july/20/underground-hansa-market-taken-over-and-shut-down.html. Accessed Mar 2018.

FBI. (2017, July 20). *Darknet takedown: Authorities shutter online criminal market AlphaBay, Federal Bureau of Investigation announcement.* Available at https://www.fbi.gov/news/stories/alphabay-takedown. Accessed Mar 2018.

Federal Rules of Criminal Procedure. Rule 41: Search and seizure. Available at https://www.law.cornell.edu/rules/frcrmp/rule_41. Accessed Mar 2018.

Greenberg, A. (2015, April). Silk road boss' first murder-for-hire was his mentor's idea. *Wired Magazine.* Available at https://www.wired.com/2015/04/silk-road-boss-first-murder-attempt-mentors-idea. Accessed Mar 2018.

Greenberg, A. (2018, March). Operation bayonet: Inside the sting that hijacked an entire dark web drug market. *Wired Magazine.* Available at https://www.wired.com/story/hansa-dutch-police-sting-operation. Accessed Mar 2018.

Hern, A. (2013, October 3). Five stupid things dread pirate Roberts did go get arrested, *The Guardian.* Available at https://www.theguardian.com/technology/2013/oct/03/five-stupid-things-dread-pirate-roberts-did-to-get-arrested. Accessed Feb 2018.

Hintz, A. (2013). Fingerprinting websites using traffic analysis. In *Proceedings of the 2nd international conference on Privacy Enhancing Technologies (PET)*, 2003.

Johnson, A., Chris, W., Jansen, R., Sherr, M., & Syverson, P. (2013, November). Users get routed: Traffic correlation on tor by realistic adversaries. In *Proceedings of the 2013 ACM SIGSAC conference on computer & communications security.*

Judge Katherine Forrest. (2015, May 29). *United States of America v Ross Ulbricht, 14 Cr. 68 (KBF) (sentencing hearing).* Available at https://freeross.org/wp-content/uploads/2015/05/Sentencing_2015-May-29.pdf. Accessed Feb 2018.

Judge Richard Voorhees. (2017, April 19). *U.S. v. Chase, Amended Preliminary Order of Forfeiture* (Docket No. 5:15cr15). Available at https://www.leagle.com/decision/infdco20170420c76. Accessed Apr 2018.

Krebs, B. (2014). Silk road lawyers poke holes in FBI's story. *KrebsonSecurity.* Available at https://krebsonsecurity.com/2014/10/silk-road-lawyers-poke-holes-in-fbis-story. Accessed Mar 2018.

Krebs, B. (2015). Dark web's "evolution market" Vanishes. *KrebsonSecurity.* Available at https://krebsonsecurity.com/2015/03/dark-webs-evolution-market-vanishes. Accessed Mar 2018.

Krebs, B. (2017). Exclusive: Dutch cops on AlphaBay "Refuges". *KrebsonSecurity.* Available at https://krebsonsecurity.com/2017/07/exclusive-dutch-cops-on-alphabay-refugees. Accessed Mar 2018.

Levine, Y. (2014, July 16). *Almost everyone involved in developing Tor was (or is) funded by the US government.* Pando.com. Available at https://pando.com/2014/07/16/tor-spooks. Accessed Mar 2018.

Newman, L. H. (2017, March 7). The feds would rather drop a child porn case than give up a Tor exploit. *Wired Magazine.* Available at https://www.wired.com/2017/03/feds-rather-drop-child-porn-case-give-exploit. Accessed Mar 2018.

O'Neill, P. H. (2014, November 7). The police campaign to scare everyone off Tor. *The Daily Dot.* Available at https://www.dailydot.com/layer8/tor-crisis-of-confidence/. Accessed Mar 2018.

Popper, N. (2015, December 25). The tax sleuth who took down a drug lord. *The New York Times*. Available at https://www.nytimes.com/2015/12/27/business/dealbook/the-unsung-tax-agent-who-put-a-face-on-the-silk-road.html. Accessed Mar 2018.

Popper, N (2017, July 18). Hansa market, a dark web marketplace, bans the sale of fentanyl. *The New York Times*. Available at https://www.nytimes.com/2017/07/18/business/dealbook/hansa-market-a-dark-web-marketplace-bans-the-sale-of-fentanyl.html. Accessed Mar 2018.

Rumold, M. (2016, September 15). *Playpen: The story of the FBI's unprecedented and illegal hacking operation*. Electronic Frontier Foundation. Available at https://www.eff.org/deeplinks/2016/09/playpen-story-fbis-unprecedented-and-illegal-hacking-operation. Accessed Feb 2018.

Syverson, P., David Goldschlag, D., & Michael Reed, M. (1997, May). Anonymous connections and onion routing. In *Proceedings of the 1997 IEEE symposium on security and privacy*.

Syverson, P., Tsudik, G., Reed, M., & Landwehr, C. (2000, July). *Towards an analysis of onion routing security, international workshop on designing privacy enhancing technologies: design issues in anonymity and unobservability* (pp. 96–114). Berkeley/New York: Springer.

Tarbell, C. (2014, September 5). *Declaration of Christopher Tarbell, United States v Ross Ulbricht, US District Court, Southern District of New York* (S1 14 Cr. 68 (KBF)). Available at https://freeross.org/wp-content/uploads/2018/01/140905-Tarbell-Declaration.pdf. Accessed Mar 2018.

The Tor Project, Exit Notice. (undated). *This is a tor exit router*. https://gitweb.torproject.org/tor.git/plain/contrib/operator-tools/tor-exit-notice.html. Accessed Feb 2018.

The Tor Project, Inc. (undated). *Tor: Onion Service Protocol*. https://www.torproject.org/docs/onion-services.html.en. Accessed Feb 2018.

The Tor Project, Metrics. (2018). *Welcome to tor metrics*. https://metrics.torproject.org. Accessed Apr 2018.

Ulbricht, R. (by assumption) (2013). https://stackoverflow.com/questions/15445285/how-can-i-connect-to-a-tor-hidden-service-using-curl-in-php, March 2013. Accessed Mar 2018.

Ulbricht, R. (2015 [estimated]). https://www.linkedin.com/in/rossulbricht. Accessed Mar 2018.

Wikipedia. (undated). *Operation onymous*. https://en.wikipedia.org/wiki/Operation_Onymous. Accessed Feb 2018.

Tor Black Markets: Economics, Characterization and Investigation Technique

Gianluigi Me and Liberato Pesticcio

1 Introduction

The introduction of tools ensuring a high level of anonymity and confidentiality to communications represent an amplification factor to spread illegal conducts, including the firearms trafficking, drugs or prescriptions trading, money laundering, child pornography or to spin-off new ones as cyber attacks targeting. These criminal behaviors are part of the underground economy or black market, defined by Smith (1994), as market based production of goods and services, whether legal or illegal, that escapes detection in the official estimates of gross domestic product. Hence, black markets have definitively online counterpart consisting of darknet markets (Europol 2016), where vendors and buyers can safely trade illegal goods and commodities, with heavily mitigating risks, as shown, e.g. in Lewis (2016).

The activities carried out in the Dark Web, and thus the transactions taking place on the dark markets, are supported by different technologies, whose combination provide the basic substrate for their achievement.

Therefore, these technologies represent an enabling infrastructure as a necessary prerequisite to the existence of the phenomenon itself, such as the ability to surf the net in an anonymous way, host web services in a hidden way or pay by anonymous and decentralised currency.

G. Me (✉)
LUISS University "Guido Carli", Rome, Italy
e-mail: gme@luiss.it

L. Pesticcio
Independent researcher, Rome, Italy

In particular, the widespread diffusion of new marketplaces, as black markets hosting platforms, has been enabled by the combination of the Tor network (hidden services) and decentralized electronic payment systems (e.g. Bitcoin, Monero, Zcash).

Based on the success of the first well-known dark market Silk Road, several anonymous marketplaces were built: when Silk Road was shut down in 2013 by law enforcement, the phenomenon was not defeated but simply sellers and buyers have moved on other marketplaces (Soska and Christin 2015) as a *rebirth*. Moreover, other marketplaces unexpectedly closed due to a fraud called *exit scam*, where the marketplace owner stops shipping orders while continuing to receive payment for new orders. Both phenomena represent an invariant of darknet marketplaces with a non-negligible frequency: at time of writing of this paper, less than ten marketplaces are active (*Hidden marketplace list changelog* 2017) with respect to more than 20 marketplaces active in April 2017 (e.g. take down of Alphabay and Hansa markets in Operation Bayonet, July 2017).

Therefore, aside the technological aspects, the real strength is the business model provided by the markets: this approach mitigates the critical issues of illegal transactions, encouraging participation in the transaction by removing human fears. In fact, this model eliminates the physical harm, prevents (mitigates) the intrusion of law enforcement, provides an escrow system to prevent financial risk and finally provides a quality control system based on feedbacks. Although it is not possible to know the overall revenues of the Tor marketplaces, according to T. Economist (2016) and *Global drug survey* (2016), Tor markets still account for a small share of illicit drug sales: they are growing fast, sellers are competing on price and quality, and seeking to build reputable brands. Turnover has risen from an estimated $15–$17m in 2012 to $150–$180m in 2015. According to *Spending on illegal drugs* (2017) the total illicit drug industry would have a realistic expected turnover of around $400 billion per annum: if only 1% of the expected drug revenues would pass through Tor network it would in the order of magnitude of a small-medium country GDP. Analogous consideration can be applied to other categories, such as Digital goods (e.g. fraud data, botnet rental). Finally, these activities could be managed by one or few criminals in order to fund serious organized crime (e.g. terrorism): this is the main reason why investigating Tor marketplaces cannot have a negligible part in the overall crime investigations.

2 Related Works

Tor is a P2P network for promoting traffic anonimity of various forms of internet communications for millions of users worldwide. In most cases, Tor users are very unlikely to become the target of an adversary, as they are namely shielded via Tor against opportunistic local hackers, local censorship authorities and hostile destinations.

Hence, Tor markets can be analyzed in order to detect rationals of criminal phenomena (descriptive analysis), to identify possible organizations or alliances and then target the deanonymization of single criminal. Law enforcement agencies (LEAs) have been utilizing a myriad of exploits to deanonymize some of Tor users, involving exploits of human errors, in addition to complicated mathematical methods that can take advantage of software flaws. Moreover, operational security (OPSEC) failures, which are usually related to mistakes committed by users, can facilitate deanonymization. Apart from techniques based on online behavior correlation (as happened for Ross Ulbricht, the Silk Road founder), identity exploitation can happen via typical attacks at application layer as exploitation of bugs (Firefox/Tor browser, FBI, August 2013), attacks on hidden services (e.g. via SSH) or through social engineering techniques (Sameeh 2017a, b; Berte et al. 2009).

The former descriptive analysis paper is by Soska and Christin (2015), where authors present an analysis of the anonymous marketplace ecosystem evolution. Their study is a long-term measurement analysis: in more than 2 years, they collected data from 16 different marketplaces, without focusing on a specific products' category. With respect to this work, our study focuses on illicit drugs trade, setting a short-term analysis on a reduced set of marketplaces. The results of their study suggest that marketplaces are quite resilient to law enforcement takedowns and large-scale frauds. They also evidence that the majority of the products being sold belongs to the drugs category. Several research works about anonymous marketplaces focuses on Silk Road, in particular on the drug selling (Celestini and Me 2016): in Van Hout and Bingham (2013) where the authors present the Silk Road user's experiences: in both cases the analysis concerned drug purchasing. In Van Hout and Bingham (2013) they monitored and observed the market's forum for 4 months and collected anonymous on-line interviews of adult 'Silk Road' users.

Activities on marketplaces imply relationships between actors, namely vendors and buyers: even if transactions cannot be tracked outside the marketplaces, links between participants can be analyzed due to the openness of adopted technologies (e.g. reputation, PGP signatures). Hence, a participants graph can be sketched in order to better analyse the phenomena.

In particular, Social Network Analysis (SNA) has been an active area of research in psychology, sociology, anthropology, and political science for many years. Research topics include isolation and popularity, prestige, power, and influence, social cohesion, subgroups and cliques, status and roles within organizations, balance and reciprocity, marketplace relationships, and measures of centrality and connectedness. Therefore, social networks can be useful in making predictions because we can expect that a our attitudes and behavior are affected by the people we know. Detection of clusters, information communities, core groups, and cliques is an important area of research in SNA (Zubcsek et al. 2014). Social network analysis often involve multidimensional scaling, hierarchical cluster analysis, log linear modeling, and a variety of specialized methods. Such as review a variety of statistical methods in network science (Kolaczyk 2009; Schnettler 2009; Borgatti et al. 2013).

Warren et al. (2006) have applied SNA to PGP keys using time-stamps to track the evolution of the social network. Moreover, they have shown how to build a social network with publicly available information retrieved from Keyservers (2016), in order to understand how the network may change in a dynamic way against an event which concerns the participants to the network (such as conferences).

Finally, several recent works focused on the study of Internet organised crime (Europol 2016; Thomas et al. 2015), in particular the study of on-line drugs marketplaces (Celestini and Me 2016; Laura and Me 2017) has become quite popular, where novel techniques can be applied, e.g. criminal networks analysis with graphs, as in Firmani et al. (2014).

3 Economic Characterization

Studies with a social approach towards the underground markets (Allodi et al. 2016) showed that criminals prefer trading in a more secured and hierarchical system to further increase trading efficiency and stability of the market. In particular, running an efficient underground economy in which criminals trade goods and services often lead to scammers and, consequently, to market failure (Herley and Florencio 2009). In particular, in Herley and Florencio (2009), several key features have been identified to be responsible for market failure:

- Users could join the market freely and with an arbitrary identity. Feedback mechanisms, as reputation, on the reliability of the users are not effective;
- There is no history of transactions available, so it is impossible to look back to user's trades or community provided feedbacks;
- The community is largely unregulated and no assurance for the buyer or the seller exists, e.g. there's no way to identify a legitimate trader and not a scammer before the transaction.

Hence, the aforementioned mechanisms could lead to well known Akerlof's lemon's market, where undesired results occur when buyers and sellers have asymmetric information driving the "low quality" products or services to be the only possible deal on the market (therefore the need of signaling systems as reputation). Hence, the asymmetric information generates imbalance of power in Darknet Marketplaces transactions (and, generally, in underground economy), which can sometimes cause the transactions to go away or scams to appear, and represent a kind of market failure in the worst case. In fact, unlike buyers in traditional settings, online shoppers are often physically unable to inspect the products for sale and typically must rely on pictures and descriptions provided by the seller (information asymmetry). Any time buyers cannot determine the quality of a product until after the purchase has been made, sellers have less incentive to provide high quality products (e.g. lemons market). In case of Tor marketplaces illegal trade of goods/services reinforces information asymmetry, due to the poor reliability of the criminal activity of the vendor, increasing the fraud risk. This creates an imbalance of power in transactions, which can sometimes cause the transactions to go away, a kind of market failure in the worst case. Hence, the Tor Marketplaces replicate the reputation mechanisms, reinforced by offering the escrow service and, sometimes, by denying the Finalized Early (pay the goods before shipping) option. Many examples show that many vendors, even after very positive feedback, disappear after a period of fair trading. This is the well known phenomenon named *exit scam*. Finally, Tor Marketplaces present many features of two-sided markets, with effects of interdependence or externality between groups of agents (vendors and buyers) that the Marketplace owner (the intermediary) serves.

4 The Criminal Impact of the Marketplaces

As shown in Europol (2017), darknets are a key facilitator for various criminal activities including the trade in illicit drugs, illegal firearms and malware. Hence, Darknet marketplaces represent an evidence of enablers for Crime as a Service, because they provide goods and services typically found in physical black markets. In particular, illicit online markets, both on the surface web and Darknet, provide criminal vendors the opportunity to purvey all manner of illicit commodities, with those of a more serious nature typically found deeper in the Darknet. Many of these illicit goods, such as cybercrime toolkits or fake documents, are enablers for further criminal activities. Sale of illicit goods to dedicated criminal websites and markets hosted on anonymising networks such as Tor, I2P and Freenet, although such activity appears to be mainly concentrated on the Tor network, which is increasingly decentralised (more than 2.2 million directly connecting users and almost 70,000 unique *.onion* domains in April 2018).

The drugs market is undoubtedly the largest criminal market on the Darknet, offering almost every class of drug for worldwide dispatch. As of June 2017, AlphaBay, one of the largest Darknet markets, had over 250,000 separate listings for drugs, accounting for almost 68% of all listings (with 30% of the drugs. Thirty percent of the drugs listings related to Class A drugs). While it is assessed that the majority of vendors are lone offenders, dealing in small amounts, it is reported that many of the top sellers are likely organised crime groups earning significant profits. Some studies suggest that the total monthly drugs revenue of the top 8 Darknet markets ranges between EUR 10.6 million and EUR 18.7 million when prescription drugs, alcohol and tobacco are excluded. Thanks to the related changes of business models for drug trafficking and organized crime the new crime opportunities can generate revenues on top of current ones, between a fifth and a third of the income of transnational organized crime where 60–70% of global drug proceeds may be laundered (UNODC 2017). Moreover, as shown in Europol (2017), increasingly sophisticated security features protecting documents against forgery as well as improved technical control measures have compelled Organized Crime Groups to improve the quality of fraudulent documents and suppliers of raw materials now primarily rely on Darknet marketplaces to sell their products.

Finally, Infringements of intellectual property rights (IPR) are a widespread and ever-increasing worldwide phenomenon. In 2013, the international trade in counterfeit products represented up to 2.5% of world trade. The impact of counterfeiting is even higher in the European Union, with counterfeit and pirated products amounting to up to 5% of imports. As discussed earlier, most counterfeit products can more readily be sold on the surface web, being presented as, or mixed with, genuine products. Consequently, counterfeit products only account for between 1.5% and 2.5% of listings on Darknet markets. Moreover, the most commonly listed counterfeit products are those which are obviously illegal counterfeit bank notes and fake ID documents, which account for almost one third and almost one quarter of counterfeit listings respectively. The majority of reported law enforcement investigations in the EU relating to counterfeit goods on the Darknet relate to counterfeit bank notes.

5 Investigating the Marketplaces

5.1 System Description

Crawling and scraping are fundamental to web and network data science. In fact, modeling and analysis begin with data, whose massive store is represented by the web. and the web is a massive store of data. Extracting relevant data in an efficient manner, preserving the data quality, is an essential skill of data science. In this paper, due to the format of target pages it is not needed to turn to sophisticated techniques based on semantics (Laura and Me 2017), while, after issuing a focused crawl, a raw web scraping with XPath regular expressions for text parsing has needed. The environment used for the analysis is a Linux box with Tor (Dingledine et al. 2004) on-board, in particular, Whonix (a desktop operating system designed for advanced security and privacy), and custom script for scraping every marketplace. The scraping software has been implemented relying on top of Scrapy (http://scrapy.org), an open source framework written in Python specialized for web crawling, enhanced by parsing capabilities (through CSS or Xpath selector), on-the-fly database population with parsing results and HTTP metadata access.

Furthermore, the spiders implemented an additional feature in order to overcome anti-DDoS systems; in fact, a delay was applied via an adaptive tuning on the timing requests. In addition to the delay parameters, we needed to provide a session cookie that we obtain by manually logging into the marketplace, after solving a CAPTCHA. Recently, new open-source web crawlers have been unveiled to serve the Tor community, e.g. Fresh Onions (https://github.com/dirtyfilthy/freshonions-torscraper) promising to potentially become one of the best Tor indexing tools out there, it is designed for indexing hidden services on the Tor network and comes out with many features. The analysis was aimed to acquire the PGP keys items in category Digital Goods, both software and fraud from five different markets. In particular, We extracted the information regarding the vendors and buyers from each item's page of marketplaces in Table 1.

Table 1 Markets URLs

Market	URL
Alphabay	http://pwoah7foa6au2pul.onion
Dream	http://lchudifyeqm4ldjj.onion
Hansa	http://hansamkt2rr6nfg3.onion
Outlaw	http://outfor6jwcztwbpd.onion
Valhalla	http://valhallaxmn3fydu.onion

5.2 Analysis

Data mining can be defined as the process of mining for implicit previously unknown, and potentially useful information from very large database by efficient knowledge discovery techniques. Data mining techniques are widely applied to the issues of cyber security and crime: many LEAs today are faced with large volume of date that must be processed and transformed into useful information. This science can greatly improve crime analysis and aid in reducing and preventing crime. As D.E. Brown stated in 2003, "*no field is in greater need of data mining technology than law enforcement*" crime analysis and prevention can dramaticaly benefit of data mining, and, more in general, data science techniques.

The typical structure of anonymous marketplaces offer a list of products and their individual pages where authorized vendors can set up a virtual shop and place listings. Items for sale are organized in categories and subcategories, the organization vary from market to market, with large coincidence of main product categories (e.g. drugs, weapons, frauds). Generally, it is possible to search products for sale both by product categories and by keywords, but the last option is not always available. Buyers and sellers are able to leave feedback about their transactions, typically including a rating (e.g., good/bad or a value between 0 and 5), a comment and the obfuscated user' nickname who leaves the feedback. Such information are used to construct users' reputations inside the market both as sellers and as buyers. The main difference with surface market is the regulation of market access: users accounts of anonymous marketplaces are needed not only to carry out transactions, but they are required to access the market itself, which is not true for surface markets Every marketplace vendor has a PGP public key shown in his/her home page, to be used to provide confidentiality to the negotiation and purchasing phase between vendor and buyer in a Tor marketplace. The strength of PGP (and its open-source OpenPGP-compliant programs) is based on the decentralized model (Web of Trust, WOT), used to establish the authenticity of the binding between a public key and its owner. The PGP WOT can be analyzed as a graph representing a social network, exploiting the signature of keys: in fact, signing a key represents a relationship between the PGP owners, e.g. vendor/buyer. The resulting social network is a structure consisting of nodes (actors) and the links between them, that identify a kind of relationship, such as friendship, professional cooperation, and in our case the signature of the PGP identity. As mentioned earlier, SNA relies on algorithms and tools used in the analysis of Graph Theory. Formally, a graph is a pair $G = V, E$

where V is a set of nodes and E is a set of edges which are pairs of nodes (i.e. E = e = (x, y) : x V and y V). An undirected graph is a graph without edge orientation, thus the edge (x,y) is identical to the edge (y,x). Whereas directed graph is a graph with a set of edge E of ordered pairs of vertices: this is the case of analysis of PGP signatures in this paper.

In particular, as hypothesis, we assume that if a node x signs a node y then node x has bought goods/services from node y. This general hypothesis can hold because of the need of good reputation of vendors, stated by more received signatures. We applied SNA to this graph, with analogous reasoning in crime, where data refers to the relationship element, namely the information on a social network made up of actors with their attributes (nodes), and the links between the nodes.

The main criminal graph metrics to consider are:

1. *Homophily*, defined as a network where criminals are more likely to be connected to other criminals, and legitimate people are more likely to be connected to other legitimate people. Generally, the network homophily relies on crime type: criminal nodes are not uniquely connected to other criminal nodes, but connect to legitimate nodes as well. A network is homophilic if criminal nodes are significantly more connected to other criminal nodes, and as a consequence, legitimate nodes connect significantly more to other legitimate nodes. Further useful metrics are dyadicity and heterophilicity:
2. *Centrality metrics*, quantifying the importance of an individual in a social network (Borgatti et al. 2013). Centrality metrics are typically extracted based on the whole network structure, or a subgraph. The main metrics are:

   - *Closeness*, the average distance of a node to all other nodes in the network;
   - *Betweenness*, the number of times a node or connection lies on the shortest path between any two nodes in the network;
   - *Graph theoretic center*, the node with the smallest maximum distance to all other nodes in the network;
   - *Eigenvector centrality (EC)* which is a measure of a vertex's centrality which often reflects its importance based on the graph's structure. Using EC, a vertex is considered important if it has many neighbors, a few important neighbors, or both. More formally, the eigenvector centrality xi for a vertex i in a graph G is defined in Eq. (1)

$$x_i = K_i^{-1} \sum_j A_{ij} x_j \qquad (1)$$

where A is the adjacency matrix of G, K_1 is its largest eigenvalue, $0 < x_i < 1$ and xj are i's neighbors eigenvector centralities. The EC is a useful metric for identifying important vertices in a graph independent of the underlying data being represented. We will use this to help determine a strategy that attempts to maximize damage to a criminal network. Removing important vertices targets portions of the criminal network that are used both frequently and collectively to host the operations of multiple criminals.

3. Neighborhood metrics characterize the target of interest based on its direct associates. The n-order neighborhood around a node consists of the nodes that are n hops apart from that node;
   - *Degree* summarizes how many neighbors the node has. In crime, it is often useful to distinguish between the number of criminal and legitimate neighbors;
   - *Density* derives how closely connected the group is. A high density might correspond to an intensive information flow between the instances, which might indicate that the nodes extensively influence each other ($d = 2M/N(N-1)$, with M and N the number of edges and nodes in the network respectively);
   - *Triangles*, number of fully connected sub-graphs consisting of three nodes. In an egonet a triangle includes the ego and two alters. If the two alters are both criminal (legitimate), we say that the triangle is criminal (legitimate);
   - *Modularity* class, which is used to identify clusters or communities, as clusters are popularly called in the networks world, in a given graph. Modularity results in grouping of nodes that are far more strongly connected than they would have been in a random graph.
4. *Collective inference algorithms*, as Page rank algorithm, used as a propagation of page influence through the network. The same reasoning can be used to propagate crime through the network. That is, we personalize the ranking algorithm by crime. Instead of web pages, the adjacency matrix represents a crime network (e.g., a people-to-people network). The final ranking assigned to each node should be interpreted as a ranking and not as a score. The top-ranked nodes are the most influenced by crime.

6 Investigation Analysis on Marketplaces

Social networks are an important element in the analysis of crime, which is often committed through illegal setups with many accomplices. When traditional analytical techniques fail to detect crime due to a lack of evidence, SNA might give new insights by investigating how people influence each other. These are the so called guilt-by-associations, where we assume that criminal influences run through the network. Hence, SNA is the process of investigating social structure through the use of network and graph theories. This methodology to analyze social relationships is applied in different sciences, such as sociology, psychology, economic, and criminal investigations. For example in the publication Europol Review, Europol has adopted state-of-the-art SNA as an innovative way to conduct intelligence analysis and support major investigations on organized crime and terrorism. As a straightforward intuition result, funneling effect plays an important role in criminal analysis, thus resulting network analysis can provide original results with no negligible impact. Moreover, e.g., multi-edge nodes (two nodes are connected by more than one edge) can represent multiple purchases of a buyer to the same vendor, or hyper-edge nodes (an edge that connects more than one node in the network) can represent different purchases to the same vendor. When analyzing Marketplaces WOT networks, we integrate the vendor label of the nodes into the network: a node can be vendor or buyer, depending on the condition of the object it represents. I.e., a Marketplaces WOT network can represent the vendor and the buyer people by white- and black-colored nodes, respectively.

For the scope of this paper, the actors are represented by the PGP identities and the directed edges between nodes, that is to say the relationship, will be the signature by an actor on the identity of another actor, according to the principle of the web of trust. The WOT is a concept used by PGP to establish the authenticity of the pair user-public key, based on reputation, following the intention as a cryptographic tool for the masses, there are no central authorities globally trusted. There may be many independent networks of trust, where every user can be part of it and act as a link between many of them. An identity certificate can be signed electronically by other users, who in this way, attest to the objective association of that public key to that particular user.

6.1 Drugs Vendors and Items

Analyzing one of the most important market, the drugs items represent aroud 44% of the entire marketplace, out of 25,360 distinct items observed throughout the year 2015 for market 11,057 are in the drug category. In particular, the types of substances make up the market as shown in the Fig. 1. As can be seen, the stimulants, ectasy, prescription along with light drugs, cannabis and hashish, cover about 60% of the market.

The drug category is extremely fragmented, with the top 10 sellers not exceeding the 11% of share and the top vendor holding a quota of 1.99 Fig. 2.

6.2 Identity Detection

The identity category, representing around the 15%, includes many *items* that can interweave the cybernetic with less technological threats such as terrorism or more traditional crimes; e.g. you can easily buy a new passport, driver license to cross borders or deceive customs officers. In Fig. 3 the subdivision of sub-categories. In particular, the identity category is less fragmented while the top vendor holds about 8.50% of the category (Fig. 4).

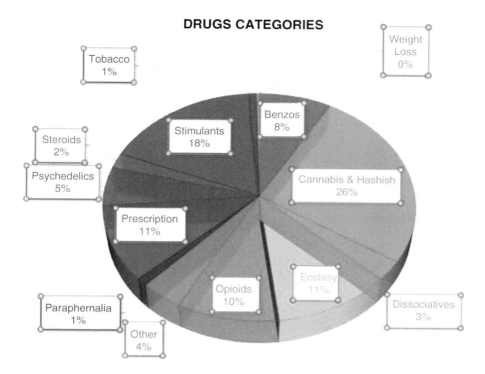

Fig. 1 Drugs categories

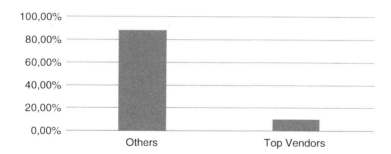

Fig. 2 Drug category quotas

6.3 Digital Goods Check

The items in the "digital good" category represent around 14% and include technology threats such as botnets, exploits, and predominantly those items threatening business profits, such as DRM violations or credit card codes (Fig. 5). The digital goods category is fragmented like identity, but the top vendor holds about only 2.40% of the category (Fig. 6).

Fig. 3 Identity categories

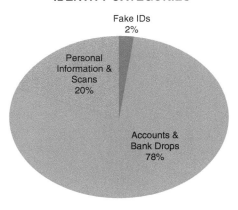

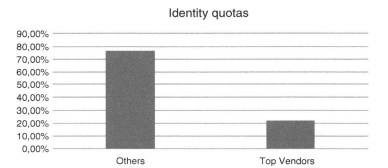

Fig. 4 Identity categories quotas

Fig. 5 Digital goods categories

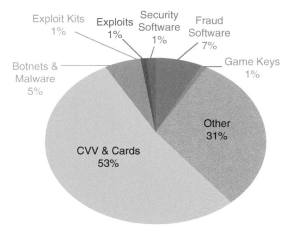

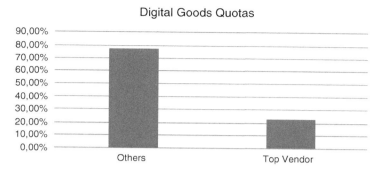

Fig. 6 Digital goods categories quotas

Table 2 PGP identity fields

UserID	Commonly represented by the owner's e-mail address, 'name <userid@domain>
Public Key	The public key associated with identity
KeyID	'0x0123456789ABCDEF'
Signatures	"Which are signatures affixed by other PGP

6.4 The Importance of Digital Identity

The core of this work is represented by the analysis of the information embedded in PGP identities, contains different information, for our purposes the most important are whose relevant subset is shown in Table 2.

Because UserID can be arbitrarily assigned to a public key, it is a widely accepted practice within the PGPcommunity to refer to a public key by its key-ID rather than its UserID. The keyservers are publicly available for data mining, therefore they are free to be openly accessed (the downside of WOT), allowing to track the relationship between PGP identities, thus enabling additional identity information retrieval and related relationships. effect allowing to retrieve additional information on likely relationships of a user. Based on the abovementioned considerations, the intuition is to retrieve the PGP identity from a vendor profile on the marketplace and extract from it the KeyID to be submitted to the key servers. The aim is to discover who has signed that identity, and thus iteratively proceed also for the signer. The goal is to build, gradually, the social network made up of the relationships between the public keys.

The underlying methodology of the analysis consists of several steps:

1. Vendors list: the initial operation is to retrieve the keys of vendor profiles on marketplaces; this task is split into two steps, the first one is to set up a spider that browses the site by extracting the whole list of active vendors (then those sellers who currently offer at least one item), in the second step, we identify the sections in the DOM of the HTML page where the vendor's PGP key is present, so the key is retrieved.

2. PGP keys retrieval: extraction of information in PGP keys from a vendors repository based only on the information retrieved from the marketplaces.
3. Data update: using spider functionalites again, keys are checked against the PGP keyserver by comparing information with those in the repository as the information on the keyserver is latest. It is not feasible to start directly with the requests to the keyserver, given that the keys may not be uploaded on keyserver.
4. Extraction of the signed keys: the signing keyIDs are recursively extracted from the signatures field to verify their presence on the keyservers.

At the end of the process we get a hybrid database containing both the keys on the keyserver and the keys recovered from the market and not uploaded. The information about vendors within the repository are as follows:

- Market
- Categories
- Alias
- UserID
- KeyID
- The date when the key was created
- Key size
- Signatures update after the entire process
- The date of revocation, if present
- Possible new UserID
- A flag indicating whether the key is on the keyserver
- A flag indicating whether the signatures are different between the version in store and that on the keyserver
- The old signatures at the end of the process

This container can be used to carry out deeper analysis on the keys, discussed in the next paragraphs. However, in order to successfully apply SNA, a further step has needed, consisting of the signatures conversion in a list mapping the sources and targets, and edges of the graph. So the outcome of the whole process is a directed graph, signed and signer: the calculation of SNA indexes has been carried out with Gephi, an open-source network analysis and visualization software. Looking at the degree distribution, the Tor marketplaces users graph shows feature of the preferential attachment model and follows from the work in Barabasi and Albert (1999). This model adds new links to a random graph in a manner that gives preference to nodes that are already well connected (nodes with higher degree centrality).

The preferential attachment model, as shown for our case, is the underlying model of the tipping markets, where 'The rich get richer, and the poor get poorer.' The distribution of the degree, in fact, has the long tail network shape (scale-free).

Moreover, the overall graph including only vendors confirms the long tail shape, which is in contrast, e.g., with the Real-Life Network of Social Security Fraud, whose degree distribution follows a power law.

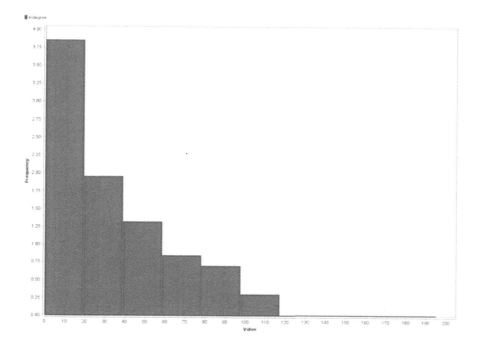

Fig. 7 In-degree distribution

Table 3 Giant component

	Whole	Giant	Giant%
Nodes	7301	6902	94.53
Edges	11,538	11,287	97.87
AVG degree	1.58	1.635	

Furthermore, as expected, the top vendors (those with the highest in-degree) result in the highest eigenvector centrality (top influencer or best seller) and in the lowest betweenness and closeness centrality, as they could represent separate markets with different customers (in general). Finally, the top vendors hold the highest page rank values, suggesting to investigate on possible inter-relations, because their geodesic distance is two (Fig. 7).

In the overall graph there is a macro-network, a connected component of the graph that contains a constant fraction of the entire graph's vertices (Giant Component), that includes most of the nodes, and it is the network on which we carried out further analysis. The table shows the measurements for the Giant Component.

Gephi provides a set of algorithms to identify, e.g. the Giant Component data as shown in Table 3 for measuring indexes of the graphs and the SNA, such as the PageRank algorithm for detecting the Authorities.

Id	Label	Authority
0x	Alphabay – DREAM	0.888024
0x	Alphabay – DREAM	0.179341
0x	Alphabay	0.142529
0x	Alphabay – DREAM	0.132848
0x	Alphabay – Dream – O...	0.079979
0x	Alphabay	0.077786
0x	Alphabay – Dream – Va...	0.075057
0x	Alphabay	0.067587
0x	Alphabay	0.064441
0x	Dream – Valhalla	0.063917
0x	Alphabay	0.058393
0x	Alphabay	0.056427
0x	Alphabay	0.05541
0x	Alphabay	0.055195
0x	Alphabay – Dream – Va...	0.052176
0x	Alphabay	0.051403
0x	Alphabay	0.048057
0x	Alphabay – DREAM	0.047381
0x	Alphabay	0.044705
0x	Dream – Valhalla	0.044602
0x	Dream	0.04331
0x	Alphabay	0.042555
0x	Alphabay – DREAM	0.041457
0x	Alphabay	0.041243

Fig. 8 Authorities degree

In the Fig. 8, the results of applying algorithms for Authorities and hubs.

The Fig. 9 shows the Giant Component with top authorities labels showing the markets where they are active. Using the values of Authority, we focused the analysis on the identified node, this entity with graph of depth 1 (then with a single hop) creates a Social Network of 196 nodes: taking the same entity but selecting a graph with depth 2, it creates a network consisting of 569 nodes and 943 edges, (Fig. 10) and reaches all 25 top authorities.

From the analysis of the top 25 of the authority nodes, we retrieved some additional information about the nature of "top vendors" such as ship-to, ship-from, markets, and categories, as shown in Table 4.

About the types of items sold are all highly specialized (e.g., only pills, drugs, cannabis) except for one which is cross-category. The Table 5 shows the categories

Tor Black Markets: Economics, Characterization and Investigation Technique 135

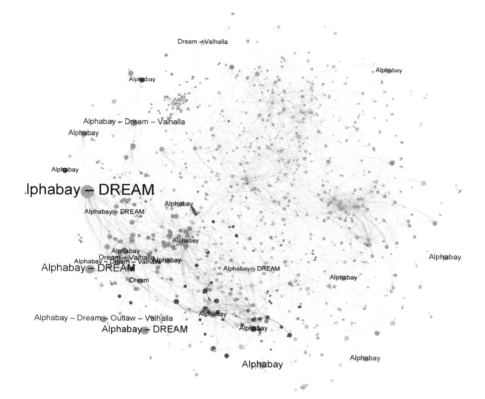

Fig. 9 SNA graph

of items handled by the top authorities. Several authorities operate on more markets, but some are monomarket in particular 12 are only present on Alphabay and 1 on Dream market.

In general, we can point out that the authorities have only input edges, only one has an outbound edge to an identity that has in turn signed it.

Finally, Fig. 11 shows there is one dominant class with more than 1100 nodes embedding the third top vendor, while the next two classes contain the other two top vendors. The suggested intuition is to firstly investigate on these communities which potentially hide a criminal structure. In particular the bigger classes contain even the most signed vendors, suggesting exactly a possible criminal chain with different roles. Moreover, the biggest class embeds even the node with highest closeness centrality, suggesting to further investigate it as a bridge between different criminal groups. Finally, the presence of a mid-sized class embedding the nodes with the highest eigenvector centrality, whose distribution is in Fig. 12, and betweenness centrality infer to investigate the group as a link between different vendors groups.

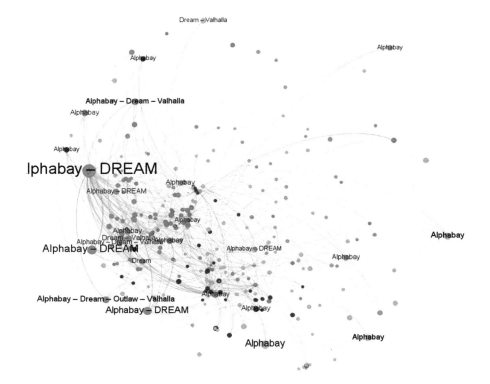

Fig. 10 Authority two hop

Table 4 Top vendors shipments

Country	Ship-from	Ship-to
USA	18	17
WW	4	7
GERMANY	1	
EU		1
UK		1
CAN		1
MEX		1
ARG		1
BRA		1
N/A	1	

7 Conclusions

It is possible to analyze the relationship between the various identities, and we can detect the authorities, node that have zero out-degree, thus receiving signatures only. We have shown that in an anonymous environment, a disclosed link between

Table 5 Top vendors categories

Category	N
Drug	20
Pharma	2
CVV/Digital	1
Multi	1
N/A	1

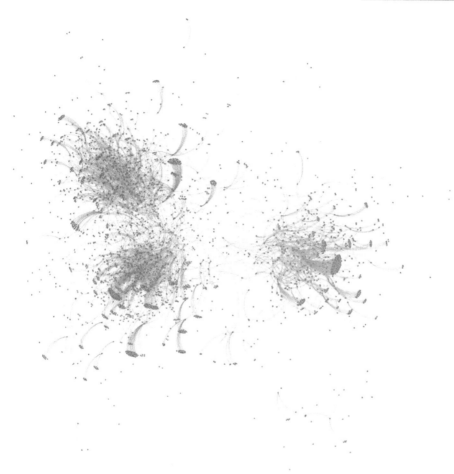

Fig. 11 Top three modularity classes

dark web and surface can be found in web of trust system identities, through the information leakage exploitation given by PGP.

By integrating publicly accessible information on the surface (keyservers) with information accessible on Darknet (marketplaces, stores, website, forums, etc.),

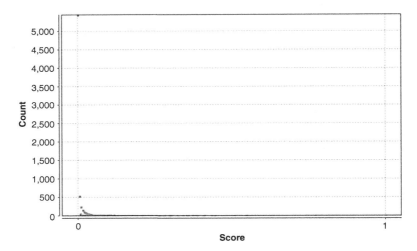

Fig. 12 Eigenvector centrality

you can attempt to attribute some KeyID to UserID (names, alias, email), so this information can be used as a starting point for more in-depth analysis with specific tools (e.g. Maltego or a Cyber Intelligence platform) and content enrichment services.

Furthermore, leveraging this technique, there is some past exploitation of the context in which identity has been found, simply by looking at the built relationships. For example, a KeyID that signed five other keys related to drug vendors. Moreover, analyzing the items offered by these vendors, we can get additional information, the intersection between the items offered (i.e., in this paper, the cannabis). Therefore, we can assume that digital identity is associated with a cannabis consumer or any person interested in that category of goods. In addition, the key discussed above was not found on the keyserver, but the assumptions we have made are only due to signatures analysis, so even if the key is not found, it is possible to infer a minimum of additional information.

The resulting repository allowed to quickly and easily analyze the behavior of vendors on various marketplaces: i.e., several sellers use different aliases on different markets while maintaining (perhaps for simplicity of management) the same PGP key. Consequently, multiple different aliases were disclosed leading to the same vendor. Finally, using the repository enables the deanonymization of some relationships: in fact, if an identity is present on the keyserver so we can retrieve the information about the signed keys, but the same keys are not present on the keyserver, so if the keys that signed the identity you are looking for are not uploaded the analysis would stop.

By leveraging the hybrid repository we can deanonymize the identity by attributing, i.e., to the keyID 0x1EXXXXXXXXXXXXXX, market, vendor alias, categories of items sold, UserID and others information on PGP identity.

This leakage allows the building of a social network among the entities belonging to underground world, where to apply standard SNA metrics, indicating possible priorities in investigation paths, as well hypothesis (to be verified) on possible structures in case of organized crime or relationships between criminals (e.g. dealer). The future direction of this work is both enlarging the set of the vendor, either conduct periodic checks to capture the change in the graph and the possible rise of new relationships and the links between subnets previously unrelated, applying further metrics able to identify new possible investigation paths to prioritize the criminal analysis.

References

Allodi, L., Corradin, M., & Massacci, F. (2016). Then and now: On the maturity of the cybercrime markets the lesson that black-hat marketeers learned. *IEEE Transactions on Emerging Topics in Computing, 4*(1), 35–46.

Barabasi, A., & Albert, L. R. (1999). Analyzing social networks. In *Emergence of scaling in random networks* (p. 509512).

Berte, R., Lentini, A., Me, G., et al. (2009). Fast smartphones forensic analysis results through mobile internal acquisition tool and forensic farm. *International Journal of Electronic Security and Digital Forensics (IJESDF), 2.* online.

Borgatti, S. P., Everett, M. G., & Johnson, J. C. (2013). *Analyzing social networks.* Thousand Oaks: SAGE.

Celestini, A., & Me, G. (2016). *Tor marketplaces exploratory data analysis: The drugs case* (J. Hamid, C. Alex, E. David, H.-F. Amin, B. Guy, S. Graham, J. Arshad, Eds.), (pp. 218–229).

Dingledine, R., Mathewson, N., & Syverson, P. (2004). *Tor: The secondgeneration onion router* (Technical report, DTIC document).

Europol. (2016). *The internet organised crime threat assessment* [Online]. Available: https://www.europol.europa.eu/activities-services/main-reports/internet-organised-crime-threat-assessment-iocta-2016.

Europol. (2017). *Serious organized crime threat assessment* [Online]. Available: https://www.europol.europa.eu.

Firmani, D., Italiano, G. F., & Laura, L. (2014). The (not so) critical nodes of criminal networks. In *International conference on social informatics* (pp. 87–96). Springer.

freshonions. [Online]. Available: https://github.com/dirtyfilthy/freshonions-torscraper.

Global drug survey. (2016). [Online]. Available: https://www.globaldrugsurvey.com/.

Herley, C., & Florencio, D. A. F. (2009). Nobody sells gold for the price of silver: Dishonesty, uncertainty and the underground economy. In *Proceedings (online) of the Workshop on Economics of Information Security (WEIS)*.

Hidden marketplace list changelog. (2017). [Online]. Available: https://www.deepdotweb.com/hidden-marketplace-list-changelog/.

Keyservers. (2016). *dsadsa* [Online]. Available: https://skskeyservers.net/status/.

Kolaczyk, E. (2009). *Statistical analysis of network data: Methods andmodels* (Springer Series in Statistics, p. 386).

Laura, L., & Me, G. (2017). Searching the web for illegal content: the anatomy of a semantic search engine. *Soft Computing, 21*(5), 1245–1252. https://doi.org/10.1007/s00500-015-1857-4.

Lewis, S. (2016). Onionscan report June 2016-snapshots of the dark web. Hentet fra https://mascherari.press/onionscan-report-june-2016.

Sameeh, T. (2017a). *An overview of modern tor deanonymization attacks* [Online]. Avaible: https://www.deepdotweb.com/2017/09/12/overview-modern-tordeanonymization-attacks/.

Sameeh, T. (2017b). *Targeting adversaries and deanonymization attacks against tor users* [Online]. Available: https://www.deepdotweb.com/2017/08/21/targeting-adversariesdeanonymization-attacks-tor-users.

Schnettler, S. (2009). A structured overview of 50 years of small-world research. *Social Networks, 31*(3), 165–178.

Scrapy. [Online]. Available: http://scrapy.org.

Smith, P. (1994). Assessing the size of the underground economy: the canadian statistical perspectives. *Canadian Economic Observer, 3*, 16–33 Catalogue No. 11-010.

Soska, K. & Christin, N. (2015). *Measuring the longitudinal evolution of the online anonymous marketplace ecosystem.*

Spending on illegal drugs. 2017. [Online]. Available: http://www.worldometers.info/drugs/.

The Economist. (2016). *Shedding light on the dark web* [Online]. Available: https://www.economist.com/news/international/21702176-drug-trade-moving-street-online-cryptomarkets-forced-compete.

Thomas, K., Yuxing, D., David, H., Elie, W., Grier, B. C., Holt, T. J., Kruegel, C., Mccoy, D., Savage, S., & Vigna, G. (2015). Framing dependencies introduced by underground commoditization.In *IProceedings (online) of the Workshop on Economics of Information Security (WEIS)*.

UNODC. (2017). World drug report 2017.

Van Hout, M. C., & Bingham, T. (2013). Surfing the silk road: A study of users experiences. *International Journal of Drug Policy, 24*(6), 524–529.

Warren, R., Wilkinson, D., & Warnecke M. (2006). Empirical analysis of a dynamic social network built from pgp keyrings. In *iCML'06 Proceedings of the 2006 conference on statistical network analysis* (pp. 158–171).

Zubcsek, P. P., Chowdhury, I., & Katona, Z. (2014). Information communities: the network structure of communication. *Social Networks, 38*, 50–62.

A New Scalable Botnet Detection Method in the Frequency Domain

Giovanni Bottazzi, Giuseppe F. Italiano, and Giuseppe G. Rutigliano

1 Introduction

One of the most insidious threat in the cyber domain is currently represented by the diffusion of botnets, which are networks of infected computers (called bots or zombies), typically propagated through malware. The manager of a botnet, a.k.a. the botmaster, controls the activities of the entire network giving orders to every single zombie through various communication channels and some Command-and-Control servers (C2).

Botnets are very common in various cybercriminal contexts, because they are able to provide a very efficient and distributed platform that can be used for several malicious activities, such as Distributed Denial of Service, click fraud, cyber extortion or crypto currency mining.

Over the years, two main approaches to botnet detection have been widely deployed. The first approach can be summarized as the application of active countermeasures for the identification of the specific malicious agents and/or the communication protocols. The second approach, usually labeled as passive countermeasures, is carried out essentially through traffic analysis.

A preliminary version of this chapter was presented at the 9th International Conference on Security of Information and Networks (SIN 2016) (Bottazzi et al. 2016).

G. Bottazzi (✉) · G. F. Italiano
Department of Civil Engineering and Computer Science, University of Rome "Tor Vergata", Rome, Italy
e-mail: gbottazzi@luiss.it; giuseppe.italiano@uniroma2.it

G. G. Rutigliano (✉)
Department of Electronics Engineering, University of Rome "Tor Vergata", Rome, Italy
e-mail: rutigliano@ing.uniroma2.it

This second approach is particularly weakened by the fast-paced level of malware sophistication, by the use, ever more frequent, of obfuscation/encryption techniques of both malware and related traffic and by the "Internet of Things". Furthermore, the passive analysis of traffic must be carried out on a sufficiently large data set: the larger the set of available data, the greater the capability for analysis.

Moreover, the exploitation of new vulnerabilities and the evolution of attack techniques do not make easy the detection of botnets through traditional security solutions such as firewalls, intrusion detection/prevention systems and antiviruses.

In this context, there has been much attention on the use of behavioral analysis for botnet detection. In particular, several approaches have been used, looking for patterns able to find in advance the action of botnets, before they could be identified through their footprints. Those patterns, in order to be used effectively, should be able to identify common botnet behaviors, regardless of architectures, payloads, protocols and infection techniques used.

Indeed, many botnet architectures follow a common pattern of interactions between clients (zombies) and servers, essential for bot synchronization and command deployment (towards the C2 servers) and for storing exfiltrated data (towards the so-called *Drop* servers).

The identification of a network behavior as suspicious, regardless of the particular technique used, should be:

- Tested on a dataset as real and/or wide as possible, with both positive and negative samples;
- Scalable to large set of real-world scenarios such as service providers, enterprise networks, common workstations and mobile devices (increasingly subject to infections);
- Not directly linked with the specific threat;
- Fast in finding new threats, given the large number of different variants of the same malware and the never-ending presence of software vulnerabilities (many of which are Zero day).

In this chapter, we consider the traffic analysis between the malicious agents and the C2 servers (post-infection stage a.k.a. the *Rallying* stage, considered the weakest phase in the botnet lifecycle).

In fact, the communication between zombies and C2 servers, often started by bots, outlines characteristics of periodicity and timeliness (Gu et al. 2008a) and allows the botmasters to (see, e.g., the Citadel configuration file in Sood and Bansal 2014):

- Get a "complete picture of the situation" almost in real time;
- Update bot executables;
- Collect the information exfiltrated from victims;
- Plan the subsequent actions to be commissioned to the bots.

Consider also that, from the botmasters' point of view, a frequent update of the agents' "vitality" is already threatened by a critical variable that is not easily predictable: the availability of the device hosting the malicious agent (e.g., when the infected machine is switched off or disconnected).

The periodical communication analysis also appears to be particularly interesting because it is completely independent from agents, protocols, architectures, C2 addressing (hard-coded IP addresses, DGA, Fast Flux, etc.) and from encryption techniques, mainly when it considers only the timestamp of connections, but used in the frequency domain.

In fact, we believe that other parameters such as the time taken to exchange information or the bytes exchanged, do not have the same independence, but are strongly influenced by possible network congestions or by the specific protocols used by botnets.

What we propose is a method able to detect the periodical communication between zombies and C2 servers, without using the well-known Fast Fourier Transform (FFT).

The application of the FFT needs a number of interventions closely related to the best fitting of a function of continuous signals to a bursty dataset, while the method we propose considers only the data available. In order to confirm our insights, we:

- Developed an ad-hoc agent for testing, able to contact an Internet domain with either a fixed timing and a random timing within a fixed range;
- Applied the proposed method to a set of workstations (using the network logs of the hosts of a large corporate network), detecting a number of so-called Possible Unwanted Programs (PUPs) and real malicious agents.

The advantages of the proposed method, reside mainly in the less computational effort required, because of:

- The dataset used, considerably lower than the one necessary for the FFT;
- The calculations implemented that do not require any pre and/or post-processing.

For instance, the efficient application of the FFT requires an amount of samples that must be a power of two. Moreover, the signal must be:

- Analyzed during all the time interval used for the observation;
- Properly described and recorded among two positive observations – when we have connections – creating "acceptable" rising and falling edges – where we have no connections.

We defined also a Signal-to-Noise Ratio (SNR), as the one used in the signal theory, able to separate the signal from noise, that in our case are respectively the traffic generated by bots (signal) and the traffic generated by humans (noise).

Moreover, it makes it possible to find what are the specific connections outlining a periodic behavior, and thus suspicious, in completely "blind conditions", without any previously acquired knowledge about the bots' activity. Finally, our implementation can be easily deployed on any real-world scenario.

In the following, we will describe in detail the method implemented and tested on a /8 corporate network, much wider than the test performed in (Bottazzi et al. 2016), and on real malware samples.

2 Related Work

The frequency domain analysis is widely used in signal processing. The continuous signals are in fact representable as a sum of "simple" sinusoidal signals of different frequencies. The set – and the amount – of all the frequencies contained in a signal is defined as the frequency spectrum of the signal.

Further, the frequency domain analysis highlights features that are not easily observable through the time domain analysis, especially in complex situations in which several signals overlap.

Transforming a discrete time series into a discrete frequency series is a matter of computing the Discrete Fourier Transform (DFT), shown in Eq. (1), through its efficient implementation commonly identified as Fast Fourier Transform – FFT – (Heideman et al. 1984) or Sparse Fast Fourier Transform – SFFT – (MIT Staff 2012).

$$X(k) = \sum_{n=0}^{N-1} x(n) e^{-i\frac{2\pi kn}{N}} \; k \in [0, N-1] \qquad (1)$$

Where, *X(k)* are the frequency domain samples,

- *x(n)* are the time series samples,
- *e* is the Napier number,
- *i* is the imaginary unit.

Recently, many researchers proposed botnet detection methods, feeling that the pre-programmed botnet activities related to C2 traffic could highlight spatial-temporal correlations and similarities (Gu et al. 2008a). Exploiting these similarities, could result in a framework whose operation is independent from protocols, architectures and payloads used, just because it can exploit correlated communication flows (Gu et al. 2008b).

Due to these insights, the analysis of the communication flows between bots and C2 servers disclose a certain degree of regularity detectable by means of the FFT, widely used in signal theory (AsSadhan et al. 2009a).

In fact, in (Tegeler et al. 2012), three main features were used as the basis for the extraction of network traces (to be understood as a sequence of chronologically ordered flows between two network endpoints):

- The average time interval between the start times of two subsequent flows in the trace;
- The average duration of connections;
- The average number of source bytes and destination bytes per flow.

Other approaches (Paul et al. 2014; Thaker 2015; Balram and Wilscy 2014; Zhou and Lang 2004; Zhou and Lang 2003; Tsuge and HidemaTanaka 2016; Yu et al. 2010; Zhao et al. 2013; AsSadhan and Moura 2014; Chimetseren et al. 2014; AsSadhan et al. 2009b; Eslahi et al. 2015) exploited the frequency analysis of suspicious flows, based on tuples of information, composed again by source and destination addresses, bytes exchanged, average packet size, TCP/UDP ports, etc.. The main goal was to evaluate the effectiveness of a specific methodology tested in a controlled environment, but especially with a limited number of samples collected in a restricted observation period (Bartlett et al. 2011). Finally, we mention the work done in (Kwon et al. 2014, 2016), which makes use of frequency analysis based on FFT, but applied to DNS queries.

All the aforementioned approaches base their results on the use of the FFT applied to a certain type of network flows, often extracted using NetFlow, or to a set of DNS queries. Unfortunately, all the data used so far (log files, pcap files, etc.) reflect the usual operation of computer networks, characterized by bursty traffic and inactivity periods, that is quite different from the definition of a continuous time-variant system. The bursty traffic involves the need to handle large amounts of data many of which, especially for the inactivity periods, do not provide useful information to the analysis.

Therefore, there is a need to pre-process the data presenting impulsive behaviors, in order to make them more similar to a continuous function.

Finally, starting with a large number of samples in the time domain, the FFT calculates, again, a large number of samples in the frequency domain, and this makes it more difficult to analyze the frequency spectrum, both methodologically and computationally, looking for anomalies.

Even the SFFT cannot be useful, given the lack of knowledge of the initial assumptions required by this technique (the data source must highlight some characteristics of *sparsity*).

3 The Botnet Life Cycle

The first phase of the botnet life cycle (Fig. 1), is the *Infection*, wherein a host is infected and becomes a potential bot. This phase is characterized by a regular computer infection procedure, which may be carried out in different ways as a typical virus infection would be, for instance, through unwanted downloads of malware from websites, infected files attached to email messages, infected removable disks, etc.

This first phase is not used to install the malware used by the botnet, but only to establish a first communication channel with the victim. The malicious codes used in this phase, are sometimes identified as *droppers*.

In the second phase, the *Secondary Injection*, the infected host runs a program that searches for malware binaries in a given network database. When downloaded and executed, these binaries make the host behave as a real bot (or zombie). The download of bot binaries is usually performed through FTP, HTTP(s) or P2P protocols.

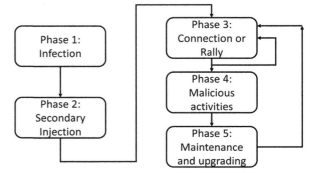

Fig. 1 The botnet life cycle

At some point in time the new zombie must contact a C2 server to receive instructions or updates, through the third phase called the *Rallying* phase. This procedure is sometimes also called the *Connection* phase.

This phase is scheduled periodically by the zombie, at least every time the host is restarted, in order to ensure the botmaster that the bot is taking part in the botnet and is able to receive commands and perform malicious activities. Therefore, the connection phase is likely to occur several times during the bot life cycle.

Because they must contact C2 servers, bots may be vulnerable during this phase. Bots often establish connections with C2 servers by default, allowing mechanisms to be created to identify traffic patterns and hence identify the components of the botnet or even of the C2 server.

In order to find the C2 server, the malware installed during the first two phases should contain the address of the machines to be contacted. As previously said, there are several ways for addressing C2 serves (hard-coded IP addresses, DGA, Fast Flux, etc.).

After establishing the command-and-control communication channel, the bot waits for commands to perform malicious activities. Thus, the bot passes into phase 4 and is ready to perform an attack that can be as wide ranging as information theft, performing DDoS attacks, spreading malware, extortion, stealing computer resources, monitoring network traffic, searching for vulnerable and unprotected computers, spamming, phishing, identity theft, manipulating games and surveys, etc.

The last phase of the bot life cycle is the maintenance and updating of the malware. Maintenance is necessary if the botmaster wants to keep his army of zombies. It may be necessary to update codes for many reasons, including evading detection techniques, adding new features or migrating to another C2 server. After bots are updated, they need to establish, again, connections with the C2 infrastructure.

As previously mentioned, the rallying stage, is one of the most critical phase during the botnet lifecycle and occurs after bots have been successfully recruited into the bot army. They are rallied back to a central C2 unit which could either be administered centrally (by the botmaster) or in a peer-to-peer manner (by other bots in the botnet), but usually remotely via the Internet. The bots maintain synchronization with the C2 unit at all times in order to receive new commands, infiltration parameters and takeover specifications, which they readily execute (Ogu et al. 2016). Thus, Synchronization and Rallying are possible because the process of the bot code installation includes (also) a "rendez-vous" mechanism with C2 servers.

The high level of sophistication achieved in the deployment process of botnets, further worsened by an innovative business model (long supply-chain strongly oriented to outsourcing), produces, however, some weaknesses mainly concentrated in the rallying stage. In particular:

- The commoditization of skills, products and services has, as a side effect, a software reuse tendency. This is probably due both to the efforts (technical and economic) needed to develop innovative payloads and because the black market continues to be very inclined to "Lemon Markets" and so, despite the many attempts to build real post-sales supports (e.g., Citadel botnet), the creation of affiliates or trust-based middlemen is still strongly present (Herley and Florencio 2009). In fact, the black market reveals a perfect situation where sellers have better information than buyers do, about the quality of their wares and "the bad drives out the good". Choosing the specific example of used-car markets (a.k.a. Lemon Markets) where only the seller knows whether the car is a lemon or not, the buyer will consider the average likelihood of getting a lemon into the price.
- Regardless of the particular architecture implemented and used by the specific malware, bots must periodically synchronize with their C2 servers.

Our approach tries to take advantage of the previous considerations, in order to identify bots manifesting a certain periodicity when contacting the Internet trying to synchronize with the C2 servers.

4 Our Proposed Method

We propose a method that does not use the Fourier Transform. We prefer, instead, an approach based on the underlying principles of the histograms applied directly to the data to be examined, to minimize the requirements in terms of memory and computational cost.

The main idea is to properly mine the logs for extracting knowledge about the periodic behavior of bots, concealed behind the connection timing. Hence, each connection trace stored in the logs, referred in the following as **hit**, was used to compute the time difference, referred in the following as **period**, existing between two subsequent hits, previously sorted on the time field.

In case of logs related to multiple workstations, the logs must be previously sorted on both the workstation source IP address and the time. Hence, the first period of the logs, or the first period of each source IP address, must be set to zero (Fig. 2).

Fig. 2 Calculation of periods

Algorithm: calculation of periods

Input: log table ordered on the source IP address and time containing, for each hit, the source ip address (c-ip) and the time of connection.

Output: the same log table with a further column containing the differential timing values (period).

period(1) ← 0
for i = 2 **to** (end of logs) **do**
   if c-ip(i) = c-ip(i-1) **then**
     period(i) ← time(i) − time(i-1)
   else
     period(i) ← 0
   end if
end for

The "differential" values thus obtained – the periods (in Fig. 3 an extract of proxy logs where the fields *time* and *period* are expressed in "hh:mm:ss"), must be clustered (Fig. 4), in order to find anomalous peaks compared to the shape of the resulting distributions, a.k.a. **periodograms**, for every single workstation (after clustering).

In Fig. 4 we plotted the periodogram of few proxy logs of a single workstation, denoting a typical non-periodic distribution where almost all the hits collapse in the "small" periods (0, 2 and 3 s). The y-axis shows how many times the workstation has actually had a period of t seconds of inactivity (x-axis).

In a periodogram without any periodic activity, most of the occurrences tend to collapse in the small periodicities of the series, denoting also the typical human usage of the Internet. In fact, when someone (human) activates a hyperlink, by clicking or typing directly the URL, a series of events almost concurrent are triggered (all the simultaneous events, seen in the frequency domain, fall within the same period, zero seconds). Thus, the time spent by the user (human) to read the content of an Internet page, causes the workstation inactivity (the reason why we observed low values in the high periodicities). Bots (or agents) instead act, by definition, with time periodicities.

The effectiveness of this method makes it possible to "isolate" these periodicities because human behavior has numerous periods of inactivity where many of the actions of an agent can fall. Hence the periodical agent's activity can be highlighted just looking at the anomalous peaks of the distribution.

In order to properly test our method, we developed and injected within a corporate network infrastructure, an ad-hoc agent able to contact two different

time	URL	period
00:13:02	http://go.microsoft.com/fwlink/?LinkId=121315	00:00:00
00:13:02	http://www.msn.com/it-it?devicegroup=downlevel.mobile&webslice=ieslice	00:00:00
00:23:14	http://crl.microsoft.com/pki/crl/products/WinPCA.crl	00:10:12
00:23:16	http://crl.microsoft.com/pki/crl/products/MicrosoftTimeStampPCA.crl	00:00:02
00:23:18	http://crl.microsoft.com/pki/crl/products/MicCodSigPCA_08-31-2010.crl	00:00:02
00:23:21	http://crl.microsoft.com/pki/crl/products/MicWinHarComPCA_2010-11-01.crl	00:00:03
00:23:23	http://www.microsoft.com/pkiops/crl/MicSecSerCA2011_2011-10-18.crl	00:00:02
00:23:26	http://crl.microsoft.com/pki/crl/products/MicRooCerAut2011_2011_03_22.crl	00:00:03
00:23:28	http://crl.microsoft.com/pki/crl/products/MicTimStaPCA_2010-07-01.crl	00:00:02
00:23:30	http://crl.microsoft.com/pki/crl/products/microsoftrootcert.crl	00:00:02
00:23:33	http://crl.microsoft.com/pki/crl/products/MicRooCerAut_2010-06-23.crl	00:00:03
00:23:46	vortex-win.data.microsoft.com:443	00:00:13
00:23:46	settings-win.data.microsoft.com:443	00:00:00
00:28:15	http://www.msn.com/it-it?devicegroup=downlevel.mobile&webslice=ieslice	00:04:29
00:28:15	http://go.microsoft.com/fwlink/?LinkId=121315	00:00:00
00:36:08	ent-shasta-rrs.symantec.com:443	00:07:53
00:43:31	http://www.msn.com/it-it?devicegroup=downlevel.mobile&webslice=ieslice	00:07:23
00:43:31	http://go.microsoft.com/fwlink/?LinkId=121315	00:00:00
00:58:50	http://www.msn.com/it-it?devicegroup=downlevel.mobile&webslice=ieslice	00:15:19
00:58:50	http://go.microsoft.com/fwlink/?LinkId=121315	00:00:00

Fig. 3 Example of periods

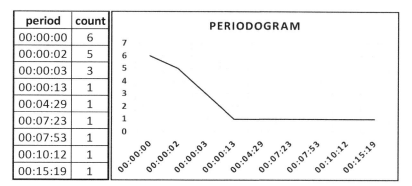

Fig. 4 Clusters of periods

Internet URLs with either a fixed timing and a random timing within a fixed range, trying to mimic the behavior of a real malicious agent (Bottazzi et al. 2016). Within this experiment, we collected the proxy logs for one day.

The results revealed that our method is able to highlight the periodical actions made by the ad-hoc agent, in presence of both fixed periodicity (Fig. 5) and random periodicity (Fig. 6). Moreover, the test described in Fig. 5 allowed us to find the presence of another bot (period of 75 s), corresponding to the URL "nexus.officeapps.live.com", the service used by some Microsoft software as Office

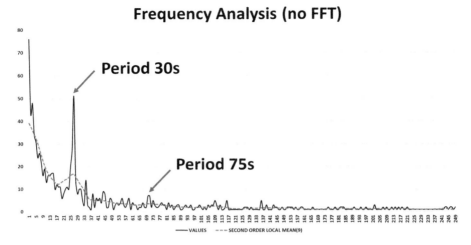

Fig. 5 Ad-hoc agent with fixed periodicity (no FFT)

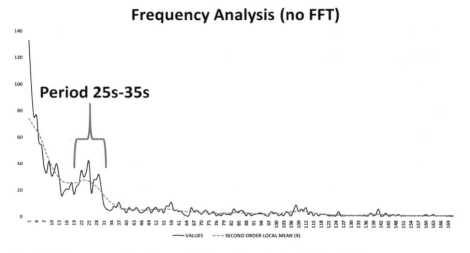

Fig. 6 Ad-hoc agent with random periodicity within a fixed period (no FFT)

365, used to report periodically the software health (the idea was to inject the agent on the workstation while the user continues to do normal actions).

The feature exploited is able to "resist" also to a second order of randomness. To highlight this, we deleted intentionally some agent's logs, depicting a behavior that skips randomly some hits, still highlighting periodicity partially with its "*original behavior*" and partially with its "*harmonic behavior*" (Fig. 7).

The periodicities plotted in the previous figures are related to the logs of just one workstation. In order to confirm the efficiency of the proposed method, we

A New Scalable Botnet Detection Method in the Frequency Domain

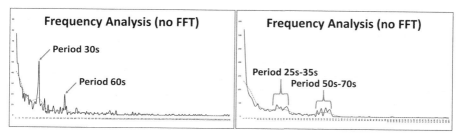

Fig. 7 Normal and harmonic behavior of the ad-hoc agent (no FFT)

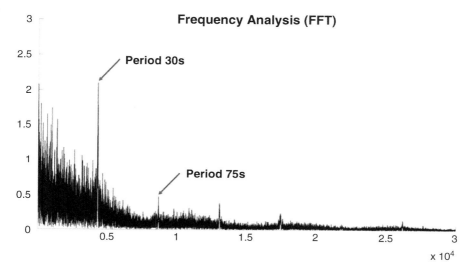

Fig. 8 Ad-hoc agent with fixed periodicity (FFT)

used the same proxy logs, related to only one workstation, to compute the same periodograms, but with the FFT method (in Figs. 8 and 9 the two distributions related to the agent's action with fixed and random periodicity).

As we can see in the two previous pictures, the same data treated with the FFT resulted in a much noisier distribution. This is mainly due to the high number of frequency samples (as many as those in the time domain), although we:

- Built a timeline able to cover all the observation periods, whose elements is a power of two;
- Created "acceptable" rising and falling edges, related to periods of inactivity.

The described interventions are closely related to the best fitting of a function of continuous signals to a bursty dataset.

Leaving aside all the possible benefits resulting from the processing of logs with a DBMS, related to the optimizations of the sorting and aggregation operators, the

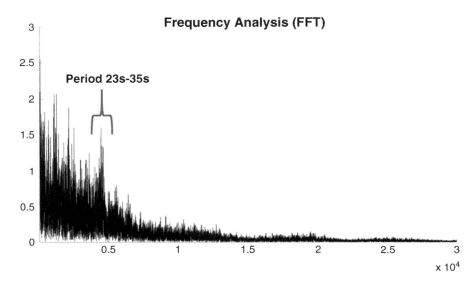

Fig. 9 Ad-hoc agent with random periodicity within a fixed period (no FFT)

main advantage of the proposed method, compared to the FFT, fully lies on the exploitation of a considerably reduced dataset.

In fact, the efficient application of the FFT on the logs of one day for a single workstation requires a dataset containing the smallest power of two greater than the dataset within the observed period. Thus, having a timeline expressed in seconds, we need to consider the smallest power of two greater than 86,400, the seconds in a day, equal to 131,072 (2^{17}), although in many cases there has been no activity. Moreover, the FFT treats, in the frequency domain, the same amount of data of the time domain.

Our method considers, as shown in the examples in Figs. 5 and 6, only the data observed (equal to 11,276 periods), that when clustered are reduced to about 250 different items. Moreover, the data treated with the proposed method can be further reduced, after clustering, by deleting all the occurrences with small periods, which gives no contribute to frequency analysis.

In the example shown in Fig. 5, the hits related to the period of 0 s are 8,959, almost the 80% of total hits, and the periods longer than 10 s are only the 7% of total hits (Fig. 10).

The visual analysis of anomalous peaks can be used only to analyze the periodograms of single workstations. For multiple workstations we must use a method, similarly to the signal theory, which is able to detect signal from noise, that in our case are respectively the traffic generated by bots and by humans. In order to automatically identify the peaks contained within the series, similarly to the signal theory, we propose a method, able to separate signal from noise.

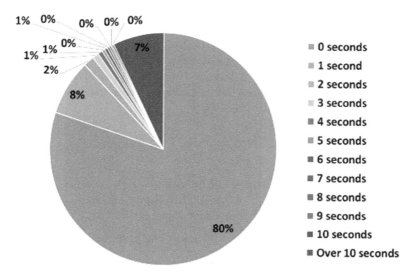

Fig. 10 Distribution of periodicities of a single workstation

In the examples shown in Figs. 5 and 6, the continued line series is the traffic generated by bots (the signal) and the dotted line series is the traffic generated by humans (the noise).

The dotted line series, the noise, has been considered to be as the local average of the second order (the average of the average), computed on a neighborhood of 9 values of the continued line series, the signal (2). The local average of the first order is the one shown in (3). As we can see, we excluded the value of x_i in the first order average, $Avg_9(x_i)$, in order to better highlight anomalous peaks.

$$\forall i, Noise(x_i) = \frac{\sum_{j=i-4}^{j=i+4} Avg_9(x_j)}{9} \qquad (2)$$

$$\forall i, Avg_9(x_i) = \frac{\sum_{j=i-4}^{j=i+4} x_j}{8} \quad i \neq j \qquad (3)$$

Finally, we assumed as anomalous peaks (Signal to Noise Ratio) all the signal values greater than 1.5 times the noise values (4). Obviously, this threshold is a critical parameter in order to minimize false positives and false negatives.

$$\forall i, \; SNR(x_i) = \begin{cases} Peak, & x_i \geq 1.5\, Noise(x_i) \\ NoPeak, & otherwise \end{cases} \qquad (4)$$

We applied our method to a wide set of workstations (Bottazzi et al. 2016), using the proxy logs of the hosts of a /16 network. The corporate network considered deals with critical information and it is thus equipped with its own security organization and infrastructure, which is composed by managed perimeter security systems and endpoint protection systems. Moreover, the current corporate policies forbid the interaction with the Internet with protocols other than http and https. In this case, we were able to find a number of so-called *"Potentially Unwanted Programs"* and the actions of a real malicious agent installed on a single workstation.

5 Discussion

The method proposed in this chapter highlights the same issue raised by any other method of frequency analysis, the so-called *"frequency scattering"*. Looking at the peak with a period of 30s in Fig. 5, some hits were partially "scattered" because:

- Fifty-one hits out of 75, made by the ad-hoc agent, have been logged in the right period;
- Eleven hits out of 51 are not related to the ad-hoc agent's activity.

This mean that only the 53% of the logged hits started during an inactivity period of the workstation, and so they were logged correctly within the 30s-period, while the others were "hidden" over other periods. This phenomenon becomes more prominent when the agent's frequency increases.

Stressing the concept of extended periodicity, we made in (Bottazzi et al. 2016) an experiment, by injecting our ad-hoc agent with a periodicity of 30 min, making 15 hits in total. As expected, the agent's actions were completely distributed over other periods. In fact, as it can be seen in Fig. 11, no peaks are related to a period of 30 min.

Instead, removing from the proxy logs the hits related to just two well-known legitimate Internet domains (those related to the operating system and the antivirus), we had a great enhancement of agent's visibility (Fig. 12) compared to the previous periodogram.

Hence, the "success" factor, from the botmaster's point of view, is neither the randomness of the periods nor the number of URLs contacted (the ad-hoc agent was able to contact two different URLs) but the number of "useful hits" during the observation period. The useful hits must be understood as those connections whose periodicity is obfuscated by normal traffic that, if reduced in amount, can be completely "hidden" within other periods.

Thus, the method we implemented exploits the agent's hits, whose periodical behavior fits completely within the inactivity periods of the workstation.

Implementing an agent with a high inactivity time, on one hand increases the probability of being detected, because its actions fall within an area where there are few hits (little activity). However, on the other hand, it increases the probability that its extended periodical behavior is "obfuscated" by other traffic, even other agent's traffic with a shorter periodicity, making it invisible.

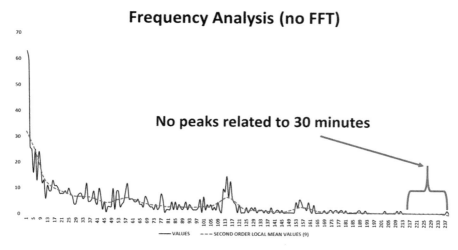

Fig. 11 Periodogram of the ad-hoc agent's activity with a periodicity of 30 min

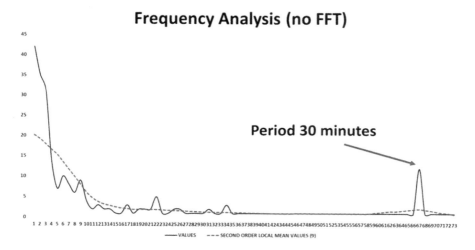

Fig. 12 Periodogram of the ad-hoc agent's activity with a periodicity of 30 min, removing the logs of some well-known licit Internet domains

Moreover, developing a malicious agent trying to contact the C2 servers few times on a daily basis, e.g., twice an hour, could be particularly disadvantageous for the botmasters. In fact, the prompt knowledge about the zombie army, can be undermined by the possible and unpredictable decrease of active agents just because the infected hosts are switched-off, unconnected or cleaned-up by antivirus. Of course, the "prompt knowledge" about the number of active zombies must be also related to the purpose for which the botnet has been created (much more critical for Spam and DDoS botnets). Hence, both attackers and defenders have to balance the scalability and agility of the botnet to deploy/detect, with the increasing risk of detection associated with an increasing number of C2 server connections.

For the reasons mentioned above, the application of some white lists is of fundamental importance both for reducing the amount of data to be inspected and for deleting the periodical actions made by a wide set of Internet services.

As mentioned earlier, the hits contained in every single peak may have occurrences that do not concern any agent, but the great majority must necessarily refer to a periodic action, mostly if the period is particularly extended. On the other hand, the agent's hits can be distributed over other periods, mostly, again, when the periodicity of actions is particularly extended.

Therefore, the cited white lists must be used for finding and deleting all the occurrences (periodical or not) related to known licit domains, in order to isolate malicious agents as if they were acting alone (without background noise). In this way, while introducing an overhead, it is possible to mitigate the limit posed by all the frequency analysis techniques.

Since the use of white lists is crucial for isolating the malicious agent's behavior, the public white lists should be further enlarged by adding legitimate Internet domains:

- Always related to ALEXA-like lists, but not referred to the main web sites (e.g., *.gstatic.com and *.ggpht.com are Google-related websites but not included in ALEXA white list);
- Not included in the ALEXA-like lists.

However, the massive use of white lists could expose this kind of analysis to compromised licit Internet domains, used for hosting C2 servers.

6 Double Blind Experiment

The experiment presented in this paragraph has the purpose to confirm the scalability of the proposed method, by applying it to a set of workstations belonging to a /8 corporate network. We labeled this experiment as "*double blind*" only because no users were warned in advance and no knowledge was acquired prior to apply the method.

We collected and mined the proxy logs of one day of more than 60,000 workstations, stored in huge text files in W3C format (more than 100 GBytes). In order to mine log files, we used the free tool Log Parser (Giuseppini et al. 2015) to implement a script formatted in a SQL-like language, in order to extract the data stored in the W3C files and upload the results to a database.

It was possible to store, effectively and efficiently (e.g., in few minutes using a common desktop machine), all the data associated with our experiment. It is important to note that the same log extraction method can be easily scaled to bigger size logs as already demonstrated in a previous work (Bottazzi and Italiano 2015)

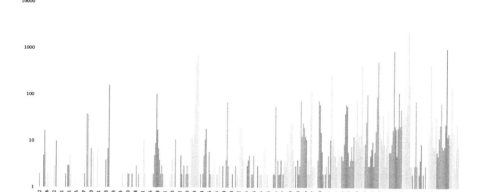

Fig. 13 Distribution of peak periodicities over the workstations

where, although in a very different context, a starting dataset of almost 3 TBytes has been used. In fact, since the mentioned SQL-like script can be applied also to portion of daily logs, uploading the partial results to a data base, the global extraction time, using the same hardware, is thus a mere multiplication factor.

Hence, starting from more than 75 million rows, we were able to reduce the dataset to less than 400 periodical peaks by:

- Deleting all the domains included in the ALEXA top 500 web sites and all the domains related to public web mails, online newspapers, online weather forecasts, etc.;
- Not considering the periods related to 0, 1, 2 and 3 s;
- Ranking the peaks, according to the distribution of periods on the various workstations involved. In other words, we verified how many source IP addresses were involved in a given frequency, by clustering the periodicities and counting the source IP addresses. As we can see in Fig. 13, many periods involve a high number of source IP addresses (displayed in logarithmic scale).

Now, in order to give operators (e.g., those of a Security Operation Center) a treatment priority to the resulting periodicities, we decided to start from those affecting up to 20 distinct source IP addresses. This assumption is justified by the context, in which a mass distribution of a specific frequency over a huge number of source IP addresses would mean a mass infection (rare in case of highly managed corporate networks). With these premises, we were able to find, again, a number of "Possible Unwanted Programs" (PUPs) and some suspect agents.

7 Malware Samples Tested in a Controlled Environment

The experiments presented in this paragraph have the purpose to verify the effectiveness of the proposed method, applied to a set of real malware. The idea is based on the deployment of a controlled network composed of one or more workstations based on Windows 7, installed on both physical hardware and virtualized environment. The need arises from the fact that some malware samples do not carry out any activity when "noticing" to run on virtualized environment and/or in debugging mode. For the same reason the interception proxy module has been installed on the same network, but on a physically different workstation. (lab architecture in Fig. 14).

All the tests performed, considered a time span no longer than 30 min and a set of malware samples resumed in Table 1.

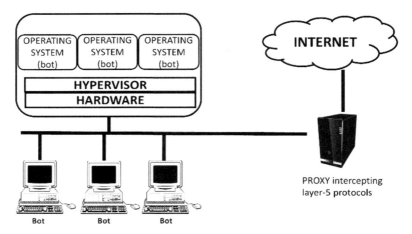

Fig. 14 Lab test environment

Table 1 Malware samples tested

Malware	Description
Andromeda	Andromeda is a modular bot. The original bot simply consists of a loader, which downloads modules and updates from its C&C server during execution
DriveDDOS	http DDOS. Different GET and POST http flooding
NjRAT Bladabindi	A R.A.T. malware, developed in .NET, allowing attackers to take complete control of an infected device
Kuaibpy	Generic Trojan
Katrina	It's a POS (Windows connected) credential stealer
NgrBOT	A worm stealing FTP and browser passwords causing also a DoS by flooding. It could download updates and more malware

The tests performed on real malware revealed that, depending on the purpose for which the botnet has been developed, some application protocols could be particularly useless in tracking periodic behaviors, regardless of the frequency analysis method used. In fact, exploiting the method, always the same, selectively on more than one application protocol, allowed us to identify the periodic behaviors we were looking for.

For instance, in case of http-flooding botnets, e.g. DriveDDOS, the frequency analysis of the http protocol does not produce any useful result, simply because of the massive use of http, saturating all the inactivity periods of a workstation.

Instead, the alternative application protocol that often provided useful information about the periodic behavior of the bots, always using the same method, is the DNS, used by zombies to resolve domain names of the C2 servers.

We resumed the test performed in Fig. 15 displaying, for each malware sample, the periodograms of both DNS and HTTP protocols. For almost all malware we had the need to infect physical workstations, given their capability of recognizing the presence of a hypervisor installed, except for DriveDDOS. Each subpicture contains both the signal and the second order local mean as described in previous paragraphs.

While all the samples tested gave us results on the DNS protocol, the HTTP protocol test failed in three cases (DriveDDOS, NgrBOT and NjRAT). As already said, DriveDDOS makes http-flooding, saturating thus all possible inactivity periods. NgrBOT and NjRAT, however, although not reporting any periodic peak in HTTP (NjRAT did not have any http traffic), show evident spikes, considering only the TCP-PSH packets, that can be related to a customized layer-5 protocol (Figs. 16 and 17).

8 Sality Botnet Traffic Capture

In this experiment we used the traffic capture (.pcap file) made available by the "Stratosphere IPS Project" (https://stratosphereips.org/), related to the well-known P2P botnet, called Sality. The P2P version of Sality first appeared in early 2008 and is a variant of the centralized Sality malware downloader. Sality uses a pull-based unstructured P2P network to spread URLs, where payloads are to be downloaded. Peers regularly contact their neighbors to exchange new URLs (Stratosphere IPS Project "https://stratosphereips.org/"; Rossow et al. 2013).

The capture, performed in the 2014, was intentionally made on a real network environment in order to have mixed logs related to botnet, normal and background traffic (Falliere 2011). As you can see in this case we did not exploit the proxy logs, but instead the raw traffic logs. This could represent a great advantage, demonstrating that the method we propose can be adopted regardless of the technology used for capturing layer-5 traffic.

The mentioned sample (downloaded from Falliere 2011) is related to a 15-days traffic capture but, in line with the previous experiments, we used the logs of just one day. In particular we extracted the logs related to the protocols DNS and HTTP.

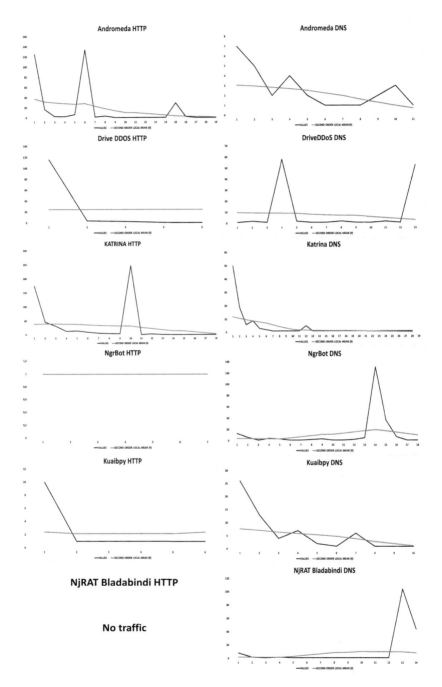

Fig. 15 Periodograms of malware samples for HTTP and DNS protocols

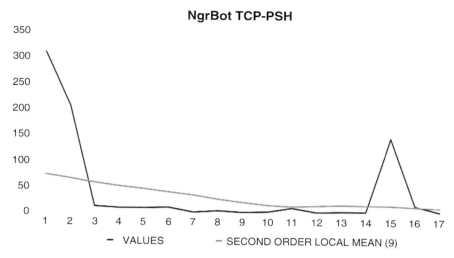

Fig. 16 NgrBot periodogram considering the TCP-PSH packets

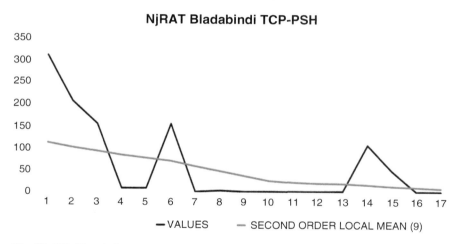

Fig. 17 NjRAT periodogram considering the TCP-PSH packets

Starting with the DNS protocol, the application of our method, excluding the periods of 0, 1, 2 and 3 s, resulted in the periodogram shown in Fig. 18.

As we can see the periodogram is very noisy with many frequency peaks distributed all along the periods. Inspecting the logs in detail, we noticed a huge number of DNS requests for the domain "msftncsi.com", that is the "Network Awareness" feature provided by Microsoft and used for determining the state of the network connectivity. Deleting these logs (thus applying a whitelist), we have as a result the periodogram in Fig. 19.

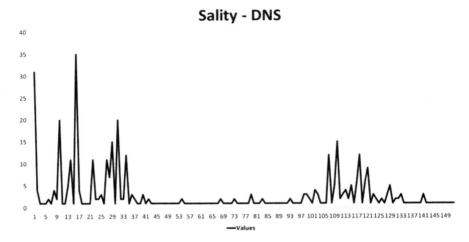

Fig. 18 Sality periodogram considering DNS protocol

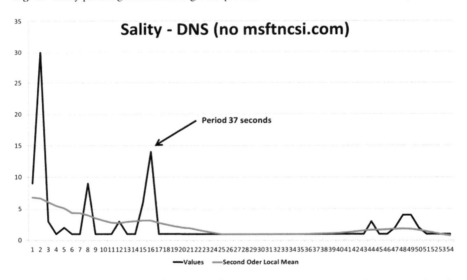

Fig. 19 Sality periodogram considering DNS protocol, without the logs related to the domain "msftncsi.com"

The periodogram is now much less noisy and the frequency peak highlighted with a 37 s period corresponds to the "*gardapalace.it*" domain, that is easily identifiable as belonging to the Sality network through any common search engine. Moreover, exploiting the method over the HTTP protocol (in this case there was no need to apply any white-list), we obtained the periodogram in Fig. 20, excluding again the periods of 0, 1, 2 and 3 s.

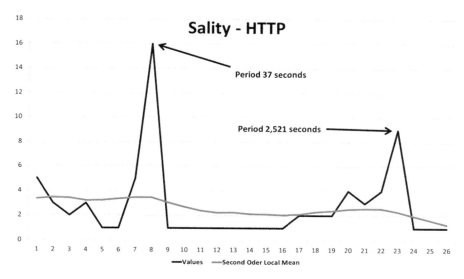

Fig. 20 Sality periodogram considering HTTP protocol

The two peaks shown in Fig. 20 are both related to URLs used by the Sality protocol in the so called "Pack Exchange" command (Falliere 2011). In particular the peak with period of 37 s corresponds to the IP resolved by the DNS with the same frequency in Fig. 19.

9 Conclusions and Future Work

We stress that our method is able to find the periodical actions of bots, similarly to what has been already done by previous approaches, but offers additional advantages since:

- It does not require any data preprocessing (flows extraction, average packet length, average duration of the connections, etc.);
- It does not need any previously acquired knowledge on bot's behavior;
- It is completely independent from the number of periods composing the observation, while the FFT treats the whole frequency spectrum;
- It is completely independent from the technology used for capturing the traffic of layer-5 protocols.

Moreover, the exploitation of the method on proxy logs allowed us to build a bot detection tool that can be implemented efficiently and effectively in almost every practical scenario.

Once the traffic have been extracted, the application of some white lists is of fundamental importance both for reducing the amount of data to be inspected and for deleting the common refreshing actions made by a wide set of Internet services. In this way, while introducing an overhead, it is possible to overcome the limit posed by all the frequency analysis techniques. On the other hand, white lists cannot resolve the issue related to legitimate (and compromised) web services, used for hosting C2 servers.

Furthermore, the frequency analysis described in this chapter is particularly:

- Effective, given its absolute independence from agents, protocols, architectures and from the Internet domains contacted, and does not need any previous knowledge about the periodical behavior of malicious agents. We demonstrated also that the frequency analysis is applicable on every layer-5 protocol, even if it is customized and thus unknown.
- Efficient, because it can fulfill the frequency analysis of the logs directly on the observed traffic with a high level of scalability, while the FFT-based approaches, for the same purpose, require pre and post processing activities. The data observed are the only information we need.

The time required for uploading the logs into a data base depends only on the initial availability in W3C format, prior to the use of the methodology, that can be optimized easily by writing the logs directly into a data base or using, for example, those uploaded to a Security Information and Event Management (SIEM) system.

The exploitation of proxy logs, moreover, has an additional high benefit, given that they:

- Are natively interposed to communications between client and server;
- Are widely used within enterprise networks and their analysis is particularly useful for a Security Operation Center (SOC);
- Can be used both to identify a suspicious Internet domain or to block the connections to it. Of course, the timely intervention and the observation period must be correctly balanced. A longer observation period allows to detect the actions of an agent with greater accuracy, but increases the intervention time;
- Can be deployed also in consumer environments such as laptops and mobile (Bottazzi et al. 2015) – high versatility – requiring on one hand a lower computational complexity, and allowing, on the other hand, the identification of the specific processes originating the observed traffic (for which has been found a frequency peak).

Hence, the methodology just described allowed us "to put in place" a highly scalable detection tool in almost every real-world scenario (the computational cost is proportional to the amount of data to be processed). Of course, the vastness of the adversary model (e.g. vectors of infection, architectures, protocols, etc.), poses some limits to the framework, in terms of false positives.

However, false positives must be understood as clearly identified periodic behaviors that are not malicious, for which no frequency analysis method can establish their intention. The only possibility for removing this sort of "ambiguity", is a detailed analysis of the processes underlying the periodic behaviors that, while being a much more time-consuming activity especially for attacks not yet known as such, is out of the scope of this chapter.

The proposed method, instead, can be extremely useful for heavily-managed computer networks, where the workstations should have a software stack little dynamic and well-known, and where the white-lists can heavily reduce the noise within the periodograms.

For the future, we plan to implement the method as an online IDS tool for laptops and mobiles.

References

AsSadhan, B., & Moura, J. M. F. (2014). An efficient method to detect periodic behavior in botnet traffic by analyzing control plane traffic. *Journal of Advanced Research, 5*, 435–448.

AsSadhan, B., et al. (2009a). *Detecting botnets using command and control traffic*. Network Computing and Applications, 2009. NCA 2009. 8th IEEE International symposium on. IEEE.

AsSadhan, B., Moura, J. M. F., & Lapsley, D. (2009b, November 30–December 4). Periodic behavior in botnet command and control channels traffic. In *Proceedings of IEEE Global Communications conference (IEEE GLOBECOM 2009)*. Honolulu.

Balram, S., & Wilscy, M. (2014). User traffic profile for traffic reduction and effective bot C&C detection. *International Journal of Network Security, 16*(1), 46–52.

Bartlett, G. et al. (2011, April 10–15). Low-rate, flow-level periodicity detection. In *Proceedings of the 30th IEEE International Conference on Computer Communications (IEEE INFOCOM 2011)*, Shanghai.

Bottazzi, G., & Italiano, G. F. (2015). Fast mining of large-scale logs for botnet detection: A field study. In *Proceedings of the 3rd IEEE international workshop on Cybercrimes and Emerging Web Environments, in conjunction with the 13th IEEE international conference on dependable, autonomic and secure computing*, At Liverpool, UK.

Bottazzi, G., et al. (2015). MP-shield: A framework for phishing detection in mobile devices. In *Proceedings of the 3rd IEEE international workshop on Cybercrimes and Emerging Web Environments*, Liverpool, UK.

Bottazzi, G., Italiano, G. F., & Rutigliano, G. G. (2016, July 20–22). Frequency domain analysis of large-scale proxy logs for botnet traffic detection. In *Proceedings of the 9th international conference on Security of Information and Networks (SIN '16)*, Rutgers University, New Jersey.

Chimetseren, E., Iwai, K., Tanaka, H., & Kurokawa, T. (2014, October 15–17). A study of IDS using discrete Fourier transform. In *Proceedings of international conference on Advanced Technologies for Communications, ATC*, Hanoi.

Eslahi, M., et al. (2015). Periodicity classification of HTTP traffic to detect HTTP botnets. In *Proceedings IEEE Symposium on Computer Applications & Industrial Electronics (ISCAIE 2015)*, Langkawi.

Falliere, N. (2011). *Sality: Story of a peer-to-peer viral network* (Technical Report by Symantec Labs).

Giuseppini, G., Burnett, M., Faircloth, J., & Kleiman, D. (2015). Microsoft log parser toolkit: A complete toolkit for Microsoft's undocumented log analysis tool. ISBN-13: 978-1932266528.

Gu, G., Zhang, J., & Lee, W. (2008a). *Botsniffer: Detecting botnet command and control channels in network traffic*. NDSS.

Gu, G., Perdisci, R., Zhang, J., Lee, W., et al. (2008b). *Botminer: Clustering analysis of network traffic for protocol-and structure-independent botnet detection* (USENIX Security Symposium, pp 139–154).

Heideman, M. T., Don, H., & Johnson, C. (1984). Sidney Burrus, Gauss and the History of the Fast Fourier Transform. *IEEE ASSP Magazine*.

Herley, C., & Florencio, D. (2009). *Nobody sells gold for the price of silver: Dishonesty, uncertainty and the underground economy* (Microsoft TechReport).

Kwon, J., Kim, J., Lee, J., Lee, H., & Perrig, A. (2014). PsyBoG: Power spectral density analysis for detecting botnet groups. In *Proceedings of the 9th IEEE international conference on Malicious and Unwanted Software, MALCON*.

Kwon, J., Kim, J., Lee, J., Lee, H., & Perrig, A. (2016). PsyBoG: A scalable botnet detection method for large-scale DNS traffic. *Computer Networks, 97*, 48–73.

MIT Staff. (2012). *SFFT: Sparse fast Fourier transform*. http://groups.csail.mit.edu/netmit/sFFT/index.html.

Ogu, E. C., Vrakas, N., Chiemela, O., & Ajose-Ismail, B. M. (2016). On the internal workings of botnets: A review. *International Journal of Computer Applications, 138*(4).

Paul, T., et al. (2014). *Fast-flux botnet detection from network traffic*. India Conference (INDICON), 2014 annual IEEE. IEEE.

Rossow, C., et al. (2013). P2PWNED: Modeling and evaluating the resilience of peer-to-peer botnets. In *Proceedings of the 2013 IEEE symposium on Security and Privacy (SP 2013)*, San Francisco.

Sood, A. K., & Bansal, R. (2014). Prosecting the citadel botnet – Revealing the dominance of the Zeus descendent, Kaspersky Virus Bulletin.

Stratosphere IPS Project. https://stratosphereips.org/.

Tegeler, F., Xiaoming, F., Vigna, G., & Kruegel, C. (2012). BotFinder: Finding bots in network traffic without deep packet inspection. In *Proceedings of the 8th international conference on Emerging Networking Experiments and Technologies (CoNEXT '12)*.

Thaker, K. S. (2015). Modelling and detection of camouflaging worm at an advance level. *International Journal of Advanced Research in Computer Science and Software Engineering, 5*(10), 758–762.

Tsuge, Y., & HidemaTanaka. (2016). Intrusion detection system using discrete Fourier Transform with window function. *International Journal of Network Security & Its Applications (IJNSA), 8*(2), 23–34.

Yu, X., Dong, X., Yu, G., Qin, Y., Yue, D., & Zhao, Y. (2010). Online botnet detection based on incremental discrete Fourier transform. *Journal of Networks, 5*(5), 568–576.

Zhao, D., Traore, I., Sayed, B., Lu, W., Saad, S., Ghorbani, A., & Garant, D. (2013). Botnet detection based on traffic behavior analysis and flow intervals. *Computers and Security, 39*, 2–16.

Zhou, M., & Lang, S.-D. (2003). Mining frequency content of network traffic for intrusion detection. In *Proceedings of the IASTED international conference on communication, network, and information security*.

Zhou, M., & Lang, S.-D. (2004). A frequency-based approach to intrusion detection. *Journal of Systemics, Cybernetics and Informatics, 2*(3), 52–56.

Part III
Cybercrime Detection

Predicting the Cyber Attackers; A Comparison of Different Classification Techniques

Sina Pournouri, Shahrzad Zargari, and Babak Akhgar

1 Introduction

The process of identifying cyber attackers can be highly complicated and involve different steps. Authorities and cyber security experts need to identify and profile cyber-attacks to prevent further attacks and persecute the perpetrators. Once a cyber-attack reported, law enforcement agencies start to carry out the investigation process, however, time and accuracy play significant role in efficiency of the investigation. By wasting time chasing wrong cyber attacker, the real attacker can carry out more attacks and create more damages to the victims. Therefore there is an urgent need for an accurate and reliable predictor to identify cyber attackers. The accurate prediction and comprehension of current and past situation in terms of cyber-attacks not only can lead to persecute the cyber criminals but also can help cyber security specialists to block further attacks and damages.

One of the common ways of identifying cyber attackers is to follow and investigate their footprints made by attackers unintentionally. These footprints can be technical indications such as attack's origin IP address left by hackers in their attack, however, hackers tend to cover up and clean their footprints or fabricate false footprints to deviate security experts and make it more difficult and impossible to get caught. Another way of detecting cyber attackers to look at the history of similar cyber-attacks and find correlation between different factors within the attacks. What the similar attack type was, what the similar target type was and where was the similar target located in terms of country and etc. are different questions that can be asked and profile cyber-attacks and predict future and past unknown threats as learning from past failures can improve the prevention methods. In our research we

---

S. Pournouri (✉) · S. Zargari · B. Akhgar
Department of Computing, Sheffield Hallam University, Sheffield, UK
e-mail: s.pournouri@shu.ac.uk

© Springer Nature Switzerland AG 2018
H. Jahankhani (ed.), *Cyber Criminology*, Advanced Sciences and Technologies for Security Applications, https://doi.org/10.1007/978-3-319-97181-0_8

aim to utilize classification techniques in order to profile cyber attackers leading to prediction and identification of them.

Classification as a Data mining method has been used in different fields and knowledge. Classification refers to a set of techniques which assign different objects to specific predefined classes and the result can be interpreted as prediction (Han et al. 2011). Al-Janabi (2011) proposed a model using Decision Tree as a classification algorithm to analyse a crime dataset and making prediction about behaviours of different criminals based on their marital status, income and etc. The model was based on five different blocks as follows:

1. Data collection: The dataset is provided by Austin Police department about past crimes happen in their related territory.
2. Data pre-processing: This stage includes data reduction through normalization and aggregation, dealing with missing values and removing outliers.
3. Data storage: The data will be stored in a data warehouse for further analysis.
4. Applying classification technique: Decision tree has been used in order to train a classifier and make prediction.
5. Interpretation of the result: The result needs to be presented to decision makers and it should understandable for them to plan their strategy to tackle future crimes in Austin area.

In the field of education, Bhardwaj and Pal (2011) used classification techniques to predict future performance by different students based on variety of elements. They used Naïve Bayes algorithms in their method for training their classifiers and then based on that they measured the probability of each element having effect on students' performance. In field of cyber security, most of the researches have been carried out in more technical way such as identification and prediction of cyber-attacks in technical manners such as Lin et al. (2015) proposing model for improving performance of Intrusion Detection System based on K nearest neighbour algorithm.

In this study we will make and compare different classifiers in the form of predictors based on their accuracy to see to what extent classifiers can help law enforcement agencies and cyber experts in their investigations to identify cyber attackers.

2 Method

Our proposed method is based Open Source Intelligence data. The reasons of choosing OSINT as data resource are its cost effectiveness and accessibility. Our data includes past cyber-attacks where attackers were known or they claimed the responsibility of a particular cyber breach. There are different websites and blogs that report cyber-attacks around the world such as "hackread.com", "Darkread.com" and etc. Also there is a blog called hackmageddon made by Paolo Passeri which monitors OSINT and records cyber activities day by day (Passeri n.d.).

In order to build the classifiers we need to have a training set and a test set for validation purposes. Our training set consists of 1432 attacks happened from 2013 to the end of 2015 linked to known cyber attackers and the test set is made by 484 attacked taking place from 2016 to the end of first quarter of 2017. In order to make our data ready for analysis we structure the data in form of a table including nine columns as follows:

1. Date of incident: The time of cyber-attack incident.
2. Cyber attacker: who was behind of the cyber-attack. For example the Anonymous is one of the cyber attacker groups.
3. Cyber threat: it refers to type of threat that the cyber-attack poses to its victim such Denial of Service and etc.
4. Type of target: it describes the nature of target(s) of the cyber-attack such as University of Maryland as Educational organization and etc.
5. Target: it describes the name of target, for example Mitsubishi
6. Targeted Country: it refers to the origin of target(s) of the cyber-attack such as Mitsubishi in Japan.
7. Type of Cyber-attack: This feature explains type of cyber-attack in terms of type of activity such as Hacktivism and etc.
8. Description: this column describes how the cyber-attack happened based on Open source intelligence available for it.
9. OSINT resource: It refers to reference of OSINT related to the cyber-attacks.

Table 1 shows an example of a cyber-attack structured in our proposed dataset.

As this study aims to train models based on classification algorithm the data needs to be restructured. Therefore Date column will be eliminated because Time series analysis in not the area of interest of our study, Targets specifically are not relevant and can make complexity, the Target column will also be removed from data set, however, the Type of Target column will remain. Also the description column and OSINT resource will be eliminated because they are not involved in the predictive model.

After restructuring the dataset and forming the attributes into five different columns, preparation and pre-processing the values will begin. In this research it has been decided to employ a very powerful tool which is developed by Google initially in Java language called Open Refine (Verborgh and De Wilde 2013). Open refine includes different functions for data cleansing and transformation. The data will be fed into Open refine and pre-processing method will be divided into following steps:

1. Removing doubles and irrelevant records: This step includes identifying and eliminating of records which happen to appear in our data set twice and also those records that they do not have enough information to be a part of our data set.
2. Integration and Categorization: In order make the analysis simpler and less complicated the records need to be categorized and integrated. This stage applies

Table 1 Example of the dataset

Date	Cyber attacker	Cyber threat (type of threat)	Type of target	Target	Targeted country	Type of cyber attack	Description	OSINT resource
2015/1/31	Team system DZ	Defacement	NGO	Women resources center	UK	Hacktivism	The website of the Woman Resource Centre (wrc.org.uk) is defaced by Team System DZ, a group of supporters of ISIS and Jihad	http://www.thirdsector.co.uk/womens-resource-centre-website-hacked-people-claiming-support-isis/communications/article/1331684

Table 2 Type of target categorization

Acronym	Type of target	Example
BP	Broadcast and publishing	Including publisher companies and magazines
ED	Education	Including colleges, schools and universities
EN	Entertainment	Including music and video game companies and etc.
ES	Energy section	Companies and sectors operating in oil, power and etc.
FB	Finance and banks	Institutions with financing and banking functionality
GO	Government	Including states and their related departments
HC	Healthcare	Including health care providers such as hospitals and clinics
HT	Hospitality and tourism	Including hotels, restaurants and etc.
IO	Internet and online services	Including chat rooms and forums
MD	Military and defence section	Companies operating in military equipment manufacturing
MU	Multiple	Several targets
NN	NGO and no profit	Including non-profit and charity sectors
RT	Retail	Including retail shops
SI	Single individual	Publicly known figures
SN	Social network	Including Facebook, Twitter, Instagram and etc.
TC	Telecommunication	Sectors providing telecommunication lines such as internet and telephone
THS	Technology hardware and software	Companies and business providing hardware or software products
TM	Terrorism	Terrorist groups
TP	Transportation	Including traffic lights and etc.

to Type of target, Type of Threat and Type of cyber-attack. Tables 2, 3, and 4 Show the categorization of mentioned attributes.

After the pre-processing stage, we start the analysis with classification algorithms. Classification algorithms that have the ability to build a predictor which have been used in this study are as follows:

Table 3 Type of threat categorization

Acronym	Type of threat	Definition
AH	Account hijacking	Any online account such as email, social media and etc. associated with a person or a company hijacked by a hacker(s)
DF	Defacement	Unauthorized changing a web page by hackers through penetration to web server
DS	DDOS	Disturbing availability of victims' server by hackers through sending high volume of requests
MWV	Malware	A piece of malicious code including virus, worm, Trojan horse and etc. designed by hackers for compromising victims' system
PH	Phishing	A malicious method tries to steal sensitive information by deceiving victims through an email conversation,
SQ	SQL injection	Attackers' code try to compromise the database
TA	Targeted attack	Anonymous and un trackable attackers actively are trying to penetrate to victims' system
UA	Unauthorized access	Any unauthorized access to computer devices and software by hackers
UN	Unknown attacks	Those attacks when type of threat has not been reported in OSINT resource
CSS	XSS vulnerability (cross site scripting)	Attackers inject client side script into a webpage
ZD	0 day	Unresolved security bugs get exploited by hackers

1. Decision Tree: Decision trees is also known prediction trees and include sequence of decisions and their outcomes as consequences. The prediction process can be done through making a decision tree with nodes and their branches. Each node represents a specific input variable and each branch means a decision making process. Leaf node refers to those nodes that they do not have branch and return class labels and in some occasion they generate probability scores. Decision trees are being used in most data mining application with predictive purposes due to the fact that they are easy to implement, visualize and present. Freund and Mason (1999) divide decision tree techniques into

Table 4 Type of cyber attack categorization

Acronym	Type of cyber attack	Definition
HA	Hacktivism	A cyber attack based political or social reasons
CW	Cyber war	State sponsored attacks to damage critical infrastructure
CC	Cyber crime	A crime which a computer device is involved
CE	Cyber espionage	A malicious attempt to gain access to secret or confidential information held by governments or critical firms

three main categories; C4.5, Random Forest and Recursive partitioning. C4.5 is a decision tree algorithm which was developed by Ros Quinlan (1993) and it is an extension of ID3 algorithm based on information entropy. Recursive Partitioning is a method of decision trees based on greedy algorithms to classify members of population correctly based on independent variables. It mainly suits categorical variables and does not perform well on continues variables (Friedman 1976). Random Forest is another important decision tree algorithm developed by Breiman and Cutler (2007). This algorithm builds decision tree by a process called bagging which means combination of learning trees to increase classification accuracy,

2. K nearest neighbour: According to Larose (2005) K Nearest Neighbour (KNN) is an algorithm used in both classification and regression problems. As Input the algorithm gets K nearest training example and produces the output as class label. The object will be classified based on majority vote of its neighbour. In KNN algorithm, the training set is the set of vectors in multidimensional space assigned to different classes. Then in classification stage, K will be defined by user and the object will be assigned to relevant classed based on maximum frequency of that class.

3. Naïve Bayes: Naïve Bayes is one of the most important data mining methods which leads to classification and prediction. One of the most common application of Naïve Bayes is spam filtering and text categorization (Murphy 2006). Naïve Bayes assumes that all variables in the dataset are independent from each other and calculate the probability of different events based on the variables, however, some authors like (Ray 2015) consider this as a disadvantage because it is almost impossible to obtain data set that its features are completely irrelevant in real life.

4. Support Vector Machine: Support Vector Machine (SVM) is a supervised classification technique that is being used for both classification and regression problems. SVM has been used in different application such as text classification and image processing due to high level accuracy compared to other algorithms. Initially SVM algorithm was developed by Cortes and Vapnik (1995) and it is based on choosing the optimal hyperplane which separates two or more classes

Table 5 Models accuracy

Classification models	Accuracy (%)
C4.5	61.25
Random forest	59.51
Recursive partitioning	60.48
K nearest neighbour	61.27
Naïve Bayes	58.17
Artificial neural network	59.71
Support vector machine	61.34

with the maximum distance called Margin between their closest. Support vectors refer to those objects located on the boundaries. Support vectors are the most challenging objects to classify and they play a significant role in defining and identification of the optimal hyperplane.

5. Artificial Neural Network: Artificial Neural Network (ANN) is another classification algorithm which is designed based on neural system of the brain. In initial stage, ANN classifies objects then it will compare them with actual classes and the measure the error rate. Then it will repeat the process until the error reaches to its minimum value. An ANN algorithm consists of three main layers; the first layer is input layer getting inputs, the second layer named hidden layer which can have one or more sublayers doing the process task of the algorithm and the last layer is the output layer produces the result (Agatonovic-Kustrin and Beresford 2000).

To apply these classification algorithms, we use tenfold cross validation for training method which means the training data will be divided into ten equal samples randomly and one sub sample is considered as test sample for validation and the others belong to the training set and this process keep taking place for ten times therefore the result will be averaged and an estimation will be provided. Table 5 shows the accuracy each model.

After obtaining accuracy of each model, we compare them then we conclude that Support Vector Machine model has done more accurate job in terms of prediction of cyber attackers. With some slight difference K nearest neighbour and C4.5 are placed in second and third best respectively. Figure 1 shows the comparison between different models in terms of minimum, maximum and average accuracy.

3 Discussion

This section aims to discuss and evaluate the nominated model for prediction of cyber attackers, as it was mentioned Support Vector Machine model has been chosen as the most reliable model with 61.34% accuracy. For evaluation purposes we apply the model to the test dataset which it 484 records of cyber-attacks from 2016 to the end of first quarter of 2017. As it was mentioned before one of the limitation of this

Fig. 1 Models' accuracy comparison

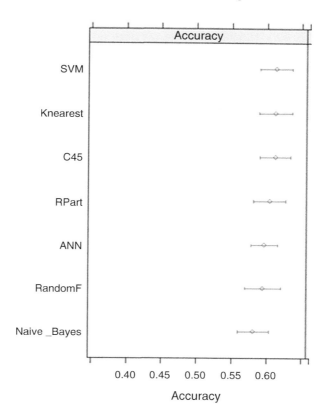

study was cyber attackers identity because they might change their identity or stop carrying out cyber-attacks, however, some known attackers such as the Anonymous they are present in the cyber space. In order to measure the success of our model for future prediction, we consider the following criteria (Fawcett 2006):

1. True Positive rate (Recall): It refers to number of instances which are classified correctly in one class, divided by total correctly classified instances and it is formulated as: TP rate = TP/(TP+FN)
2. False Positive rate: It is also called false alarm rate, explaining the number incorrectly classified of one class over the total number incorrectly classified instances. The FP rate equation is: FP rate = FP/(FP+TN)
3. Precision: It is also named as Positive Predicted Value which describes the number of instances classified correctly in one class, divided by total number of classified objects in that class. Precision is calculated as: Precision = TP/TP+FP
4. ROC area: Roc explains a two dimensional graph where X axis is labelled as False Positive rate and Y axis is named as True positive rate. The Area under Roc curve is important for measuring the success of a classifier.

By taking above criteria, following table demonstrates the result of evaluation:

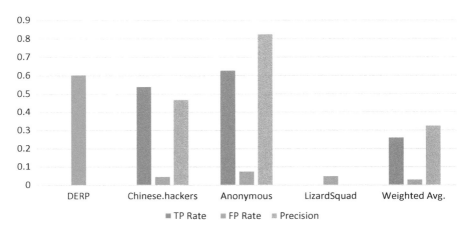

Fig. 2 TP, FP and precision comparison

The overall accuracy of this model is 25.92% which is not reliable enough for cyber experts as a single predictive method, however, by analysing this result by comparing different benchmarks following points can be highlighted:

1. It should be mentioned that there are some attackers who were not available in validation dataset which means some attacker are not active in recent years or they might have changed their alias names. Figure 2 shows the bar plot of comparison of TP rate, FP rate and precision for those classes that they are available both in the training set and the validation set. The highest TP rate which is 0.627 belongs to the Anonymous hacker group which shows the model despite its insufficient accuracy level, it has done reliable prediction about this cyber attacker group which can be concluded with contribution of significant level precision which is 0.824. The Chinese hackers are in the second rank in terms of TP rate with level of 0.538 which also demonstrates accurate task of the predictive model in terms of detection of these cyber attackers. The highest FP rate goes to DERP group which shows in the validation set there are some hackers that they have similar pattern of feature but they are not in DERP group.
2. ROC is another benchmark to evaluate the predictive model. Figure 3 Shows ROC area for each class. The Anonymous group has the highest level of ROC with 0.897 indicating they are will predicted by the model. Chines hackers are the second in terms of ROC area with amount of 0.84 which also demonstrates the model is reliable in terms of predicting them.
3. The last step of this analysis is comparing recall, precision and ROC for each class which is shown in Fig. 4. Again this step indicates the Anonymous and the Chinese hackers are the most well predicted among other cyber attackers despite unreliable overall accuracy, precision and ROC with amount of 0.26, 0.324 and 0.684 respectively.

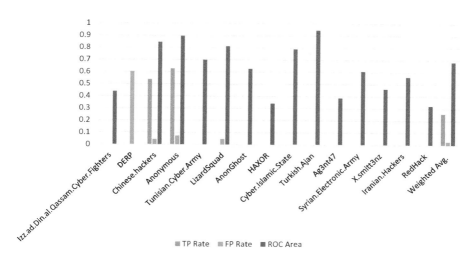

Fig. 3 TP, FP and ROC comparison

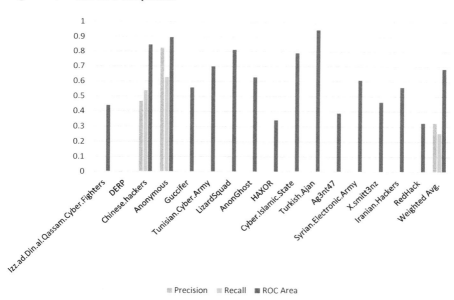

Fig. 4 Precision, recall and ROC comparison

4 Conclusion

Nowadays data mining and predictive analytic techniques are being used in daily life including cyber security. Predictive analytic techniques not only can help cyber experts to tackle future threats but also aids law enforcement agencies to identify and catch criminals before they cause more damages. As our proposed model based on SVM classifier using OSINT indicated the accuracy for current and past prediction of cyber attacker is higher than prediction of future by almost 35% and this difference can be because changing identities and motivation of cyber criminals over time, however, this method can be used as a technique with combination of other techniques to make more reliable and accurate prediction. Using OSINT in this study has its advantages and disadvantages; on one hand it is accessible and free to use for research purposes and other hand it can be noisy and incomplete which can lead to less accuracy. If OSINT combines with the data collected by Intelligence firms and organization, then the more accurate and reliable predictors can be built.

In order to see how much impact each variable has on the proposed model accuracy, we used Classifier Attribute Evaluator in Weka which is used to evaluate the value of each attribute in classification techniques. The first important attribute in cyber attacks was identified as the type of threat which means Cyber Attackers can be often identified by looking at type of threat that impose to their victims. The second significant attribute was type of target which indicates cyber criminals follows the same pattern in terms of selecting their targets. The third variable in cyber attacks is concluded type of cyber attack indicating the motivation of cyber attackers which means the motivation can be varied from one attack to another for attackers and the motivation mainly depends on different factors which can be changed such as political, economy and etc. The least important variable which does not matter in the prediction of cyber attackers is the country of their target which means the attackers do not follow a same pattern for choosing which country they want to target.

Future work in this field which can be done is to gather more data on cyber-attacks and make effort to add more variables to the collected data set as cyber-attacks are driven by many other factors than we mentioned in our study. Time series analysis can be another extension to this study leading to behavioral identification of cyber attackers time by time.

References

Agatonovic-Kustrin, S., & Beresford, R. (2000). Basic concepts of artificial neural network (ANN) modeling and its application in pharmaceutical research. *Journal of Pharmaceutical and Biomedical Analysis, 22*(5), 717–727.

Al-janabi, K. B. S. (2011). A proposed framework for analysing crime data set using decision tree and simple K-means mining. *Algorithms, 1*(3), 8–24.

Bhardwaj, B. K., & Pal, S. (2011). Data mining: A prediction for performance improvement using classification. *(IJCSIS) International Journal of Computer Science and Information Security, 9*(4), 136–140.

Cortes, C., & Vapnik, V. (1995). Support-vector networks. *Machine Learning, 20*(3), 273–297.

Fawcett, T. (2006). An introduction to ROC analysis. *Pattern Recognition Letters, 27*(8), 861–874.

Freund, Y., & Mason, L. (1999). The alternating decision tree learning algorithm. In *icml*, 99 (pp. 124–133).

Friedman, J. H. (1976). A recursive partitioning decision rule for nonparametric classification. *IEEE Transactions on Computers, 26*(SLAC-PUB-1573-REV), 404.

Han, J., Pei, J., & Kamber, M. (2011). *Data mining: Concepts and techniques*. Elsevier.

Larose, D. T. (2005). *k-nearest neighbor algorithm. discovering knowledge in data: An introduction to data mining* (pp. 90–106).

Lin, W. C., Ke, S. W., & Tsai, C. F. (2015). CANN: An intrusion detection system based on combining cluster centers and nearest neighbors. *Knowledge-Based Systems, 78*, 13–21.

Murphy, K. P. (2006). *Naive bayes classifiers*. University of British Columbia.

Passeri, P. (n.d.) HACKMAGEDDON [WWW Document]. HACKMAGEDDON. URL https://www.hackmageddon.com/. Accessed 5.24.18.

Quinlan, J. R. (1993). *C4. 5: Programming for machine learning*. Burlington: Morgan Kauffmann.

Verborgh, R., & De Wilde, M. (2013) *Using OpenRefine*. Packt Publishing Ltd. Birmingham.

Crime Data Mining, Threat Analysis and Prediction

Maryam Farsi, Alireza Daneshkhah, Amin Hosseinian-Far,
Omid Chatrabgoun, and Reza Montasari

1 Introduction

Today, many businesses benefit from cyberspace for communicating with their clients and promoting their commercial activities. Moreover, the Internet has resulted in numerous new business models. Number of individuals who have online presence has increased significantly from 2000 to 2018 by 10520% (Internet Growth Statistics 2018). Cybercrime in general is a form crime which has emerged since the emergence of the Internet as we know it today. Jahankhani and Hosseinian-Far (2014) and Jahankhani et al. (2014) classify cybercrime and provide the characteristics of the different categories. They also argue that the existing preventive measures to tackle cybercrime are ineffective and new techniques should be ensued. Some of the techniques are quite simplistic and are really a simple response to a cyber attack.

M. Farsi
School of Aerospace, Transport and Manufacturing, Cranfield University, Cranfield, UK
e-mail: Maryam.Farsi@cranfield.ac.uk

A. Daneshkhah
School of Computing, Electronics and Mathematics, Coventry University, Coventry, UK
e-mail: Ali.Daneshkhah@coventry.ac.uk

A. Hosseinian-Far (✉)
Department of Business Systems & Operations, University of Northampton, Northampton, UK
e-mail: Amin.Hosseinian-Far@northampton.ac.uk

O. Chatrabgoun
Department of Statistics Faculty of Mathematical Sciences & Statistics, Malayer University, Malayer, Iran
e-mail: O.Chatrabgoun@malayeru.ac.ir

R. Montasari
Department of Computer Science, The University of Huddersfield, Huddersfield, UK
e-mail: Reza.Montasari@bcu.ac.uk

© Springer Nature Switzerland AG 2018
H. Jahankhani (ed.), *Cyber Criminology*, Advanced Sciences and Technologies for Security Applications, https://doi.org/10.1007/978-3-319-97181-0_9

Some tend to be preventive; for instance provision of Intrusion Detection Systems (IDS's), honey-pots, firewalls, etc. There are also techniques for designing secure systems in the design and development phases (Yu et al. 2017). Other techniques are to minimise the damages that could be caused by the attacks; for instance, keeping regular backup, cold sites, etc.

The main aim of this paper is to provide another dimension to the above providing a concise review of the latest developments in data mining techniques applicable in crime analysis to explore and detect crimes, and their relationships with criminals/hackers. The paper provides a concise overview of a crime analysis framework within the context and then discusses the data mining and machine learning applications in the field of cybercrimonology.

This paper consists of four main sections. In the second section, we present a generic framework for crime analysis. Section 3 outlines machine learning based methodologies that are widely and recently used in data mining in the context of crime data analysis. Finally, Sect. 4 is devoted to conclusion and future works of the authors.

2 Crime Analysis Framework

In this section, a generic framework for crime data analysis and crime data mining is presented to clarify the relationship between crime data mining and types of data analytic, see Fig. 1.

2.1 Data Domain Specification

Crime data can be gategorised as confidential (e.g. narcotics or juvenile cases), public (e.g. burglary) or partly public information (e.g. sex offenders). Data domain and restrictions are then specified based on their defined category. There are a number of challenges regarding the crime data categorisation which are as follows: (i) free-text field: the collected data are heavily dependent on the investigator, and therefore due the subjectivity, the meaning of crime data may change. Although free-text fields provide the investigator with some level of flexibility in the data recording activity, nevertheless these free-text fields cause lower data accuracy which may lead to lower required clarity for further applications. Moreover, free-text fields are not readable by computers automatically, as they require additional data cleaning (Jabar et al. 2013); (ii) data diversity in terms of data source and nature due to modularity and variability aspects (Thongtae and Srisuk 2008); (iii) Noisy data due to subjective observation, investigation and judgement; (iv) crime trend analysis is a challenging task since crime data are highly time-dependant and any collected data around behaviour of criminals and types of crime may repeatedly change throughout the investigation period; (v) acquisition of media (e.g. pictures,

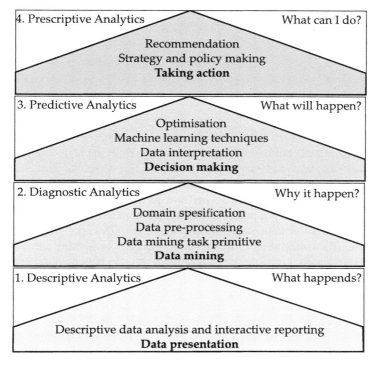

Fig. 1 Crime analysis framework to demonstrate the relationship between crime data mining and types of data analytic

videos, voice record) is a significant challenge in respect of data quality issues; and (vi) the high level of data security and privacy requirement through crime data value chain (Ghaffari et al. 2017). This latest item is vital to protect highly sensitive business and personal data for legal procedures and in secure sharing techniques.

2.2 Extracting the Target Dataset

A crime dataset has an inherent geographical feature where all data in the dataset are not randomly distributed. There are four dimensions of crime in such a dataset: (i) Legal which refers to when a law is broken; (ii) Victim refers to someone or something that has to be targeted for a cyberattack; (iii) Criminal who has committed a crime; and (iv) Spatial which refers to the fact that a crime has to occur somewhere and in a specific time (Chainey and Ratcliffe 2013). The first dimension specifies the crime domain. Whereas the other three create the Crime Triangle which is defined within the context of Criminological Theory (Gottfredson and Hirschi 1990). Identifying crime characteristics is essential for any further crime data analytics. Crime analysis involves breaking the crime problem apart and extracting the specifics and crime dimensions.

2.3 Data Pre-processing

Crime dataset as a source for both confidential and public information requires a highly regulated and complex data modelling. Moreover, the high volume of crime datasets and also the complexity of relationships between the two above-mentioned data domains have made criminology an appropriate field for the application of data mining techniques. In order to perform an effective data mining, data pre-processing is vital. The pre-processing activities transform uncertain, incomplete and inconsistent raw data to an understandable format. Data pre-processing covers data cleaning, data integration and data transformation. Considering crime data, data cleansing (or cleaning) involves language detection and entity identification for the textual sources. With regards to data value chain, pre-processing is a part of the data acquisition phase (Hosseinpournajarkolaei et al. 2014; Curry et al. 2016). Moreover, data integration includes Extract, Transform and Load (ETL) extensively and is a key to load data from diverse sources to the data warehouse. The outcome from pre-processing activities can be later fed into data analysis phase within the data value chain.

2.4 Data Mining Task Primitives

Crime data processing is challenging and complex mainly due to the diversity, noise and uncertainty attributes. Consequently, throughout data mining, several data patterns may be identified which many of them are not genuine. Therefore, a set of primitives should be set in order to communicate with the data mining system. Such primitives can be specified in the form of data mining query. Primitives that define data mining task are: task-relevant data, type of knowledge to be mined, background knowledge, pattern interestingness measurements and visualisation of discovered patterns (Han et al. 2006, 2011; Geng and Hamilton 2006). Data mining task primitives can provide more flexibility during the data mining process by making data mining more interactive and providing a visualisation interface. Data mining primitives can specify items illustrated in Fig. 2.

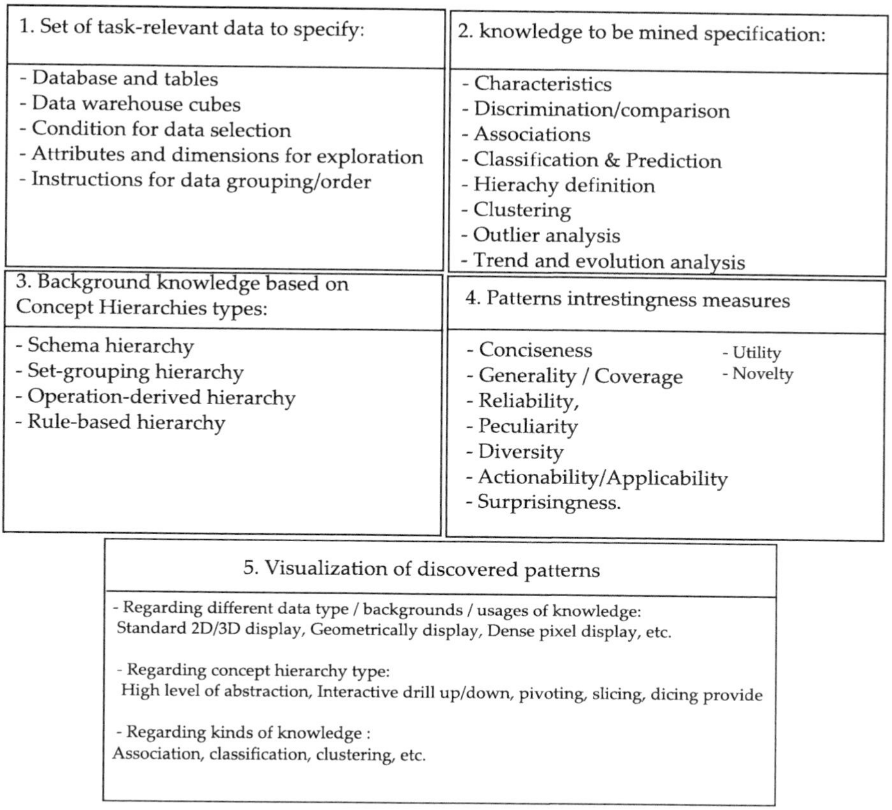

1. Set of task-relevant data to specify:	2. knowledge to be mined specification:
- Database and tables - Data warehouse cubes - Condition for data selection - Attributes and dimensions for exploration - Instructions for data grouping/order	- Characteristics - Discrimination/comparison - Associations - Classification & Prediction - Hierachy definition - Clustering - Outlier analysis - Trend and evolution analysis
3. Background knowledge based on Concept Hierarchies types:	4. Patterns intrestingness measures
- Schema hierarchy - Set-grouping hierarchy - Operation-derived hierarchy - Rule-based hierarchy	- Conciseness - Utility - Generality / Coverage - Novelty - Reliability, - Peculiarity - Diversity - Actionability/Applicability - Surprisingness.

5. Visualization of discovered patterns
- Regarding different data type / backgrounds / usages of knowledge: Standard 2D/3D display, Geometrically display, Dense pixel display, etc. - Regarding concept hierarchy type: High level of abstraction, Interactive drill up/down, pivoting, slicing, dicing provide - Regarding kinds of knowledge : Association, classification, clustering, etc.

Fig. 2 Primitives to define a data mining task

2.5 Data Mining

Data mining is a computational approach to discover the behavioural patterns in a large dataset. Considering a crime dataset, data mining determine the indicative patters among crime-triangle elements using neural network and machine learning algorithms for extracting profile of criminals and presenting the geographical map of crimes (Mena 2003; Jabar et al. 2013). Data mining application in criminology supports predictive policing with a view to decrease crime incidences and prevent future crimes. Moreover, data mining can potentially detect usual fraud patters and cyber-related crimes. Data mining has applications in both the descriptive and predictive data analysis. The descriptive data mining techniques include mining of association and correlation analysis and data clustering. Whereas, the predictive data mining applications are data classification, decision making and neural networks.

2.6 Interpretation and Using Discovered Knowledge

Data interpretation is an essential process throughout data value chain to extract valuable information and knowledge from the data mining and predictive analysis. In criminology, it is essential to consider baseline data for interpreting crime and intelligence information. Otherwise, bias error is likely to occur throughout data mining process which ultimately results in incorrect judgement and potentially distribution skewness within the analysis and findings. Moreover, quality aspects in terms of reliability, accuracy and validity of crime data is important to consider for data interpretation. Lack of quality may arise from fear, distortion, and unfaithfulness when completing the crime report. Although some of these quality issues can be addressed through the data cleansing phase, many will remain unresolved, causing a high level of uncertainty at the time of interpretations (Larose and Larose 2015).

3 Data Mining Applications in Crime

In the last decade, data mining and machine learning have become a vital part of crime analysis, detection and prevention (McClendon and Meghanathan 2015). "Data mining" is defined as a step in the knowledge discovery process and is consists of applying data analysis and discovery algorithms that, under acceptable computational efficiency limitations, produce a particular enumeration of patterns (or models) over the data (Nirkhi et al. 2012). In other words, it can be considered as an analytic (computer-aided) process constructed to examine large volumes of data (known as *Big Data*) to extract useful information, to discover hidden relationships among data using Machine learning techniques, and finally to provide predictive insights. Predictive data mining is thus the most common analytic approach with a wide range of applications such as future healthcare, business and financial case studies (e.g., customer relationship management; financial banking; corporate surveillance) and criminal investigation (e.g., intrusion detection; cyber-security). In cyber-security realm, a wide range of prediction techniques have been used to evaluate when or where a crime is going to happen. Some of these techniques and algorithms are tailor-made for the crime prediction, whereas most of them are developed based on machine learning (or probabilistic Bayesian) techniques which are widely used in other contexts. The popularity of Machine learning based Data Mining techniques are further influenced by the increasing availability of Big Data, and its ease of use for the law enforcement professionals who lack data analysis skills and statistical knowledge (Fayyad and Uthurusamy 2002). In the next subsection, we briefly discuss the applications of Machine Learning techniques in crime analysis. We then describe the most widely used techniques throughout the rest of this section.

3.1 Machine Learning Techniques in Crime Analysis

The advanced Data mining techniques required to deal with Big data, focus on both structured and unstructured data to detect the hidden patterns as originally described in Chen et al. (2004). Crime analysis can encompass a broad range of crimes including a simple burglary to the complex ones (e.g., organised gang crimes), such as cyber crimes which are usually difficult to resolve as available information on suspects are diffused spatially and could potentially take long periods of time to analyse. Detecting cyber crime can be even harder due to busy network traffic and frequent on-line transactions generating large amounts of data, whilst only a small portion would actually relate to the illegal activities (Hassani et al. 2016).

The machine learning techniques can be used for different purposes in crime data analysis. For instance, we can apply these techniques to identify malicious behaviour or malicious entities which are known as hackers, attackers, malware, unwanted behaviour, etc. They can be also used for the cyber security and cyber crime. Machine learning techniques can be generally partitioned into two major categories: supervised, and unsupervised learning. The former category is comprised of all the problems where we have "good", labelled data, whereas the latter is more suitable when we need to explore the hidden patterns of any given data without any labelling attachments. Clustering, dimensionality reduction, and association rule learning are among the most widely used unsupervised approaches. These are very efficient in making big data easier to analyse and understand. The dimensionality reduction would enable us to reduce the number of dimensions or data fields. The other two methods (i.e., clustering and association rules) would be useful in reducing the number of group records. Figure 3 illustrates an incomplete, yet a useful view of machine learning algorithms and their applications in cyber-security and crime analysis (Raffael 2016).

In this paper, we briefly introduce and discuss the machine learning techniques which are widely used in most crime applications.

3.1.1 Cluster Analysis and Trend

Clustering is a method to group data in which each group has identical characteristics without using a known structure in similar data. In other words, clustering is an appropriate data mining approach which groups a set of objects/subjects in such a way that an object in the same group are more similar than those in other groups. In criminology, an investigator uses this technique to identify suspects who conduct crimes in similar ways (or discriminate) among groups belonging to different gangs (Thongtae and Srisuk 2008). Moreover, clustering assists an investigator to find similar crime data trends in the future. Subsequently, cluster and trend analysis provide roadmap to security improvement from a holistic perspective. In this regard, assessing the criminal nature, risk (e.g. probability of accordance and severity) and duration are the main factors are taken into account.

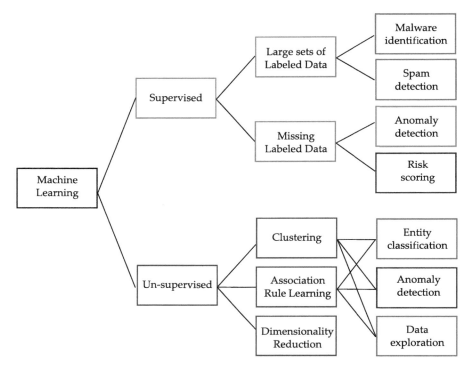

Fig. 3 A brief overview of machine learning algorithms and applications in cybercrime

There are various clustering algorithms that differ significantly in their notion, in terms of what constitutes a cluster and how to efficiently find the clusters. K-Means clustering is widely used by data scientists as the most efficient method with a wide range of applications (Dietterich et al. 2002; Pang-Ning et al. 2014). It is used by forensic science teams to evaluate their performance at crime scenes (Adderley et al. 2007). Another framework was introduced by De Bruin et al. (2006) to assess crime trends through developing a distance metric which enables the investigator to compare all suspects in terms of their profiles and then clustering them accordingly. Furthermore, clustering was at the heart of the crime analysis tool proposed in Gupta et al. (2008) which is used by the Indian police. The proposed tool is developed as an interactive questioning based interface to help the police force in their daily crime investigations. Using this tool, useful information from the big crime database can be first extracted and crime hot spots can be then found using clustering and other crime data mining techniques.

More recently, a Geographic Information System (GIS) is augmented with multivariate cluster analysis (Bello and Yelwa 2012) as a tool to evaluate spatial crime patterns (e.g., property crime in a particular city or country). A similar technique was used to determine criminals and crime hot spot (Sukanya et al. 2012). A more efficient clustering technique was proposed in Keyvanpour et al. (2011) in which self organising maps and neural networks for crime analysis and matching are combined. The RapidMiner software used with K-mean clustering techniques for analysis of the recorded crimes in England and Wales (Agarwal et al. 2013). The combination of GIS and clustering data mining techniques is adopted as a spatial-temporal predictive tool for forecasting daily criminal activities occurring in India (Vijayakumar et al. 2013).

Hierarchical clustering technique, partitioning method, Link analysis technique (including shortest path algorithms, or priority-first-search) and heuristic approach are among other clustering techniques which can be very promising for analysing vast crime datasets with a wide range of applications as discussed above (see also Hassani et al. 2016).

3.1.2 Classification Techniques and Prediction

Classification is a technique to identify in-common properties and attributes among different crime datasets and categorises them into predefined classes. Classification is incredibly powerful in many areas including medical research where one may wish to classify patients based on some medical results, e.g. "has cancer" or "does not have cancer". In criminology, classification can reduce the time required to specify crimes' characteristics. Training and testing of data is essential for a precise classification and accurate prediction. Completing the training is essential due to the high level of incompleteness in data and the missing values in crime datasets (Chen et al. 2004). In this manuscript, we briefly introduce the most widely used and suitable classification techniques including Decision Trees, Support Vector Machine (SVM), Neural network, Naïve Bayes, K-nearest neighbourhood, logistic regression, etc. One could combine these techniques for more complex cases as discussed in Hassani et al. (2016) and the references therein.

The naïve Bayes classifier is the simplest probabilistic classification technique which provides a probability distribution on the set of all classes rather than providing a single output for the given input (Sathyadevan and Gangadharan 2014). The general strategy of the naïve Bayes classifier is to calculate an a-priori probability for being in each of the classes and then classifying by 'which class produces the highest probability'. Furthermore, we can compare the highest probability of results that are in our target class. Using the classifier, one can get an answer to the following type of question: "What is the probability that a crime case B belongs to a given class A?" (Eibe and Bouckaer 2006).

The main benefit of using this classifier is that it can be performed much faster than any other classifier mentioned above, in particular the logistic regression. In comparison to SVM which requires a lot of memory, the naïve Bayes classifier takes considerably shorter time to implement which makes it distinct compared to other algorithms from the computational time perspective. The performance of SVM as the size of training set increases gets worse. The naïve Bayes classifier was used to develop a model which can appropriately classify a wide range of crimes including vandalism, robbery, burglary, sex abuse, murder, etc. (Sathyadevan and Gangadharan 2014). The naïve Bayes classifier has recently been used for detecting phishing websites (Akare et al. 2015). It was also shown that, unlike SVM, as the size of the training data increases, accuracy of classifying the new cases increases too. One of the disadvantages of using the naïve Bayes classifier is that it is constructed based on a very strong assumption of the data distribution shape, i.e. any two cases are independent given the target class (See Witten and Frank 2005 for more technical details and properties of this classifier).

A more useful classification algorithm in crime data mining is a *decision tree*. Decision Trees are a type of Supervised Machine Learning where data is continuously split according to a certain parameter. The tree can be explained by two main elements, that are decision nodes and leaves. The leaves are the decisions or the final class labels, and each node of the tree corresponds to an attribute. The general motive of using Decision Tree is to create a training model which can be used to predict class or value of target variables by learning decision rules inferred from training data. In order to construct a decision tree, one should place the best attribute of the dataset at the root of the tree, then split the training set into subsets, such that each part entails data with the same value for an attribute. The above steps should be then repeated on each subset until leaf nodes in all the branches of the tree can be reached. Figure 4 presents a decision tree illustration for prediction of a sample scenario (Tan et al. 2006). This algorithm has been widely used in a range of applications (Witten and Frank 2005), including detecting suspicious emails (Appavu and Rajaram 2007) with more than 95% accuracy in correctly classifying e-mails in a very large dateset. Later, this algorithm (and also Neural and Bayesian networks) were used in Kirkos et al. (2007) to classify auditors in

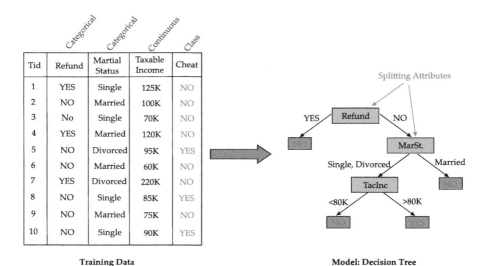

Fig. 4 A decision tree illustration for prediction of whether or not a selected person cheats

detecting firms that issue fraudulent financial statements on datasets taken from more than 70 Greek businesses (see also Gepp et al. 2012; Bhowmik 2011 for similar studies in detecting auto insurance fraud). Decision trees was then applied to detect hotspots in an urban development project dataset including 1.4 million cases, 14 independent variables and one binary response variable (Wang and Liu 2008). Another comparative study was conducted by Iqbal et al. (2012) to examine the performance of naïve Bayes classifier against decision tree to predict crime categorisation in different US states. The derived results exhibit that decision tree classifier outperforms naïve Bayes classifier by achieving 84% accuracy. In a similar study reported in Yu et al. (2011), several other classification techniques, including decision trees, neural network, naïve Bayes and support vector machines were applied and compared for forecasting crime hotspots and future trends.

Although, decision trees outperform naïve Bayes, and in some cases other classifiers, however they possess several disadvantages. First, most of the algorithms used to train this classifier require the target attribute to only have discrete values. Moreover, since decision trees use the "divide and conquer" approach, they tend to execute very well when there are few highly relevant attributes. Their performances reduce if there exist many complex interactions. The greedy characteristic of decision trees leads to another disadvantage that is their over-sensitivity to the training set, to irrelevant attributes and to noise (Maimon 2010).

Artificial Neural Networks (ANNs) are regularly used for classification in data mining and machine learning applications. Their feature vectors are put together into classes, making them desirable to input new data and detect which label fits best. This can be used to label anything, such as spam emails, phishing web-pages, etc. ANNs have proved their high resilience to noisy data. They have also demonstrated strength to classify future patterns. This has been well substantiated within recent published literature. The network is a set of artificial neurons, connected like the neurons in brain. It learns the associations through seeing numerous examples of each class, and learning the similarities and differences between them. This is thought to be a similar process to how the brain learns, with repeated exposure to a pattern forming a stronger and stronger association over time. An ANN can learn patterns in data by mimicking the structure and learning process of neurons in the brain. Using ANNs, the posterior probabilities can be easily computed which provides the basis for establishing classification rule and performing statistical analysis (Richard and Lippmann 1991; Gish 1990). The details of classification algorithm using ANNs can be found in Hassani et al. (2016), Tan et al. (2006) and Gish (1990).

In recent years, ANNs and Support Vector Machines (SVMs) have become more popular in classification of Big Data (Roy 2015). ANNs were first applied in Fuller et al. (2011) for exposing lies from a large amounts of statements associated with various kinds of crimes. ANNs were also used for crime matching in Keyvanpour et al. (2011). ANN uses a classification process by means of a Multi-Layer Perceptron neural network with back-propagation training method (see Tan et al. 2006 for the details of these technical terms). The predication accuracy of the ANNs reported to be higher than other classification techniques in a wide range of crime investigation, such as spotting smuggling vessels (Wen et al. 2012), observing phishing emails from a large set of e-mails (Pandey and Ravi 2012), etc.

It should be noted that working with ANNs has several advantages, including ability to implicitly discover complex non-linear relationships between dependent and independent variables, ability to discover all possible interactions between predictor variables, and the availability of multiple training algorithms; nevertheless they suffer from several shortcomings which on should be mindful before using them in practice. The shortcomings of ANNs are: their "black box" nature; greater computational burden; proneness to over-fitting; and the empirical nature of model development. The pro and cons of using this modelling technique are extensively discussed in Tu (1996).

Support Vector Machine (SVM) was originally proposed as a classification approach in which the objects were divided into two classes based on an optimal separating hyperplane; this minimises the classification error (Cortes and Vapnik 1995). SVM is viewed as a supervised machine learning procedure that can be used for both classification or regression tasks; Nevertheless, it is widely used for classification problems. Within the approach, each data item as a point in k-dimensional space (k stands for the number of features we have) is plotted with the value of each feature being the value of a particular coordinate. The classification will be then implemented by finding the hyper-plane that differentiates the two classes, as shown in Fig. 5 (see also Ray 2017). The technical details of hyperplanes can be found in Hassani et al. (2016) and Tan et al. (2006), and relevant computational codes written in Python and R are also available from Ray (2017).

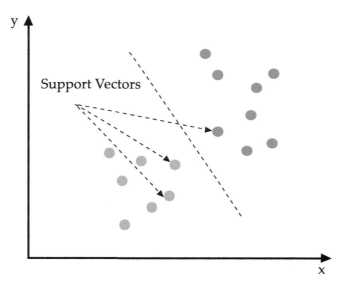

Fig. 5 A SVM illustration for classification problem

In crime data mining, the SVM classification has been originally used to detect the origins of e-mail spams in terms of the sender linguistic patterns and structural features, first in De Vel et al. (2001) and later in Chen et al. (2004). In these works, this classifier was efficiently used to help the investigator identifying crime patterns and features through mining e-mails' contents. Another application of SVM is the prediction of crime hotspots. For such an application SVMs' performance is much better than ANNs and spatial auto-regression based approaches as discussed in Kianmehr and Alhajj (2008). This classifier has been recently used in several other applications, including crime scene classification to a small database (Abu-Hana et al. 2008), detecting identity theft (Salem and Stolfo 2010), detecting advanced fee fraud activities on the Internet and comparing its performance against Random Forest classifier (Modupe et al. 2011), detecting credit card fraud by assessing real world transaction data (Bhattacharyya et al. 2011), detecting and preventing cyber-crime activities by analysing Facebook data (Deylami and Singh 2012), and forecasting real-time crime in urban areas by analysing Twitter data collected from urban sub-area in San Francisco city (Bendler et al. 2014). The modified version of SVM, known as *Support Vector Regression (SVR)* model has been combined with other time series model for forecasting property crime rates as discussed by Alwee et al. (2013a,b).

3.1.3 Association Rule Mining

Association rule mining/learning is a rule-based machine learning method for discovering the interesting relationships between variables in Big Data. The primary goal in this approach is to detect strong rules discovered in databases using some measures of interestingness (Piatetsky-Shapiro 1991). Association rule mining is originally introduced in Agrawal et al. (1993) as a new way for discovering interesting co-occurrences in supermarket data. Within this piece of research, it was argued that a typical example of frequent item-set mining is market basket analysis which examines customer buying patterns by determining associations between different items that customers put in their shopping baskets as illustrated in Fig. 6 (see also Han et al. 2011). The finding of these associations would be very useful to the retailers in evolving their marketing strategies by understanding which items are *frequently* bought by the most customers. Association rule mining has been recently adapted and used in a wide range of criminal case studies.

One of the earliest applications of Association Rule Mining is reported in Brown and Hagen (2003) where the method is applied in the law enforcement context. This technique automatically searches for similarities based on a new information theoretic measure known as *total similarity*. In this method, weights among attributes of a certain crime records data are determined in such a way that incidents possibly committed by the same or group of criminals are strongly associated to each other. This association method was then developed further (in Lin and Brown 2006) by combining it with outlier score function, and was tested on the same crime data as investigated in Brown and Hagen (2003). The new approach

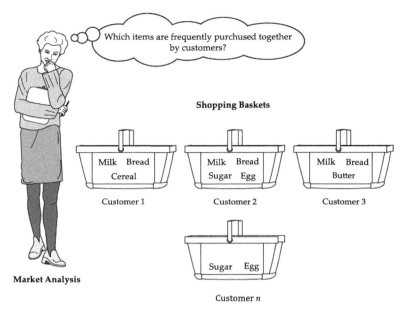

Fig. 6 Market basket analysis as illustrated in Han et al. (2011)

outperforms the similarity-based method as it provides more helpful insights in criminal investigation carried out by police forces.

Another useful association rule mining algorithm proposed by Li et al. (2005) is known as *distributed higher-order* which demonstrated promising results in analysing law enforcement data. They are very useful in overcoming the need for knowledge of a global schema; they can also assess both horizontal and vertical distributions. In order to maintain Temporal association rules with numerical attributes, an incremental algorithm was developed in Ng et al. (2007) by taking into account the negative border method. It was then used in detecting crime patterns in a district of Hong Kong. The results illustrated compelling improvements with respect to other standard methods.

The association rule method has been also used in mining cyber-crime data. For instance, this method was tested in Appavu et al. (2007) for discovering suspicious emails through detecting uncommon and fraudulent email communications through utilising Apriori Algorithm (see Han et al. 2011 for more details). The main purpose for developing this approach was to aid the investigators in efficiently extracting information and consequently taking compelling actions to decrease/prevent criminal activities. The association rule mining along with other machine learning techniques mentioned above have been applied to crime intelligence data to extract valuable information on cyber threats (Littlefield 2018). Similar to clustering methods discussed above, an association rule mining can be combined with the multi-dimensional knowledge discovery and data Mining (KDD) model (Littlefield 2018) to provide actionable knowledge on the Internet threats.

KDD model is helpful in exploring the correlation between cyber threats residing in the Internet data (e.g., Darknet dataset). This model analyses packet distributions, transports, network and application layer protocols as well as resolved domain names. Association rule mining and correlation techniques were applied on the threat that the Darknet sensors record an increasing number of TCP packets after that period (Fachkha et al. 2012). The technique extracts clusters of co-occurring malicious activities targeting certain victims, providing valuable intelligence about patterns within threats and allowing the interpretation of attack scenarios. The proposed approach is demonstrated to be very efficient in identifying three high severity threats, namely, Denial of Service (DoS) attempts, buffer overflow exploits and unsolicited VPN access (Fachkha and Debbabi 2016; Fachkha et al. 2012).

3.1.4 Social Network Analysis

Network analysis has become more popular in the past decade and has numerous applications in social sciences. Social Network Analysis (SNA) focuses on the structure of ties within a set of social actors, e.g., persons, groups, organizations, and nations, or the products of human activity or cognition such as websites, semantic concepts, and so on. It is linked to structuralism in sociology stressing the significance of relations among social actors to their behaviour, opinions, and attitudes. Social network analysis in criminology is a computational or mathematical approach to model and investigate the interactive structure of crime entities regarding the crime dataset dimensions. As presented in Mena (2003), the main structure of a social network consists of nodes that are connected to other related nodes. The link between two nodes is called an edge illustrating that two social actors participated in a certain event. The graph theory provides the measurement techniques required in SNA which contain a wide range of mathematical formulae and concepts required for studying patterns of lines (see Wasserman and Faust (1994) for further details about "Degree", "Density" and "Centrality" as the main measurement techniques needed for SNA).

Although the first application of SNA in crime analysis is reported in Sparrow (1991), nevertheless SNA has attracted considerable attention in the applications of crime analysis in the past few years. Various network structure measurements were first used in Wang and Chiu (2005) to identify the on-line auction inflated-reputation traders from regular accounts. Recently, on-line auction frauds has been detected by other supervised learning as discussed in Tsang et al. (2014). The SNA techniques used to recommend a system which can identify risks of collusion associated with an account for online auction site users. The system was then successfully (with 76% accuracy rate) detected real world blacklist data (see Wang and Chiu 2008 for further details).

Social Network Analysis (SNA) is now widely studied within the academic community for the assessment of organised criminal networks. It allows for analysing a network as a complex structure composed of actors or entities connected by links or relationships, and in which a variety of resources are subject to change. There have been several investigations using the SNA technique to study terrorist and organised criminal groups including analysing the global salafi jihad network (Qin et al. 2005), providing important intelligence to the US homeland security to be prepared by placing more effective counter-measures (Ressler 2006) (see also Chen 2008 for a similar study), analysing the core member of a terrorist group using the combination of SNA and enhanced part algorithm (Liu et al. 2007), and examining an Outlaw Motorcycle Gang operating in Canada (McNally and Alston 2006).

SNA and other web mining techniques were used in Chau and Xu (2007) to investigate characteristic of online groups which resulted in successfully detecting a selected set of hate groups. A similar methodology was used to stress the importance of investigating the rise of cyber crime groups in on-line blogs (Chau and Xu 2008). A simulation email system was developed in Qiao et al. (2008) using personality trait dimensions to model the traffic behaviour of email account users. This metric was then combined with SNA to determine crucial figures of a criminal group.

4 Conclusions and Further Work

The continuous increase in the number of on-line users and the emergence of a variety of new on-line business models require more intelligent and proactive techniques in tackling cybercrimes. In the past few years, machine learning and data mining techniques have been applied to datasets derived from different industries. In this paper, we critically discussed the existing techniques that can be used in assessing crime patterns and providing predictive and holistic insights for law enforcement units. The discussions were substantiated by literature followed by an introduction of case studies in which the techniques have been applied. Since, this chapter provides an overview and an introduction of the techniques, the authors intend to empirically apply the techniques to relevant crime related datasets with a view to compare suitable approaches in respect to their computational-efficiency, accuracy and complexity.

References

Abu-Hana, R. O., Freitas, C. O., Oliveira, L. S., & Bortolozzi, F. (2008). Crime scene classification. In *Proceedings of the 2008 ACM Symposium on Applied Computing* (pp. 419–423).
Adderley, R., Townsley, M., & Bond, J. (2007). Use of data mining techniques to model crime scene investigator performance. *Knowledge-Based Systems, 20*(2), 170–176.
Agrawal, R., Imieliński, T., & Swami, A. (1993). Mining association rules between sets of items in large databases. In *SIGMOD '93: Proceedings of the 1993 ACM SIGMOD International Conference on Management of Data*, New York (Vol. 22, pp. 207–216).
Agarwal, J., Nagpal, R., & Sehgal, R. (2013). Crime analysis using k-means clustering. *International Journal of Computer Application, 83*(4), 1–4.

Akare, P. P., Mohm, H., Maniyar, H., Thorat, T. D., & Pagar, J. K. (2015). Detecting phishing web pages using NB classifier and EMD approach. *International Journal on Recent and Innovation Trends in Computing and Communication, 3*(1), 148–151. Retrieved from http://dx.doi.org/10.17762/ijritcc2321-8169.150131.

Alwee, R., Shamsuddin, S., & Sallehuddin, R. (2013a). Economic indicators selection for property crime rates using grey relational analysis and support vector regression. In *Proceedings of the 2013 International Conference on Systems, Control, Signal Processing and Informatics* (p. 178185).

Alwee, R., Shamsuddin, S., & Sallehuddin, R. (2013b). Hybrid support vector regression and autoregressive integrated moving average models improved by particle swarm optimization for property crime rates forecasting with economic indicators. *The Scientific World Journal, 2013*, 1–11.

Appavu, S., & Rajaram, R. (2007). Suspicious e-mail detection via decision tree: A data mining approach. *Journal of Computing and Information Technology, 15*(2), 161–169.

Appavu, S., Pandian, M., & Rajaram, R. (2007). Association rule mining for suspicious email detection: A data mining approach. In *Proceedings of the IEEE international conference on intelligence and security informatics*. New Jersey, USA, pp. 316–323.

Bello, Y., & Yelwa, S. (2012). Complementing GIS with cluster analysis in assessing property crime in Katsina state, Nigeria. *American International Journal of Contemporary Research, 2*(7), 190–198.

Bendler, J., Brandt, T., Wagner, S., & Neumann, D. (2014). Investigating crime-to-twitter relationships in urban environments-facilitating a virtual neighborhood watch. In M. Avital, J. M. Leimeister, & U. Schultze (Eds.), *Ecis*. Retrieved from http://dblp.uni-trier.de/db/conf/ecis/ecis2014.html#BendlerBWFN14.

Bhattacharyya, S., Jha, S., Tharakunnel, K., & Westland, J. C. (2011). Data mining for credit card fraud: A comparative study. *Decision Support Systems, 50*(3), 602–613.

Bhowmik, R. (2011). Detecting auto insurance fraud by data mining techniques. *Journal of Emerging Trends in Computing and Information Sciences, 2*(4), 156–162.

Brown, D. E., & Hagen, S. (2003). Data association methods with applications to law enforcement. *Decision Support Systems, 34*(4), 369–378.

Chainey, S., & Ratcliffe, J. (2013). *GIS and crime mapping*. Hoboken: Wiley.

Chau, M., & Xu, J. (2007). Mining communities and their relationships in blogs: A study of online hate groups. *International Journal of Human-Computer Studies, 65*(1), 57–70.

Chau, M., & Xu, J. (2008). Using web mining and social network analysis to study the emergence of cyber communities in blogs. *Terrorism Informatics, 18*, 473–494.

Chen, H. (2008). Homeland security data mining using social network analysis. In *Isi*. IEEE. Retrieved from http://dblp.uni-trier.de/db/conf/isi/isi2008.html#Chen08.

Chen, H., Chung, W., Xu, J., Wang, G., Qin, Y., & Chau, M. (2004). Crime data mining: A general framework and some examples. *Computer, 37*(4), 50–56.

Cortes, C., & Vapnik, V. (1995). Support vector networks. *Machine Learning, 20*, 273–297.

Curry, P. A., Sen, A., & Orlov, G. (2016). Crime, apprehension and clearance rates: Panel data evidence from canadian provinces. *Canadian Journal of Economics/Revue canadienne d'économique, 49*(2), 481–514.

De Bruin, J., Cocx, T., Kosters, W., Laros, J., & Kok, J. (2006). Data mining approaches to criminal career analysis. In *Proceedings of the 6th International Conference on Data Mining* (pp. 171–177).

De Vel, O., Anderson, A., Corney, M., & Mohay, G. (2001). Mining e-mail content for author identification forensics. *ACM Sigmod Record, 30*(4), 55–64. Retrieved from http://doi.acm.org/10.1145/604264.604272.

Deylami, H.-M., & Singh, Y. P. (2012). Adaboost and SVM based cybercrime detection and prevention model. *Artificial Intelligence Research, 1*(2), 117–130. Retrieved from http://dblp.uni-trier.de/db/journals/aires/aires1.html#DeylamiS12.

Dietterich, T., Becker, S., & Ghahramani, Z. (2002). Advances in neural information processing systems. In *Proceedings of the Annual Conference on Neural Information Processing Systems*.

Eibe, F., & Bouckaer, R. (2006). Naïve Bayes for text classification with unbalanced classes. In *Proceedings of the 10th European Conference on Principle and Practice of Knowledge Discovery in Databases* (p. 503510). Berlin.

Fachkha, C., & Debbabi, M. (2016). Darknet as a source of cyber intelligence: Survey, taxonomy, and characterization. *IEEE Communications Surveys & Tutorials, 18*(2), 1197–1227.

Fachkha, C., Bou-Harb, E., Boukhtouta, A., Dinh, S., Iqbal, F., & Debbabi, M. (2012). Investigating the dark cyberspace: Profiling, threat-based analysis and correlation. In *2012 7th International Conference on Risk and Security of Internet and Systems (Crisis)* (pp. 1–8).

Fayyad, U., & Uthurusamy, R. (2002). Evolving data into mining solutions for insights. *Communications of the ACM, 45*(8), 28–31.

Fuller, C. M., Biros, D. P., & Delen, D. (2011). An investigation of data and text mining methods for real world deception detection. *Expert Systems with Applications, 38*(7), 8392–8398. Retrieved from http://dblp.uni-trier.de/db/journals/eswa/eswa38.html#FullerBD11.

Geng, L., & Hamilton, H. J. (2006). Interestingness measures for data mining: A survey. *ACM Computing Surveys (CSUR), 38*(3), 9.

Gepp, A., Wilson, J. H., Kumar, K., & Bhattacharya, S. (2012). A comparative analysis of decision trees vis-a-vis other computational data mining techniques in automotive insurance fraud detection. *Journal of Data Science, 10*, 537–561.

Ghaffari, A., Hosseinian-Far, A., & Sheikh-Akbari, A. (2017). Iris biometrics recognition in security management. In V. Chang, M. Ramachandran, R. Walters, & G. Wills (Eds.), *Enterprise Security. ES 2015. Lecture notes in Computer Science* (Vol. 10131). Cham: Springer.

Gish, H. (1990). A probabilistic approach to the understanding and training of neural network classifiers. In *Proceedings IEEE International Conference on Acoustics, Speech and Signal Processing*, Albuquerque (pp. 1361–1364).

Gottfredson, M. R., & Hirschi, T. (1990). *A general theory of crime*. Stanford: Stanford University Press.

Gupta, M., Chandra, B., & Gupta, M. (2008). Crime data mining for Indian police information system. In *Proceeding of the Computer Society of India* (pp. 389–397).

Han, J., Jian, P., & Michelin, K. (2006). *Data mining, Southeast Asia edition*. San Francisco: Elsevier Inc.

Han, J., Pei, J., & Kamber, M. (2011). *Data mining: Concepts and techniques*. Burlington: Elsevier.

Hassani, H., Huang, X., Silva, E., & Ghodsi, M. (2016). A review of data mining applications in crime. *Statistical Analysis and Data Mining, 9*(3), 139–154.

Hosseinpournajarkolaei, A., Jahankhani, H., & Hosseinian-Far, A. (2014). Vulnerability considerations for power line communications supervisory control and data acquisition. *International Journal of Electronic Security and Digital Forensics, 6*(2), 104–114.

Internet Growth Statistics. (2018). Retrieved from http://www.internetworldstats.com/emarketing.htm.

Iqbal, R., Murad, M., Mustapha, A., Panahy, P., & Khanahmadliravi, N. (2012). An experimental study of classification algorithms for crime prediction. *Indian Journal of Science and Technology, 6*(3), 4219–4225.

Jabar, E., Hashem, S., & Enas, M. (2013). Propose data mining AR-GA model to advance crime analysis. *IOSR Journal of Computer Engineering (IOSR-JCE), 14*, 38–45.

Jahankhani, H., & Hosseinian-Far, A. (2014). Digital forensics education, training and awareness. In B. Akhgar, A. Staniforth, & F. Bosco (Eds.), *Cyber crime and cyber terrorism investigator's handbook* (pp. 91–100). Syngress. Retrieved from https://www.sciencedirect.com/science/article/pii/B9780128007433000086.

Jahankhani, H., Al-Nemrat, A., & Hosseinian-Far, A. (2014). Cybercrime classification and characteristics. In *Cyber crime and cyber terrorism investigator's handbook* (pp. 149–164). Amsterdam: Elsevier.

Keyvanpour, M., Javideh, M., & Ebrahimi, M. (2011). Detecting and investigating crime by means of data mining: A general crime matching framework. *Procedia Computer Science, 3*, 872880.

Kianmehr, K., & Alhajj, R. (2008). Effectiveness of support vector machine for crime hot-spots prediction. *Applied Artificial Intelligence, 22*(5), 433–458.

Kirkos, E., Spathis, C., & Manolopoulos, Y. (2007). Data mining techniques for the detection of fraudulent financial statements. *Expert Systems with Applications, 32*(4), 9951003.

Larose, D., & Larose, C. (2015). *Data mining and predictive analytics.* Hoboken: Wiley.

Li, S., Wu, T., & Pottenger, W. M. (2005). Distributed higher order association rule mining using information extracted from textual data. *ACM SIGKDD Explorations Newsletter, 7*(1), 26–35.

Lin, S., & Brown, D. E. (2006). An outlier-based data association method for linking criminal incidents. *Decision Support Systems, 41*(3), 604–615, Elsevier.

Littlefield, R. (2018). *Cyber threat intelligence: Applying machine learning, data mining and text feature extraction to the darknet.* Retrieved from https://littlefield.co/cyber-threat-intelligence-applying-machine-learning-data-mining-and-text-feature-extraction-to-bb00c3b729bc.

Liu, Q., Tang, C., Qiao, S., Liu, Q., & Wen, F. (2007). Mining the core member of terrorist crime group based on social network analysis. In *Pacific-Asia Workshop on Intelligence and Security Informatics* (pp. 311–313).

Maimon, L., & Rokach, O. (Eds.). (2010). *Data mining and knowledge discovery handbook.* New York: Springer. Retrieved from http://public.eblib.com/choice/publicfullrecord.aspx?p=645908.

McClendon, L., & Meghanathan, N. (2015). Using machine learning algorithms to analyze crime data. *Machine Learning and Applications: An International Journal, 2*(1), 1–12.

McNally, D., & Alston, J. (2006). Use of social network analysis (SNA) in the examination of an outlaw motorcycle gang. *Journal of Gang Research, 13*(3), 1–25.

Mena, J. (2003). *Investigative data mining for security and criminal detection.* Boston: Butterworth-Heinemann.

Modupe, A., Olugbara, O. O., & Ojo, S. O. (2011). Exploring support vector machines and random forests to detect advanced fee fraud activities on internet. In *2011 IEEE 11th International Conference on Data Mining Workshops (ICDMW)* (pp. 331–335). Piscataway.

Ng, V., Chan, S., Lau, D., & Ying, C. M. (2007). Incremental mining for temporal association rules for crime pattern discoveries. In *Proceedings of the Eighteenth Conference on Australasian Database-Volume* (Vol. 63, pp. 123–132).

Nirkhi, S., Dharaskar, R., & Thakre, V. (2012). Data mining: A prospective approach for digital forensics. *International Journal of Data Mining and Knowledge Management Process, 2*(6), 41–48.

Pandey, M., & Ravi, V. (2012). Detecting phishing e-mails using text and data mining. In *Proceedings of the IEEE International Conference on Computational Intelligence & Computing Research* (pp. 1–6).

Pang-Ning, T., Steinbach, M., & Kumar, V. (2014). *Introduction to data mining.* Harlow: Pearson.

Piatetsky-Shapiro, G. (1991). Discovery, analysis, and presentation of strong rules (pp. 229–238). Menlo Park: AAI/MIT.

Qiao, S., Tang, C., Peng, J., Liu, W., Wen, F., & Qiu, J. (2008). Mining key members of crime networks based on personality trait simulation email analysis system. *Chinese Journal of Computers, 32*(10), 1795–1803.

Qin, J., Xu, J. J., Hu, D., Sageman, M., & Chen, H. (2005). Analyzing terrorist networks: A case study of the global Salafi Jihad network. In *International Conference on Intelligence and Security Informatics* (pp. 287–304).

Raffael, M. (2016). *AI and machine learning in cyber security: What zen teaches about insights.* Retrieved from https://towardsdatascience.com/ai-and-machine-learning-in-cyber-security-d6fbee480af0.

Ray, S. (2017). *Understanding support vector machine algorithm from examples (along with code).* Analytics Vidhya.

Ressler, S. (2006). Social network analysis as an approach to combat terrorism: Past, present, and future research. *Homeland Security Affairs, 2*(2), 1–10.

Richard, M. D., & Lippmann, R. P. (1991). Neural network classifiers estimate Bayesian a posteriori probabilities. *Neural Computation, 3*(4), 461–483. Retrieved from http://dblp.uni-trier.de/db/journals/neco/neco3.html#RichardL91.

Roy, A. (2015). A classification algorithm for high-dimensional data. In A. Roy, P. Angelov, A. M. Alimi, G. K. Venayagamoorthy, & T. B. Trafalis (Eds.), *Inns Conference on Big Data* (Vol. 53, pp. 345–355). Elsevier. Retrieved from http://dblp.uni-trier.de/db/conf/inns-wc/innsbd2015.html#Roy15.

Salem, M. B., & Stolfo, S. J. (2010). Detecting masqueraders: A comparison of one-class bag-of-words user behavior modeling techniques. *JoWUA, 1*(1), 3–13.

Sathyadevan, S., & Gangadharan, S. (2014). Crime analysis and prediction using data mining. In *2014 First International Conference on Networks & Soft Computing (icnsc)* (pp. 406–412).

Sparrow, M. (1991). The application of network analysis to criminal intelligence: An assessment of the prospects. *Social Networks, 13*(3), 251–274.

Sukanya, M., Kalaikumaran, T., & Karthik, S. (2012). Criminals and crime hotspot detection using data mining algorithms: Clustering and classification. *International Journal of Advanced Research in Computer Engineering and Technology, 1*(10), 225–227.

Tan, P.-N., Steinbach, M., & Kumar, V. (2006). *Introduction to data mining*. Boston: Addison Wesley.

Thongtae, P., & Srisuk, S. (2008). An analysis of data mining applications in crime domain. In *IEEE 8th International Conference on Computer and Information Technology Workshops* (pp. 122–126).

Tsang, S., Koh, Y. S., Dobbie, G., & Alam, S. (2014). Detecting online auction shilling frauds using supervised learning. *Expert Systems with Applications, 41*(6), 3027–3040.

Tu, J. (1996). Advantages and disadvantages of using artificial neural networks versus logistic regression for predicting medical outcomes. *Journal of Clinical Epidemiology, 49*(11), 1225–1231.

Vijayakumar, M., Karthick, S., & Prakash, N. (2013). The day-to-day crime forecasting analysis of using spatial-temporal clustering simulation. *International Journal of Scientific & Engineering Research, 4*(1), 1–6.

Wang, J., & Chiu, C. (2005). Detecting online auction inflated-reputation behaviors using social network analysis. In *Proceedings of the Annual Conference of the North American Association for Computational Social and or- Ganizational Science* (pp. 26–28).

Wang, J.-C., & Chiu, C.-C. (2008). Recommending trusted online auction sellers using social network analysis. *Expert Systems with Applications, 34*(3), 1666–1679. Retrieved from http://dblp.uni-trier.de/db/journals/eswa/eswa34.html#WangC08.

Wang, C., & Liu, P.-S. (2008). Data mining and hotspot detection in an urban development project. *Journal of Data Science, 32*, 389–414.

Wasserman, S., & Faust, K. (1994). *Social network analysis: Methods and applications* (Vol. 8). New York: Cambridge University Press.

Wen, C.-H., Hsu, P., Yung Wang, C., & Wu, T.-L. (2012). Identifying smuggling vessels with artificial neural network and logistics regression in criminal intelligence using vessels smuggling case data. In J.-S. Pan, S.-M. Chen, & N. T. Nguyen (Eds.), *Asian Conference on Intelligent Information and Database Systems* (Vol. 7197, pp. 539–548). Springer. Retrieved from http://dblp.uni-trier.de/db/conf/aciids/aciids2012-2.html#WenHWW12.

Witten, I. H., & Frank, E. (2005). *Data mining: Practical machine learning tools and techniques* (D. Cerra Ed.). Amsterdam: Kaufman.

Yu, C.-H., Ward, M. W., Morabito, M., & Ding, W. (2011). Crime forecasting using data mining techniques. In *Proceedings of the 11th International Conference on Data Mining Workshops* (pp. 779–786).

Yu, Y., Kaiya, H., Yoshioka, N., Hu, Z., Washizaki, H., Xiong, Y., & Hosseinian-Far, A. (2017). Goal modelling for security problem matching and pattern enforcement. *International Journal of Secure Software Engineering (IJSSE), 8*(3), 42–57.

SMERF: Social Media, Ethics and Risk Framework

Ian Mitchell, Tracey Cockerton, Sukhvinder Hara, and Carl Evans

1 Introduction

According to a study of 314 undergraduate students attending a college in US (Morgan et al. 2010), 92% of students have a Social Media (SM) presence. Independent of this study, further research in OfCom (2012) reported 91% of 16–24 year-olds and 90% of 25–34 year-olds of the population in the UK have a SM profile. Furthermore, OfCom's (2012) figures show that between 2010 and 2016 the number of people having a SM profile aged between 16 and 24 increased by 5–6% and for people aged between 25 and 34 increased by 20%. Having a SM profile does not necessarily mean that one is interacting with other user accounts. The OfCom (2012) study showed that 44% of people aged 16–24 would access their SM profile more than 10 times a day, and 41% of the same age group access their SM profile 2–10 times a day. Again, access does not equate to interaction. Whittaker and Kowalski (2015) indicated, in a study of 197 undergraduates, that 116,881 posts were downloaded during a 90 day period. Assuming a normal distribution, this equates to approx 1,300 posts made per day and approx. Six posts made per day per user. It can be argued that these statistics are unreliable since the assumption that they are normally distributed is a fragile argument, however, it does indicate that undergraduates are looking at their profiles and generating content on a regular basis. Undergraduates represents a cross-section of young adults, love or loathe SM, it is happening in your organisation and, regardless of whether or not you are a SM participant, it is highly likely your colleagues are and they are regularly checking their profiles and generating user content. This degree of online interaction in most cases is harmless and even helpful, but occasionally it causes

I. Mitchell (✉) · T. Cockerton · S. Hara · C. Evans
Middlesex University, London, UK
e-mail: I.Mitchell@mdx.ac.uk; enquiries@smerf.net

problems, such as cyberbullying, that can have a serious impact on the victim, e.g. see Weale (2015). The development of this framework was completed with respect to the context of students at Higher Education Providers (HEP), however due to SMERF's agile development there is no reason why it cannot be adapted to other domains in different organisations, so broader terms used could equally be applied to investigations, e.g. researcher and Digital Forensic Analyst, or project and investigation.

SM research is not subject constrained, and there are many examples of research across multiple subject disciplines, for example: Health Care (Korda and Itani 2013); Business (Gu and Ye 2014); Marketing (Ashley and Tuten 2015); Tourism (Munar and Jacobsen 2014); Psychology (Groth et al. 2017); Sociology (Gerbaudo and Treré 2015); Technology (Backstrom and Kleinberg 2014; Balduzzi et al. 2010; Wondracek et al. 2010); and Education (Irvin et al. 2015). It is not quite ubiquitous, but has a presence in most disciplines and a search of "social media" on a research depository (scholar.google) yielded 3.62M results. It is the welfare of the investigator or researcher that is a priority to any institution or organisation that involves SM-related activities. It is *not* the aim of SMERF to prevent SM-related research or investigations, current inertia of SM uptake makes that virtually impossible, however, it is the aim of SMERF to mitigate any risks that the investigator or researcher may be exposed to.

With increased integration of SM in projects, which offers numerous research strands in addition to easily accessible test data sets, there is a clear need to assess risks. Furthermore, such risk is not constrained merely to research projects, but can relate to any investigation that involves SM. Whilst research undergoes ethics and risk assessments, few projects undergo a risk assessment of the researcher of SM-related investigations that would be expected in other domains. This is also true of digital forensic investigations of SM-related context, First Response Teams (FRT) would complete a risk assessment of the area being searched, however there are few risk assessments relating to the individual completing the SM-related investigation or research. There is an emphasis on ethics to check that the implementation of the project is not going to cause harm or be detrimental to society, however, the converse is not always true, i.e. it does very little to safeguard the individual *from* society. There is a requirement to consider the safety of the researcher entering a potentially dangerous physical environment. Safeguarding researchers undertaking projects in a SM environment is now essential. So the motivation for this work was to introduce an integrated risk framework to enhance Research Ethics Committees (REC) for SM projects and provide some precautionary guidance to mitigate any risk the researcher may be exposed to. In addition to the above, Higher Education Providers (HEP) and organisations per se have a responsibility to protect researchers from entering environments that pose risks to their health, well-being and safety – this includes online environments – since all accidents are preventable. The proposed framework is to be integrated with existing REC frameworks and/or other existing policies in the organisation, e.g., Acceptable Use Policy (AUP), and external to the organisation, e.g. ISO17025.

1.1 Social Media Project Types

Currently, we have identified four types of social media projects, listed below:

Passive observation   potential to engage with other SM user-accounts is *discouraged* and essentially uses techniques to analyse data sets obtained from SM;

Active Experiment   potential to engage with other SM user-accounts is *encouraged*, from an individual account;

Paticipatory Action Research (PAR)   engagement with other SM user-accounts is encouraged to change or influence an outcome; and

Software Development   usually undertaken by IT-related department to develop and build SM and SM-tools (e.g. sentiment analysis), and may require user engagement to complete user testing, or de-anonymise data, i.e. using developed software to identify users (Wondracek et al. 2010), or collect data and expose data to third parties, e.g. Granville (2017) and Zimmer (2010).

The definitions for high, medium to low risk are as follows:

Low   For the completion of project outcomes there is no requirement for engaging and interacting with other account users. However, there may be some minimal risk activities to understand the workings of SM; these would be completed with necessary safeguarding and ensuring that each account was created anonymously and remains anonymous. Essentially, a low risk project would require:

- no engagement or minimal engagement under supervision in a controlled environment;
- anonymous at all times; and
- data or potential digital evidence collection

Medium   For the completion of project outcomes there would be a requirement for engagement and interaction with other SM accounts. This would be completed with necessary safeguarding and ensuring that each account remains anonymous. Engagement is observed, but at no point is the researcher engaging to influence or change outcomes at the risk of de-anonymising their SM account. Here the vulnerability is the potential to become a victim of cyberbullying and remaining unidentified. Therefore, the cyberbullying can be halted by creating new accounts.

High   For the completion of project outcomes there would be a requirement for engagement and interaction with other SM accounts to influence or capture their views. This would be completed with necessary safeguarding, but risks the de-anonymisation of account, e.g. disabling geo-location or completing some SM update that results in identity disclosure. Here the vulnerability is the potential to become a victim of cyberbullying and being identified. Therefore, the cyberbullying is difficult to halt.

A 'passive' social media project is usually minimal risk if relevant precautions are taken and has no engagement. An 'active' SM project can be medium to high risk, and as such requires precaution and is dependent on the proposal of the project. This will be considered in the subsequent sections, but first some terms of reference are presented:

1.2 Terminology

For the scope of this paper the following terms of reference are used, please note these can easily be transferred to other domains, e.g. a project can be an investigation:

Proposal A proposal can be a coursework brief or a dissertation which should then include a succinct list of aims, objectives, deliverables, planning, equipment, experiment (hypotheses), expected outcomes and brief background of problem domain.

Project Can be an in-course assessment or a dissertation. Normally, it should meet the aims, objectives and deliverables.

Supervisor A member staff supervising the research. The supervisor will be required to assess the risk and any necessary precautions applied to mitigate the risk.

Researcher(s) A researcher(s) has been assigned a supervisor and is normally the person to undertake the research.

Research Ethics Committee (REC) If a project is not low risk then it would go to the REC for individual approval, which may have some caveats.

Research Ethics Framework (REF) Existing set of questions to identify the level of risk each project undergoes.

Organisation In the case student this was a University with many undergraduate students. Could equally be a Law Enforcement Agency or a Business, that has tangible and non-tangible assets to protect.

1.3 Social Media, SM

Newman (2010, Ch.3) explains the many different types of social networks. Essentially, SM can be distilled to a graph, $G = (V, E)$, with a set of vertices, V, representing SM accounts and a set of edges, E, representing relationships between SM accounts, i.e. typically a form of internet-based connection, or link. The edges can be directed to differentiate between being 'followed' and 'following'. Furthermore the graph can be multi-modal and edges can represent messages sent between SM accounts, these edges can be weighted to represent the number of messages sent between SM accounts. Research can be on the structure of relationships between SM accounts, content of the messages, structure of the messages between SM accounts and a mixture of these. It is these relationships that are the foundation of a social network and can range from a group of users to a direct messages. Therefore, there is a wealth of publications related to SM and so it is a field that offers a lot of potential for research projects. Herein lies the problem; encouraging researchers to complete SM-related research puts them at risk.

Table 1 Including a simple question about SM at the beginning of the ethical approval procedures. Please note the vertical dots in §1 represent omitted questions relating to identifying medium to high risk projects. The "go to §2" label indicates the navigation and would not appear on online form

§1 Please check if your research involves the following?

    ☐ Non-compliance with legislation
    ⋮
    ☐ Concerns terrorism, cyber-terrorism or extreme groups
    ☐ **Social Media** (go to §2)
    ☐ None of the Above

There are many guidelines, e.g., Markham (2012), Markham and Buchanan (2012) and Rivers and Lewis (2014) with regard to SM research and these are adapted for researchers and incorporated in SMERF. Some of the guidelines require membership of the organisation and there are issues in choosing and assuaging any fears that researchers may have using SM. Despite the claim in Morgan et al. (2010) of ubiquitous use amongst students, as pointed out in Irvin et al. (2015) there may be a small percentage who are resistant or reluctant to use SM. It is recommended that organisations have a "Social Media Acceptable Use Policy (SMAUP)" which would provide succinct and clear guidelines of acceptable use whilst completing SM projects.

Given the definition above, Table 1 illustrates the first question in the framework that includes the use of SM, from here there are a number of questions that are addressed by most RECs. The problem is that almost any form of online communication could be considered SM, from email, online dating, MMORPGs to traditional micro-blog websites, e.g. Twitter, and all have their risks. Each organisation should develop a list of permitted SM services that should be reviewed annually. Developing an integrated risk assessment framework should happen in stages, and these stages are prescribed by the type of projects as follows: Quantitative Passive Observation; Qualitative Passive Observation; Quantitative Active Experiment; Qualitative Active Experiment; Participatory Action Research; and Software Development. Table 2 illustrates the question of SM usage. Typically, if this is checked then questions need to be posed to determine further elements of the research design and the potential risk, to allow reflection on actions to be taken to mitigate the potential risk.

1.4 Prevent Duty

The Counter Terrorism and Security Act (2015) introduced the Prevent strategy (Home Office 2015a,b). The framework described here could also be incorporated

Table 2 Having completed a simple question on SM, then section §2 concerns regulations and legislation. The consent of identity disclosure is important, however, the risks need to be explained to the researcher. Any project that requires identity disclosure would be considered high risk and require REC approval. These are the recommended list of de-anonymisation procedures and improvements can be ammended

§2 Social Media
§2.1 T&C and Regulations

☐ Social Media platform used:
   ☐ Facebook
   ☐ LinkedIn
   ⋮
   ☐ Second Life
   ☐ Twitter
   ☐ World of Warcraft
   ☐ Other: ....

☐ Read T&C, ToS of Social Media platform(s) identified above.
☐ Read organisation's Acceptable Use Policy (AUP).

§2.2 ID Disclosure

☐ Project or coursework requires your identity disclosure (High Risk, apply to REC)
☐ Project or coursework does not require your identity disclosure (go to §2.3)

§2.3 Anonymous User Account

☐ Do you have any reservations about using Social Media
☐ Do you require an anonymous and additional email account
☐ Please ensure you complete the following de-anonymisation protection measures:

- No identifying photos or documents
- No posts that will reveal identity
- No links to other users that may disclose identity
- No duplication of information held on other accounts
- Disable geo-location on device
- Remove any geo-location information in uploaded documents

§2.4 Project Type

☐ Passive (go to §3)
☐ Active (go to §4)
☐ Publicity (go to §5)
☐ Participatory (go to §6)
☐ Software Development (go to §7)

into the Prevent strategy of the organisation, see Universities UK (2015) for further advice, since it encourages academic freedom under some guidelines to mitigate any risks that may occur relating to radicalisation of extremist views.

1.5 Risks

Many readers may have conducted SM-related research and do not understand the need for further risk assessment. However, this is likely to be the outcome of specific approaches and design of SM research, and the time frame within which the research was conducted. The risks in SM usage can no longer be ignored and have been highlighted by results in Pyżalski (2012) stating that from a sample of 2000+ 15 year-olds, 66% reported at least one act of cyber-aggression listed in Pyzalski's typology (2012).

Many public figures have been victims of trolling.[1] or worse, for expressing a view on SM, e.g. Ms S. Creasy, MP for Waltham Forest, experienced trolling and the perpetrator was exposed, convicted and served a custodial sentence (Carter 2014). In this particular case, the perpetrator was found to have over 150 different social media accounts and was using these accounts to stalk and troll the victim. These examples are not the social norm. However, there is a SM norm that includes some experience of cyber-aggression. SM-related activities could expose the researcher to some abuse, or worse, e.g., see Weale (2015). What are the risks of cyberbullying? In Hinduja and Patchin (2010) there is a report that estimates between 10% and 25% of users have been cyberbullyied, and between 4% and 17% admitted to cyberbullying others. Whilst this survey was conducted on 11–18 year olds, it is relevant to researchers currently entering organisations. Whittaker and Kowalski (2015) conducted a study for undergraduates and noted that 18.2% of participants had experienced being cyberbullied and 12% admitted to cyberbullying others. The figures show some consistency, despite being conducted across different age groups. Study 1 in Whittaker and Kowalski (2015) makes a strong correlation between being a victim and perpetrator, with over 50% of participants witnessing cyberbullying at least once within the previous 12 months. Interestingly, the most likely perpetrator of cyberbullying was known, over that of a stranger. Study 1 in Whittaker and Kowalski (2015) was completed using a questionnaire. Study 3 used software to analyse the messages on SM. It used a list of words and modifiers (stemming[2]) to identify cyber-aggression within text and showed a difference in direct posts (38.6%) compared with indirect posts (46.1%). Furthermore, this study concluded that anonymity of

---

[1]Definition of trolling:"...intentionally posting provocative messages about sensitive subject to create conflict, upset people, and bait them..."(Zainudin et al. 2016). For further classifications on trolling, see Bishop (2012).
[2]Natural Language Processing defines stemming as identifying the same words with different affixes, see Bird et al. (2009, Ch.3.), e.g. lie and lying.

Table 3 Comparing Pyżalski (2012) to types of CA attacks. Checkmarks indicate a likelihood of this happening without volunteering information, e.g., individual characteristics and opinions are known by peers. Discs indicate a likelihood of this happening due to volunteering information, e.g., political affiliation

	AoA	AoI	IoA	IoI	Hyb
CA against peers		✓		✓	✓
CA against vulnerable	●	✓	●	✓	✓
Random CA	✓	✓	✓	✓	✓
CA against groups	●	✓	●	✓	✓
CA against celebrities		✓		✓	✓
CA against staff		✓		✓	✓

the perpertrator is a factor conducive to cyber-aggression and Facebook exhibited the lowest cyber-aggression since it has explicit ties to identity. So does anonymity of the victim help?

Given SM accounts are either Identity Disclosed (IDD) or Anonymous (Anon), then we have the following types of Cyber-Aggression (CA) attacks:

AoA Anon on Anon CA are random where all parties remain anonymous.

AoI Anon on IDD CA are more directed and where perpetrating party's identity is anonymous, however, the victim party's identity is disclosed.

IoA IDD on Anon CA, are random where perpetrating party's identity is disclosed and the victim party's identity remains anonymous.

IoI IDD on IDD CA, where all parties' identities are disclosed.

Hyb Hybrid CA, where there is a dynamic process and the CA may include a mixture of the above, e.g., AoA and IoA, this would involve several perpetrators, having identities disclosed and anonymous.

From study 1 in Whittaker and Kowalski (2015) the most common form of CA are from peers (>50%). These acts may manifest into a hybrid attack, whereby account users that maintain anonymity and account users that have disclosed their identity may perpetrate acts of cyber-aggression. Table 3 compares the classifications of cyber-aggression (Pyżalski 2012) and the types of cyber-aggression attacks in the list above. It can be seen that there is a reduction in the risk involved when the victim remains anonymous. Whilst random CA cannot be prevented, there are issues with the other two groups checked.

The motivation behind cyberbullying may help us answer this question. Intolerance, relationship rejection and relationship envy are the three main motivations behind cyberbullying (Hoff and Mitchell 2009). Intolerance of sexual orientation is most common, but disability, sexism, obesity, religion and race cannot be ruled out. A fourth motivation behind cyberbullying is exclusion, i.e. being cyber-aggressive to anyone not in your group. It is the perpetrator that uses anonymity to mask their identity; the victim has no anonymity. For this reason, and classifications shown in Table 3, one of the recommendations is to provide protection with an anonymous account with the attributes of the individual hidden for the duration of the project.

This is not a policy on cyberbullying per se, rather a guidance framework for researchers conducting projects with an SM element to reduce the probability of cyberbullying.

Providing anonymisation whilst completing SM-related research is not the end. To remain anonymous, the user needs to ensure they do not reveal their identity. Naturally, there will be projects that require a user to reveal individual attributes. This then becomes a project of medium to high risk, and would require ethical approval.

It may be infeasible for SMERF to take account of every possible situation, but it can exist as mitigation and can inform what procedures to complete when things go wrong. The development of SMERF has considered a wide range of activities that could lead to vulnerabilities being exposed in SM-related research. The list is not complete but includes the following:

Cyber-bullying and Harassment Extended arguments that result in other actions such as trolling and stalking. Repetition. Power imbalance. Completed anonymously. Victim. Publicity and humiliation. Intentional harm. Obsessive following of another user, not passive but usually aggressive;

SM CSE Child Sexual Exploitation, CSE, is illegal and needs reporting directly.

SM Pornography There are many websites for pornography and ethical guidelines will apply;

SM Revenge Whilst this is a motivation, there is more and more evidence that this results in either take-over accounts or impersonation and defamation;

SM Hatred Inciting hatred to a group of people;

Unauthorised Access Common in open office/lab environments. Unauthorised access to an account can result in misrepresentation or public distribution of information intended for a private network. This information can include personal details or in extreme cases financial details and is often referred to as "Doxxing";

Private SM There is the ability to have a private network, in fact, a trend in SM is for organisations to have internal SM. There is a risk of releasing and making public information intended for such private social networks and this can occur non-maliciously and maliciously;

Impersonation/Identity theft Results in the impersonation implemented via a non *bona-fide* account and misrepresentation of views and opinions in an attempt to defame;

De-anonymisation Passive observation; someone may wish to remain anonymous and use an anonymous account. The identification of someone, or the unauthorised access and display of personal information is a risk which is often followed by threatening behaviour. Research and reports (Balduzzi et al. 2010; Wondracek et al. 2010) show how easy it is to de-anonymise users in social networking sites (Zimmer 2010);

Multiple accounts attacks One person, or organisation can have multiple accounts. These accounts, if managed well, can result in a single attack to give the argument credibility and disguise their stalking, e.g. see Stella Creasy case, where the stalker had 150+ accounts (Carter 2014);

Group attacks Has the same effect as multiple attacks, but the user accounts belong to group, such as a political party or sports fans, and cyber-abuse and cyber-aggression are used to intimidate a minority, often referred to as "mobbing";

Baiting There is a fine line between trolling and baiting, but typically an outrageous and usually offensive statement is made (i.e. the baiting) and abuse, hatred, trolling follows. The statement can be false, and can generate advertisement revenues (Sambrook 2017);

Artificial trending bot accounts that have no known user and are paid for accounts, e.g. a *bona-fide* user can pay for 20,000 followers and artificially trend their opinion.

Extremism Any extreme material that attempts to radicalise individuals, and should be part of the Prevent strategy of HEP (CT&S 2015);

Retrospective deprecation Ever done something you regret when you were young? Whilst many readers would not dream of putting what they have done online, it is fast becoming a social norm to share this experience, e.g. Alcohol-abuse (Groth et al. 2017), this growing trend can lead to future employers making judgments; and

Exclusion Being excluded from groups and unable to communicate can lead to isolation.

This list is likely to increase over time and should be reviewed annually. Many of the social media companies have applied filters and reporting opportunities to overcome such issues and all users involved should know how to report abuse and record abuse. Legislation is catching up and provides support for organisations, e.g., in the UK the Crown Prosecution Service (CPS) introduced new laws that mean that those who create derogatory and offensive modified images may face prosecution, since this incites people to harass others online, known as "mobbing", and is among the offences included in the guidance. For REC guidance a risk assessment matrix, see Table 4, is used; the probability of identity disclosure occurrence is considered in the rows as: unlikely; occasionally; and likely. The consequences are defined as:

Minor Damage to user and perhaps some minor exposure to one-off offensive remarks, but no long term effect and easily repairable and overcome by the user.

Moderate Potential for full identity disclosure; user may experience repetitive offensive remarks and start to see a pattern emerging. Typically, this would be a result of user spending time on an anonymous account and be at risk of disclosing identity. The user may find this harder to overcome and require some support.

Major This is full identity disclosure, which may prove impossible to overcome and stop. There is the increasing possibility of user becoming withdrawn as a consequence of the abuse and in worse cases results in suicide. It can involve liability and litigation, especially if the cause of the identity disclosure is found to be due to the project and recommendations were not followed.

Table 4 Risk assessment matrix, identity disclosure (IDD) correlates to risk

IDD	Consequence		
	Minor	Moderate	Major
Unlikely	Low	Low	Medium
Occasionally	Low	Medium	High
Likely	Medium	High	High

2 Passive Observation

2.1 Quantitative

Typically used to investigate the structure of the SM, targeted techniques may be employed to transform an incoherent graph of 20 K tweets into a coherent graph, which conveys clear clusters, and shows key influencers in each cluster based on some centrality measure. There is no need to study the message content, even though this may be downloaded in the data. Such analysis usually requires that the individual has an account; the data does not need to be anonymised since the agreement of the SM service includes that the user contribution is public. However, no further attempt to de-anonymise data other than that information available should be made. For example, if an account uses a pseudonym then that pseudonym should not be de-anonymised. The T&C of the SM service may place restrictions on manipulation and storage of data. Efforts to keep publication of data anonymous can still be challenging due to unauthorised secondary use of data, it is therefore advised that data collected for all experiments is only used by authorised personnal and stored securely using encryption (Zimmer 2010).

Essentially, Quantitative Passive Observation is seen as minimal risk. However, there is some potential problem with the only interaction that is required – the downloading of the data and search term. It is therefore advised that the search term is discussed and agreed upon between supervisor and researcher, and if in doubt, then the search term should be sent to the ethics committee to ensure that there is no perceived issue with it and the downloaded data cannot be used as inculpatory evidence against researcher, e.g., if a search term for a hashtag on Twitter for "#terrorism" could result in downloading data related to terrorists. GDPR is not confined to EU, it is confined to EU citizens, so there is an issue of completing data acquisition from SM and not securely storing that data in a GDPR compliant manner. The concerns regarding SM-related research are listed below and should be taken in the context of approval from REC, e.g. compliance to legislation is checked elsewhere in REC applications.

- Pseudonyms should be kept and data should be anonymised where possible.
- Strictly no engagement with users, other than under supervision to understand the nuances of the SM platform.
- Search terms should be discussed with supervisor

- If there is doubt over a search term, then a submission should be made to REC stating concerns.
- Ensuring that data acquisition is legally obtained – The legal "Terms of Service" are pointed out in Beurskens (2014) for Twitter data acquisition.
- Reluctant users: There is minimal risk in exposing researchers' details when using anonymous accounts; there are recommendations in Lin et al. (2013) to provide further assurances by not following peers, rather a subject, or as in Twitter, a hashtag. To ensure a further level of anonymity, an additional email account can be provided. In all cases, it is recommended that all accounts are disabled after research is complete.
- Storage of dataset is local and only distributed within the confines of the "Terms and Conditions" of the use of the Social Media in question, e.g. it is pointed out in Beurskens (2014) that there is some contradiction between reproducibility of experiment and availability of data due to Twitter, Inc. restriction on storage of datasets. Access is only by authorised personnel and secondary re-use of data is restricted to REF approved projects. This is due to the risk of de-anonymisation as seen with Lewis et al. (2008) and documented in Zimmer (2010).

2.2 Qualitative

Due to the nature of reviewing content, which may be illegal, only SM services with the ability to self-regulate, review and block certain content are permitted to be used. The concerns with qualitative passive observation and SM-related research are listed below:

- Comply with all guidelines in Quantitative Passive Observation
- All applications should go through REC approval.
- The reaction to discovery of offensive material should be measured and contextualised, where it is clearly offensive and illegal, e.g., inciting hate speech, then appropriate measures should be completed and reported to SM authority.

Only the collection of data does not need Ethical approval as this is minimal risk if the above rules are completed (exceptions can apply). However, qualitative research may give rise to legal issues over privacy, copyright and offensive content that requires reporting and should undergo Ethical approval to ensure that appropriate measures are in place, e.g., compliance with data protection regulations.

2.3 SMERF: Passive Observation

3 Active Experiment

This framework does not intend to cover all experiments, but merely offer guidelines to complete experiments. The annual quality review ensures improvements and amendments are updated and the cycle continues. The following are some recommendations made for active experiment SM-related research.

3.1 Quantitative

The problem is the revealing identification via SM. Regardless of whether or not the project starts out with minimal risk on SM, this can lead to tempestuous battles and result in medium to high risk to the researcher and organisation involved. So the issue is not so much about the topic, but rather the identification of the researcher or organisation via SM. The issue of covert and overt research methods are covered elsewhere by REC. Our recommendations for Quantitative Active Experimentation are:

- Comply with all guidelines in Passive Observation, see Table 5.
- All applications should go through REC approval.
- Anonymity of researcher and supervisor – this can include temporary email accounts for the duration of the experiment to add a layer of abstraction, e.g., use of current email may exist in other domains. Expire email accounts when experiment completed.
- Repeat engagement of users should be monitored.
- No physical interaction or meeting with users; there should be no need for interviews after the experiment is completed. The analysis should satisfy the objectives of the project.
- Unless essential to the project, disable geo-location on software applications that the researcher is using to access SM.
- Ensure privacy settings to protect anonymity.
- Gaining informed consent is an issue for SM research. Overt operations should not reveal the identity of the researcher, organisation or supervisor and need only apply with applications where the ownership of the message is private, e.g., in a dating website, or a private group. Informed consent requires a contact; an anonymous email can be used that does not reveal researcher's identity but remains as a contact – revealing researcher emails could be seen as a security vulnerability.
- Use of public data whereby the terms of condition means that informed consent is not required, e.g., see Beninger et al. (2014) and Williams et al. (2017).

Table 5 SMERF Passive Observation: To understand a SM platform, it may be necessary to use it in a supervised manner. Otherwise no engagement should be required and all users can remain anonymous for the duration of the project

§3 Passive Observation
§3.1 Quantitative Passive Observation
☐ Engagement with other users is kept minimal, under supervision and only for the benefit of understanding the nuances of the chosen SM Platform. ☐ Anonymise collected data ☐ Store data in accordance with GDPR ☐ Store data in accordance with T&C of owner ☐ Search term(s) agreed by Supervisor ☐ Enter the search term(s) below:
§3.2 Qualitative Passive Observation
☐ Completed and agree with terms in §3.1 ☐ All attempts made to anonymise data. Messages, if published, should not include identification of user. ☐ No response to discovery of offensive and cyber-aggressive material. If the data has been anonymised then it is difficult to report such cyber-aggression. In this stage cyber-aggression can only be observed, and reported under the supervision. ☐ No use of software tools or programs to de-anonymise data. ☐ No unauthorised re-use of data.
All the above identified as minimal risk

- Ensuring that data acquisition is legally obtained, modified and stored – see Beurskens (2014).
- Reproducibility and Experimentation is often required to justify empirical results, therefore several temporary email accounts may be required for repeat experiments.
- Creating new accounts can be difficult, since the influence has been diminished by losing the number of relations an account has. This is problematic, since many researchers may have built up a profile that has influence in a certain domain. There is then a temptation to mimic the growth of the individual, by creating an exact copy profile and contact lots of peers in the same organisation. The experiment objectives in the Ethics application should make it clear that users, in order to remain anonymous, should avoid creating duplicate profiles.

Table 6 Risk assessment: please note this is intended to integrated into an ethics assessment, e.g,. have you read the acceptable use policy or computer misuse policy are covered elsewhere?

§4 Active Experiment
§4.1 Quantitative

- ☐ De-anonymise data
- ☐ Meetings with participant(s)
- ☐ Enable geo-location
- ☐ Unprotected anonymity of my user account
- ☐ Reveal my identity via SM
- ☐ No consent of participants, covert
- ☐ Create online relationships that can compromise your anonymity
- ☐ Create duplicate profiles
- ☐ Employ third parties to create User Generated Content or Accounts

§4.2 Qualitative

- Answer questions in §4.1
- On observing Cyberbullying I should...
- On observing Cyber-aggression I should...
- On observing repetitive messages I should...
- Including extracts of SM messages I should...
- Measures taken to remain anonymous whilst communicating via SM...

Any of the above then identified as medium risk, and if appropriate measures are taken there is no reason to allow this project to continue, but under supervision to monitor and ensure there is nothing untoward. Revealing identity would result in high risk and would require supporting evidence and background and finally approval from REC.

3.2 Qualitative

Many of the guidelines that have to be followed are covered by Quantitative research, and therefore section §4.1 in Table 6 is to be completed. There are other issues with Qualitative research that may require the inclusion of comments made from users, and these need to be kept anonymous. The biggest challenge is the interaction and engagement of other users, which increases the risk to the medium level of being exposed to cyber-aggression and cyberbullying. It needs to be clear that if researchers are engaging with other users they should not reveal their identity by accidentally providing information in a post that is irrelevant to the research, but, releases information about their identity. When completing Qualitative research the following recommendations are:

- Comply with all guidelines in Quantitative Active Experiment, Table 6.
- Engagement should be honest and it is recommended to adopt the excellent advice and guidelines from Aragon et al. (2014) and Isaacs et al. (2014).
- All data should be anonymised, but any extracts included in reports should not be able to identify the user.
- Any cyberbullying should be reported immediately, covered by HEP AUP and T&C of SM service.

3.3 SMERF: Active Experiment

4 Miscellaneous Use of Social Media

4.1 Questionnaires

It is not recommended that Social Media is used for Questionnaires, a specific website should be designed or many of the customisable questionnaire websites available employed to complete online questionnaires. SM can be used in conjunction with a website. For example, the deployment of SM to publicise the questionnaire is acceptable and low risk. Typically, SM was not designed for questionnaires although there have been developments to make this facility available. Therefore, it is appreciated that there may be a valid reason to complete the questionnaire on SM, but extreme care should be taken to ensure various controls are implemented appropriately. As with all questionnaires there are issues, which are as follows: gaining consent; controlling sample and *bona-fide* participants; anonymity; ensuring confidentiality; and data protection. If done incorrectly the SM platform could contravene all these issues and at best make the research void, or worse break GDPR (Table 7).

4.2 Participatory Action Research (PAR)

By far the biggest challenge to mitigating risk is PAR and Social Media. This type of research is common amongst Business or Prevention/Awareness campaigns, to introduce ideas and measure the influence, and is ideal for SM. As in all PAR, the participants need to give their consent; SM is not an exception, but surreptitiously gaining participants is problematic. PAR requires the control of the group; stricter guidelines are to be employed by using private groups and hence reduce the possibility of any cyber-aggression from outside the group. Any reaction and influence can still be measured. This is completed by control over group membership and ensures that the effect is generated from *bona-fide* members who have given consent, provided anonymity and privacy. Engagement generated from outside the defined group can be problematic with SM, e.g., a single-person with multiple accounts, say 100+, can significantly bias the results. Table 8 illustrates

Table 7 Questionnaire risk assessment. There are two parts to the form: publishing of a questionnaire, which if followed correctly is considered low risk; and completion of questionnaire using SM, including the collection and collation of data, which is considered high risk and referred to REC for approval

§5 Questionnaires
§5.1 Publicity
☐ Do *not* engage with other account users ☐ Disable geo-location ☐ Use anonymous user account ☐ Used only to publicise questionnaire ☐ Under no circumstances are questions asked on SM account
§5.2 Answers collected
It is strongly recommended that you do not use SM for questionnaires. This is considered as high risk and please state your reasons along with controls employed below and this will be considered by REC

Table 8 PAR risk assessment. If appropriate measures are completed satisfactorily in a controlled environment, then there is no reason why this cannot be deemed medium risk; it is only when the environment becomes uncontrolled the exposure to cyber-aggression is increased and therefore becomes high risk

§6 Participatory Action Research, PAR
§6.1 Controlled
– Describe measures to ensure participant's consent – Describe measures to ensure participant's anonymity – Describe measures employed to control group membership – During the research are there any meetings planned with participants – Describe measures to report and record acts of cyber-aggression
§6.2 Uncontrolled
It is strongly recommended that you control your group membership. Uncontrolled group membership is considered as high risk and please state your reasons along with conditions employed below and this will be considered by REC.

this and is split in two sections: controlled, recommended and medium risk; and uncontrolled, not recommended and high risk.

4.3 Social Media Software Development

Any software development is guided by Ethical Guidelines by Professional Organisations, e.g. see ACM (2017), BCS: Code of conduct (2017) and Markham (2012) and the organisation's RECs. The issue regards development of innate qualities

Table 9 Development of software recommends using Virtual Machines (VM) and containing data. At the VM stage some projects may stop and not progress to going live, i.e., being assigned a server other than VM Server, these are considered low risk. When a project goes live, these are considered medium to high risk

§7 SM Software Development
☐ SM virtual machine ☐ Web Host Management (WHM) Software ☐ Domain name for website ☐ If applicable, IP address of DNS ☐ Meets ethical approval ☐ Meets GDPR approval ☐ Security and privacy precautions ☐ Enter SM software and licenses below:

built within the software that deals with keeping data secure and only accessible by authorised personnel. There are additional legal risks, due to data collected inappropriately (Granville 2017) and allowed access to by unauthorised parties, all breaches could lead to litigation or lost at least loss of reputation. Most software development that involves an end client has a user test – there are strict guidelines for this in all software development. Initially, this looks extremely challenging, however, if standard software development methodologies and ethical guidelines are followed then the development should proceed just like any other project. Table 9 indicates some of the questions that should be included in Software Development.

5 Recommendations

The motivation for SMERF is different to Cyberbullying campaigns (Hinduja and Patchin 2010; UKCCIS 2010). Instead, it has been developed to provide an environment that mitigates potential risk for researchers, and provide a framework for RECs to consider the risk associated with conducting SM-related research. SMERF should not operate in isolation and should feed into various other internal quality mechanisms within the organisation, see Fig. 1. Here it is recommended that SMERF is integrated and annually reviewed, with any outcomes and recommendations as a result of debriefing to be reported to REC, these could include wider implications, such as, use of induction or staff training to make researchers aware of computer misuse policies and focus on acceptable use of SM, e.g., an investigation that had proven culpability that relied on data gathered illegally.

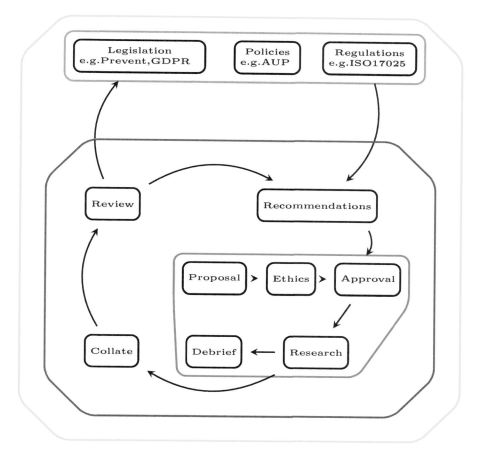

Fig. 1 Life-cycle of quality assurance for framework of risk assessment for students using social media for research. Three levels of abstraction: red: project level; blue: SMERF level; and green: prevent strategy of HEI, relevant computer misuse policies of HEI and legislation e.g. GDPR. Abbreviation: *AUP* acceptable use policy, *AM* academic misconduct, *GDPR* general data protection regulation

6 Conclusion

There is often resistance to change, and proposing something new to be integrated with an existing ethics framework may meet resistance. To summarise and justify the need for SMERF, Fig. 2 illustrates *ten* key influences, the details of which are summarised in below:

%Researchers   Prior evidence has shown that there is a high number of u/g users, approx. 90% and an even higher number of 16–24 year olds with SM presence, approx 92%, see Morgan et al. (2010) and OfCom (2012), respectively.

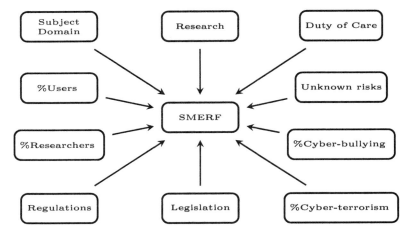

Fig. 2 Motivation: larger number of users; Widespread applications; high probability to experience cyber-aggression or cyberbullying; Organisation has a responsibility for its employees; GDPR and other Privacy laws; unconstrained research across a wide range; prevent strategies and unknown risks yet to be identified

%users Whilst there is a higher risk of cyberbulling from peer group, there is still a risk of cyberbulling from unknown perpetrators, the risk is reduce according to Pyżalski (2012).

Subject Domain Research is not constrained to a single knowledge domain, but is across subject domains.

Cyberbulling It seems inevitable that cyberbulling is going to happen, often victims become perpetrators and the cycle is difficult to stop. This has a detrimental impact on victim health and in high risk situations can damage a reputation of an organisation and in extreme cases lead to victim's suicide.

Duty of Care Organisations have a responsibility for researcher's welfare and safety, especially in projects that they are encouraged to complete.

Legislation GDPR (Council of European Union 2018) the storing and protection of data is compliant with current legislation – covered by all GDPR compliant organisations in EU.

Research Research in SM is prolific, since 2010 there have been over 3M publications relating to SM.

Cyber-Terrorism Prevent Strategies in the UK come under Counter-Terrorism Act (CT&S 2015) and require organisations to be responsible.

Unknown Risks Identified through annual monitoring reviews and integrated into SMERF.

Regulations Terms & Conditions of SM service. Organisation's own rules and regulations, e.g. AUP.

This list is useful in making change possible and gaining support of key stakeholders who are responsible for REC.

SMERF is the main contribution of the work reported here, however, there are additional unexpected outcomes. How the framework is integrated into existing quality assurance life-cycles is important and can be adapted further for more subject specialist projects that require risk and ethics, e.g. cyber-security. Integrating a set of new ethics questions in isolation can be problematic, and it is not the intention of SMERF to prescribe to researchers. SMERF is agile, see Fig. 1, and it is expected that organisations consult with participants of SMERF to introduce and update SMERF annually. SMERF focuses on identity disclosure, project type and then assessing the associated risks – there is an argument that the current project does not fit the prescribed project types. Agile design allows for the framework to adapt and be modified for the respective organisation. If new projects or modifications are identified then the annual review, see Fig. 1, can include these and being open source SMERF allows future modifications and new projects types to be reported at smerf.net, from which further details of the framework can be found and discussed. The introduction to SMERF to organisations should reduce and mitigate risks that researchers may be exposed to doing SM-related projects. SMERF is not a cyberbullying prevention programme per se, it is only intended for the duration of the project, furthermore measures would have to be deployed to prevent this from happening across the board and there are many included in SMERF that could be useful, e.g., include acceptable use of SM during induction programmes. The recommendations of this paper is the prevention of identity disclosure; SMERF for minimal risks requires anonymous users, this does not mean that projects that reveal identity cannot proceed, they can but with a higher risk. There are two levels of identity disclosure: de-anonymisation; and user permitted. If measures from the SMERF are undertaken then there is little risk of de-anonymisation and identity disclosure. Projects that require identity disclosure are high risk, and may involved reluctant researchers with good reason to remain anonymous, e.g., victims of domestic abuse or other intimidation-related crimes, and REC should consider the risk and if the proposal, especially if a brief, can be changed to accommodate anonymity. It is appreciated that some SM membership requires identity disclosure; unless absolutely necessary these SM services should be avoided and alternatives sought. Even the use of SM platforms that are private and formed to aid intra-organisation communication should be avoided. Whilst here the abuse can be closely monitored and there is some control by the organisation, however, cyberbullying and unacceptable use is restricted to one SM platform, it metastasizes to other user accounts, SM platforms and even new technology. It is the matastisization of this unacceptable use that, once started, makes it particularly difficult to stop. If the source of identity disclosure is due to the researcher undertaking the projects at the organisation, then morally and possibly legally, the organisation has a responsibility and duty of care to protect individual. Legally the organisation could be found liable for not providing a duty of care, or at the very least a risk assessment, which is why SMERF is introduced. If SM-related projects are encouraged without any risk or ethics assessment then depending on the level of interaction there is a medium to high risk that cyberbullying will occur. SMERF's intervention mitigates the risk of cyberbullying and other SM misuse.

Such mitigation and risk assessment provides some protection from subsequent litigation claims, which may not be covered by insurance premiums prior to SMERF's intervention. SMERF's mitigation also prevents any bad publicity caused by SM misuse and subsequent damaged reputation of the organisation. Finally, all organisations have a responsibility for the well-being of its staff, and ensure that conducted research causes no harm; SMERF, via ethics and risk assessment, provides a duty of care for researchers completing SM-related activities.

References

ACM. (2017). *ACM code of ethics and professional conduct*. https://www.acm.org/about-acm/acm-code-of-ethics-and-professional-conduct. Accessed June 16, 2017.

Aragon, A., AlDoubi, S., Kaminski, K., Anderson, S.K., & Isaacs, N. (2014). Social networking: Boundaries and limits part 1: Ethics. *TechTrends, 58*(2), 25.

Ashley, C., & Tuten, T. (2015). Creative strategies in social media marketing: An exploratory study of branded social content and consumer engagement. *Psychology & Marketing, 32*(1), 15–27.

Backstrom, L., & Kleinberg, J. (2014). Romantic partnerships and the dispersion of social ties: A network analysis of relationship status on facebook. In *Proceedings of the 17th ACM Conference on Computer Supported Cooperative Work & Social Computing* (pp. 831–841). ACM.

Balduzzi, M., Platzer, C., Holz, T., Kirda, E., Balzarotti, D., & Kruegel, C. (2010). Abusing social networks for automated user profiling. In *International Workshop on Recent Advances in Intrusion Detection* (pp. 422–441). Springer

BCS: Code of conduct. (2017). http://www.bcs.org/category/6030. Accessed June 16, 2017.

Beninger, K., Fry, A., Jago, N., Lepps, H., Nass, L., & Silvester, H. (2014). *Research using social media; users' views*. National Centre for Social Research.

Beurskens, M. (2014). Legal questions of twitter research. In K. Weller (Ed.), *Twitter and society* (p. 123). New York: Peter Lang.

Bird, S., Klein, E., & Loper, E. (2009). *Natural language processing with Python: Analyzing text with the natural language toolkit*. Beijing: O'Reilly Media, Inc.

Bishop, J. (2012). Tackling Internet abuse in Great Britain: Towards a framework for classifying severities of 'flame trolling'. In *Proceedings of the International Conference on Security and Management (SAM)* (p. 1). The Steering Committee of The World Congress in Computer Science, Computer Engineering and Applied Computing (WorldComp).

Carter, C. (2014). *Twitter troll jailed for 'campaign of hatred' against Stella Creasy*. The Daily Telegraph.

Council of European Union. (2018). *Council regulation (EU) no 2016/679*. http://eur-lex.europa.eu/legal-content/en/LSU/?uri=CELEX%3A32016R0679. Accessed on June 25, 2017.

CT&S. (2015). Counter Terrorism & Security Act. http://www.legislation.gov.uk/ukpga/2015/6/contents/enacted. Accessed on June 15, 2017.

Gerbaudo, P., & Treré, E. (2015). In search of the 'we' of social media activism: Introduction to the special issue on social media and protest identities. *Information, Communication & Society, 18*(8), 865–871.

Granville, K. (2018). *Facebook and cambridge analytica: What you need to know as fallout widens*. The New York Times.

Groth, G. G., Longo, L. M., & Martin, J. L. (2017). Social media and college student risk behaviors: A mini-review. *Addictive Behaviors, 65*, 87–91.

Gu, B., & Ye, Q. (2014). First step in social media: Measuring the influence of online management responses on customer satisfaction. *Production and Operations Management, 23*(4), 570–582.

Hinduja, S., & Patchin, J. W. (2010). *Cyberbullying: Identification, prevention, and response.* Cyberbullying Research Center. US.

Hoff, D. L., & Mitchell, S. N. (2009). Cyberbullying: Causes, effects, and remedies. *Journal of Educational Administration, 47*(5), 652–665.

Home Office. (2015a). *Prevent duty guidance: For higher education institutions in England and Wales.* UK Govt.

Home Office. (2015b). *Prevent duty guidance: For higher education institutions in Scotland.* UK Govt.

Irvin, E., Taper, C., Igoe, L., & Pastore, R. S. (2015). Using Twitter in an undergraduate setting: Five recommendations from a foreign language class. eLearn 11.

Isaacs, N., Kaminski, K., Aragon, A., & Anderson, S. K. (2014). Social networking: Boundaries and limitations part 2: Policy. *TechTrends 58*(3), 10.

Korda, H., & Itani, Z. (2013). Harnessing social media for health promotion and behavior change. *Health Promotion Practice, 14*(1), 15–23.

Lewis, K., Kaufman, J., Gonzalez, M., Wimmer, A., & Christakis, N. (2008). Tastes, ties and time: A new social network dataset using Facebook. *Social Networks, 30*(4), 330–342.

Lin, M. F. G., Hoffman, E. S., & Borengasser, C. (2013). Is social media too social for class? A case study of Twitter use. *TechTrends, 57*(2), 39.

Markham, A. (2012). *AOIR guidelines: Ethical decision making and Internet research ethics* (Technical report). Association of Internet Research.

Markham, A., & Buchanan, E. (2012). *Ethical decision-making and Internet research* (Technical report). Association of Internet Research (AoIR).

Morgan, E. M., Snelson, C., & Elison-Bowers, P. (2010) Image and video disclosure of substance use on social media websites. *Computers in Human Behavior, 26*, 1405–1411.

Munar, A. M., & Jacobsen, J. K. S. (2014). Motivations for sharing tourism experiences through social media. *Tourism Management, 43*, 46–54.

Newman, M. (2010). *Networks: An introduction.* Oxford: Oxford University Press.

OfCom. (2016). *Adults' media use and attitudes.* Office of Communications (April 2016).

Pyżalski, J. (2012). From cyberbullying to electronic aggression: Typology of the phenomenon. *Emotional and Behavioural Difficulties, 17*(3–4), 305–317.

Rivers, C. M., & Lewis, B. L. (2014). Ethical research standards in a world of big data. *F1000Research, 3*, 1–12.

Sambrook, R. (2017). Taking the bait: The quest for instant gratification online is seriously compromising news reporting. *Index on Censorship 46*(1), 16–17.

UKCCIS. (2010). *Child safety online: A practical guide for parent and carers whose children are using social media.* UK Govt.

Universities UK. (2012). *Oversight of seucrity-sensitive research material in UK Universities* (Technical report). UUK (October 2012).

Weale, S. (2018). Suicide is a sector-wide issue, says Bristol university vice-chancellor. The Guardian (Feb 2018).

Whittaker, E., & Kowalski, R. M. (2015). Cyberbullying via social media. *Journal of School Violence 14*(1), 11–29.

Williams, M. L., Burnap, P., & Sloan, L. (2017). Towards an ethical framework for publishing twitter data in social research: Taking into account users' views, online context and algorithmic estimation. *Sociology, 51*, 1–20.

Wondracek, G., Holz, T., Kirda, E., & Kruegel, C. (2010). A practical attack to de-anonymize social network users. In *2010 IEEE Symposium on Security and Privacy (SP)* (pp. 223–238). IEEE.

Zainudin, N. M., Zainal, K. H., Hasbullah, N. A., Wahab, N. A., & Ramli, S. (2016). A review on cyberbullying in Malaysia from digital forensic perspective. In *International Conference on Information and Communication Technology (ICICTM)* (pp. 246–250). IEEE.

Zimmer, M. (2010). "But data is already public": On the ethics of research in Facebook. *Ethics Information Technology, 12*, 313–325.

Understanding the Cyber-Victimisation of People with Long Term Conditions and the Need for Collaborative Forensics-Enabled Disease Management Programmes

Zhraa A. Alhaboby, Doaa Alhaboby, Haider M. Al-Khateeb,
Gregory Epiphaniou, Dhouha Kbaier Ben Ismail, Hamid Jahankhani,
and Prashant Pillai

1 Introduction

Victimisation can be described as an unwanted attention or negative behaviour over time, it can be performed by an individual or a group, against the victim, and sometimes multiple victims are targeted (Kouwenberg et al. 2012). The offline victimisation of people with long term conditions and disabilities is widely documented in various communities such as Canada (Hamiwka et al. 2009), Ireland and France (Sentenac et al. 2011a), the United States (Taylor et al. 2010; Chen and Schwartz 2012) and Netherlands (Kouwenberg et al. 2012).

The widespread of electronic communications such as email, phone messages, blogs or social networking websites/apps (including Facebook, Twitter, Instagram, YouTube and others) brought numerous benefits for people with long term conditions by facilitating networking for social purposes or to get health information or support (Algtewi et al. 2015). Electronic communication had empowered people with disabilities by the sense of identity, belonging and activism (Seale and Chadwick 2017). However, in the literature, there is a significant association between having a chronic condition, disability, and being a victim of harassment

---

Z. A. Alhaboby
Institute for Health Research, University of Bedfordshire, Luton, UK

D. Alhaboby
Faculty of Medicine, University of Duisburg-Essen, Duisburg, Germany

H. M. Al-Khateeb (✉) · G. Epiphaniou · D. K. B. Ismail · P. Pillai
Wolverhampton Cyber Research Institute, University of Wolverhampton, Wolverhampton, UK
e-mail: H.Al-Khateeb@wlv.ac.uk

H. Jahankhani
QAHE and Northumbria University, Northumbria University London, London, UK
e-mail: hamid.jahankhani@northumbria.ac.uk

© Springer Nature Switzerland AG 2018
H. Jahankhani (ed.), *Cyber Criminology*, Advanced Sciences and Technologies for Security Applications, https://doi.org/10.1007/978-3-319-97181-0_11

(Sentenac et al. 2011b), and this was escalated using online communications. Hence, this virtual environment has become available for the offenders too, leading to the risk of online discrimination, or what is known as 'cyber-victimisation'.

Cyber-victimisation is an umbrella term covering a range of cyber offences such as cyberharassment, cyberbullying, cyberstalking, cyber-disability hate incidents/crimes or cyber sexual exploitation. Each of these terms has its own definition that could vary between disciplines, however, they share the criteria of being an antisocial behaviour by the 'offender' towards the 'victim' via electronic communication causing fear and distress (Alhaboby et al. 2016a). This is achieved by sending harassing content, insults, creating false profiles, spreading lies or contacting the social network of the victim.

Cyberharassment includes negative attitudes or intimidating behaviours towards the victim involving the use of the Internet and/or cell phone. An example of a study that looked at cyber-victimisation of disabled people and used the term cyberharassment is the work by Fridh, Lindström (Fridh et al. 2015). This cross-sectional public health study in Sweden sampled 8544 people, of which, 762 individuals had disabilities. Participants were aged 12, 15 and 17 years with self-reported impaired hearing, impaired vision, reading/writing disorders, dyslexia and ADHD. Cyberharassment in this study was defined as a violation or harassment over the past 12 months, involving cell phones or the Internet such as email, Facebook, and text messages. Male participants reported a frequency of cyberharassment of 32.1% (one incident) to 41.5% (several incidents), while female participants reported 28% and 35% frequencies respectively. The impact upon victims was mainly subjective health complaints.

When the intimidation in harassment is associated with power imbalances, the perceived unequal power relation between the victim and the offender is described as 'cyberbullying'. Such experiences are common in schools and workplace due to the nature of relationships between the involved parties. A public health study in Sweden (Annerbäck et al. 2014) looked at 413 participants aged 13–15 years, drawn from a sample of 5248 participants. The participants had a variety of chronic conditions and disabilities including impaired hearing or vision, limited motor function, dyslexia, ADHD, asthma, diabetes, epilepsy and intestinal diseases. Cyberbullying was defined as an indirect form of bullying, indicating harassment via the Internet or mobile phones in the past 2 months and involving the use of power to control others or cause distress. The impact reported was poor health, mental health consequences and self-harm (Annerbäck et al. 2014).

Another cyber-offence is 'cyberstalking', which involves repeated unwanted contact triggering fear and distress, however, it is also characterised by fixation. Hence, scholars identify cyberstalking cases by the repetition of ten harassment incidents over a period of 4 weeks (Sheridan and Grant 2007). Cyberstalking can be regarded as a phenomenon by itself or an evolution to stalking by giving offenders new relatively easy methods to target the victim (Bocij and McFarlane 2003). There is a growing body of literature covering stalking as an ancient crime and with the surge of using technology in everyday life, cyberstalking literature has increased (Bocij and McFarlane 2003). Two types of studies emerged on review of the literature, studies that discuss stalking, introducing electronic means as new methods of stalking, referred to as cases of combined stalking and cyberstalking (Davis et al. 2002) and more recently, cyberstalking was addressed in studies purely focusing on this phenomenon (Dreßing et al. 2014). In both cases, authors tended to introduce the topic by discussing offline offences first. In a study of cyberstalking victims, the main target population was not people with chronic conditions (Sheridan and Grant 2007), however, 11.9% of pure cyberstalking cases were against people with disabilities. Additionally, more than 10 years ago in criminology, cyberstalking was defined as harassing or threatening a person or a group more than once using the Internet or electronic communication (Bocij and McFarlane 2003). Hence, it shares the same building blocks of offline victimisation definition and adding to it, electronic communication.

In the remaining part of this chapter, long term conditions such as chronic diseases and disabilities are defined in Sect. 2, in addition to a state-of-the-art review to demonstrate the inconsistency in defining cyber-victimisation in existing literature. Section 3 covers the impact of victimisation and discusses available support. In Sect. 4 we provide a comprehensive review of how DMPs developed over time. Section 5 extends the discussion towards the introduction of forensics-enabled Electronic Coaches to mitigate against cyber-victimisation. Finally, we conclude our findings and recommendation in Sect. 6.

2 Defining Chronic Conditions and Cyber-Victimisation

2.1 Chronic Conditions, Disabilities, and Vulnerability

Vulnerability within the context of research ethics is a term used to describe an individual or a group of people who require protection (Levine et al. 2004). Vulnerability connection to chronic conditions and disability status is multifaceted. The term 'chronic' is derived from the Greek word 'khronos' which means 'time', the key feature of a chronic condition. Further, the Oxford dictionary explain it as illness persisting for a long time or with a recurring nature (Oxford 2015). In medicine, 'chronic' is a term referring to a group of diseases characterised by long duration, frequent recurrence and slow progression (Webster 2015). These diagnoses, which are also known as long term conditions, have an impact on individual's life and this requires a full commitment to be taken by the person to administer medications, adopt a certain lifestyle, and make everyday decisions in order to reach the best possible quality of life (Greenhalgh 2009). Long term conditions overlap largely with disabilities. Pre-existing chronic condition can result in disability, and vice versa (Krahn et al. 2014). For example, 25% of people with chronic conditions have disabilities, and 80–90% of people with disabilities have chronic conditions (Gulley et al. 2011). However, there are variations in identifying what constitutes a disability and how it is different from the physical illness. In this study, the focus is mainly on people who are coping with both chronic conditions and disabilities.

The way in which disability is conceptualised by people impacts their understanding and subsequently influences their language, expectations, and interactions in society (Haegele and Hodge 2016). Disability discourse in relation to chronic conditions is usually addressed by employing the biomedical or the social models of disability. The medical model views disability as an impairment in a body's functions due to disease or injury that mostly requires clinical treatment (Haegele and Hodge 2016; Humpage 2007; Forhan 2009). The social model addresses disability as a construct that is imposed on the impairment. Hence, it is the society's responsibility to be more inclusive towards people with disabilities (Anastasiou and Kauffman 2013). In the UK, disability is constructed as a long term physical or mental impairment in the Equality Act (2010), and hence the legal definition is similar to the medical perspective. Another facet is what happens when one acquires a disability status and gain disability benefits (Briant et al. 2013). Emerson and Roulstone (Emerson and Roulstone 2014) argue that such compensations and disability labelling has led to a systematic error in institutions by consistently attaching negative value judgments to disability.

It could also be argued that disability is a unique element of human diversity when compared to other elements such as ethnicity. This is due to the underlying biomedical factor which influences people's lives and choices as a consequence of living with a chronic condition. Thus, people with long term conditions benefit from addressing both the biomedical and social dimensions (Anastasiou and Kauffman 2013). Hence, the medical and social models are incorporated in this study. The medical model is important in managing chronic conditions with disabilities and preventing complications, while the social model relates to vulnerability and victimisation.

2.2 Inconsistency in Defining Cyber-Victimisation

The definitions discussed in the introduction of this paper are inconsistent in the literature, they overlap and vary among disciplines and individual studies. For example, online harassment or cyberharassment, may also be referred to as trolling, or cyberstalking. Both cyberstalking and cyberharassment involve receiving online offending comments, spreading lies, insults or threats, frequently causing a significant negative impact on 'victims' (Short et al. 2014). Additionally, in UK legislation, the Crown Prosecution Service (CPS) identifies cyberstalking as a type of harassment taking place online (CPS 2016) and they are covered under the same legislation depending on the details related to each specific case.

There are numerous issues surrounding the definitions above. Firstly, when looking at online experiences, it is difficult to identify a threshold for the number of incidents, for instance, whether each email or Facebook comment is an incident, or whether each platform e.g., Facebook or Twitter is an incident. Secondly, the duration to identify a victimisation experience also varies, some researchers use a lifetime approach (Mueller-Johnson et al. 2014), others look at weekly, monthly or yearly experiences (Didden et al. 2009). Thirdly, when cyber-victimisation is perceived to be a result of hostility or prejudice, any of these offences could also be labelled as a cyber-disability hate crime, which has only been recognised recently (Alhaboby 2016b). Fourthly, people who experience cyber-victimisation do not necessarily identify themselves as victims.

Regarding cyberstalking, researchers (Dreßing et al. 2014) argue that variations in the definition of cyberstalking is reflected through the wide range of documented cyberstalking prevalence. Internationally, the prevalence of cyberstalking ranges between 3.2% and 82%, with studies in the United States reporting 3.2% (Fisher et al. 2002), 3.7% (Alexy et al. 2005) and up to 40.8% (Reyns et al. 2012). Moreover, stalking definitions show differences between specialities, as well as between practitioners and researchers (Sheridan et al. 2003); these differences are related to details in the description rather than the big picture.

When comparing offline and cyber-victimisation in clinical psychology literature, offline stalking is described as abnormal behaviour and characterised by persistence, that is, abnormal, persistent, and unwanted attention (Kamphuis et al. 2005). While it is a challenge to define what is abnormal, the two other criteria, persistent and unwanted, are consistent with definitions in other specialities in the literature. In forensic psychiatry definitions, stalking is considered as a pattern of behaviour characterised by fixated threats and intrusions, triggering fear and anxiety (McEwan et al. 2012). In law, stalking is regarded as a type of violence differing from other types in duration, which can be months or up to years, and the fear it causes, especially when this distressing conduct is seen as harmless by others (Kropp et al. 2011). In Canada, there was an attempt to develop guidelines to assess victims' vulnerability, nature of stalking and preparatory risk factor, stalking was defined as an unwanted repeated contact or conduct that deliberately or recklessly affects people resulting in experiencing fear or safety concerns of self or others (Kropp et al. 2011). Probably because violence is closely related to criminology literature, the definition adopted in criminology and clinical practice shares some similarities to the approach in law (Davis et al. 2002). In the United States, a national survey to study the effects of stalking defined a stalking case as having one or more incidents associated with any degree of fear (Davis et al. 2002). Hence, fear and distress resulting from victimisation may have a bigger impact on health than physical violence, which is an important issue in the case of cyber-victimisation as will be discussed in the next section. Despite these differences, Sheridan et al.(2003) described stalking as 'chronic, consisting of a number of nuisance behaviours that appear consistent over countries and samples' (Sheridan et al. 2003, P. 148).

Based on all these issues in defining the offence, its duration and number of incidents, the prevalence of cyber-victimisation against people with long term conditions is not clearly determined; it may range between 2% (Didden et al. 2009) and 41.5% (Fridh et al. 2015). Despite variations, it could be assumed that all of these cyber-victimisation experiences are potentially more devastating than their counter-traditional ones (Anderson et al. 2014). In fact, cyber-victimisation is further complicated by international cross-border offences where the offenders are overseas and the Police face difficulties in following up such cases (Sheridan and Grant 2007).

Differences in definitions are accompanied by using limited methods to assess the phenomenon such as using online surveys, which can not be generalised to the whole population (Boynton and Greenhalgh 2004). It must be acknowledged that the advantages of an online survey made it the method of choice to contact a relatively unreachable population due to their physical and social constraints, probably resulting from the impact of being a victim. The other factor is that these studies did not have a focused population, when the focus was attempted, it was either based on gender, age group or college context (Reyns et al. 2012; King-Ries 2010). Limiting research to a young age group is questionable, since the Office for National Statistics in the UK reported that surprisingly Internet use was 84% by all age groups in 2014 (ONS 2014). With regard to context, colleges may not reflect the whole aspect of cyber-victimisation phenomenon, furthermore, social research college students are considered an easily accessible population (Boynton and Greenhalgh 2004). Accordingly, there were few, if any, studies considering other population groups, such as people living chronic diseases who comprise 30% of the UK population (DH 2012) and already are living with compromised health (WHO 2015), and at risk of cyber-victimisation. To elaborate further, Table 1 is adopted from one of our related studies, a systematic review focusing on the cyber-victimisation of people with long term conditions. The table illustrates the different approaches and terminologies adopted to identify cyber-victimisation against this specific group.

3 Impact of Victimisation and Available Support

3.1 The Impact of Victimisation and Cyber-Victimisation

The documented impact of offline victimisation includes short and long term consequences. Psychological complications involve low self-esteem, anxiety and depression, social isolation, suicide, and unemployment (Sheridan and Grant 2007; Hugh-Jones and Smith 1999). In addition, health complications include physical health complaints (Sentenac et al. 2013), exacerbation of illness (Zinner et al. 2012) and disruption of health management (Sentenac et al. 2011b). Offline victimisation also causes financial burden, not only on a personal level, but also on national levels. In the United States, the Centre for Diseases Control (CDC) estimated that stalking has a financial burden of 342 million US dollars due to the cost of treating mental health complications (CDC 2003). This was similar to the UK as stalking resulted in financial loss due to covering therapy, legal costs and repair (Sheridan 2005). This might also have an impact on people with chronic conditions who are already have the burden of coping with impairments. Hence, offline victimisation experiences against people with long term conditions are devastating, and the introduction of the Internet in everyday communication has added to the complexity of the issue.

Table 1 Examples of the different approaches and terminologies used to describe the cyber-victimisation of people with long term conditions

Study	Definition used	Terminology used
Sheridan and Grant (2007)	Harassed via the Internet or received unsolicited emails	Harassment via the internet
	Stalking that originated online and remained solely online for a minimum of 4 weeks. Stalking is identified by repetition (10 occasions or more) and persistence (minimum 4 weeks or more)	Cyberstalking
Didden et al. (2009)	Electronic form of bullying using electronic means of communication (cell phone or the internet)	Cyberbullying
	Bullying is an aggressive act by an individual or a group that is repeated over time and intentional against victims who can not defend themselves easily	
Kowalski and Fedina (2011)	Bullying through email, instant messaging, chartrooms, webpages, receiving digital images or messages to phone	Cyberbullying
	(Recognising bullying act as a verbal, physical or socially hurtful things that is repeated over time, with power imbalance, and on purpose)	Or electronic bullying
Mueller-Johnson et al. (2014)	Subtype of non-contact sexual victimisation	Cyber-victimisation
	Clear sexual harassment or molested during an online communication (chatting, MSN, Netlog)	
Wells and Mitchell (2014)	Being a target of online harassing behaviour in the past year, if someone used the internet to threaten, embarrass or post online messages about the victim, or the victim reporting feeling worried because of someone bothering him/her online	Online victimisation
		Online harassment
	Unwanted requests for sexual information or acts, or talking about sex online	Sexual solicitation
Yen et al. (2014)	Bullying using electronic venues (Email, blog, Facebook, twitter, Plurk)	Cyberbullying
	Posting mean or harmful things, pictures or videos or spreading rumours online	
Gibson-Young et al. (2014)	Electronically bullied in the past 12 months	Cyberbullying
	(Bullying is an aggressive, intentional, electronic contact, repeated, victim can not defend self)	Electronic bullying
		Bullying in cyberspace
Annerbäck et al. (2014)	Harassment or violation via the internet or mobile phones, self-reported in past 2 months	Cyberbullying
	A form of indirect bullying	Cyberharrasment
	(Bullying -also known as mobbing- is identified as the use of power to control others or cause distress)	
Fridh et al. (2015)	Violation or harassment involving cell phone or the internet such as Facebook, email, MSN, text messages in the past 12 months	Cyberharrasment
		Cyber-victimisation

Adopted from Alhaboby et al. (2017a)

Researchers (Dreßing et al. 2014) found that offline and cyber-victimisation have comparable effects. Distress, which is prolonged stress, is an important consequence of cyber-victimisation. Stress leads to neurohormonal changes in the blood, increasing cortisol, catecholamines and insulin secretion resulting in increased blood glucose, heartbeat, blood pressure, urination and other changes (Pinel 2009). Thus, the stress caused by cyber-victimisation has a potential impact on people with chronic conditions, because it interferes directly on the changes in their bodies or indirectly via behavioural changes. Mental health consequences were studied in literature and showed subjective reactions to this experience, taking the form of fear, anger, depression, irritation and loss of control of one's life. It is argued that there is an underestimation in reporting mental health issues due to cultural influences (Davis et al. 2002).

However, quantitative studies have dominated cyber-victimisation literature (Dreßing et al. 2014; Alexy et al. 2005; Maple et al. 2011), and this does not reflect the lived-experience of the victims. One of the few qualitative studies was an online survey of 100 self-identified cyberstalking victims aged 15–68 years which thematically analysed the participants' narratives. Five overarching themes emerged: control and intimidation, determined offender, development of harassment, negative consequences and lack of support (Short et al. 2014). Negative consequences of cyberstalking identified were psychological including PTSD, panic attacks and flashbacks, physical effects and social impact. Some participants expressed being anxious, very ill, depressed, as well as long term health effects. One participant stated that she had a miscarriage as a result of the stress she experienced due to cyberstalking (Short et al. 2014). Cyberstalking differs from offline stalking in the type of invasion, in cyberstalking it is technical, while there is a greater risk of physical violence with offline stalking. The other difference observed was in the victim-stalker relationship, which was found to be more intimate in offline stalking, while acquaintance is the most common relationship in cyberstalking. Finally, the majority of stalking preparators were males, but this was unclear in the case of cyberstalking (Short et al. 2014).

A more recent systematic review focused mainly at the impact of cyber-victimisation on people with long term conditions and disabilities (Alhaboby et al. 2017a). In the ten included studies, the impact of cyber-victimisation was measured using a predetermined set of questions that focused mainly on psychological complications. The most commonly documented issue was depression, followed by anxiety and suicide or self-harm. Relatively less common problems were low self-esteem, behavioural issues and substance abuse. It is worth noting that distress was statistically significant in cyber-victimisation cases.

Two studies in the review (Alhaboby et al. 2017a) reported more detailed physical and mental health-related variables. Annerbäck et al.(2014) used a comprehensive list of health indicators, which included poor general health, physical health problems (headache, migraine, stomach ache, tinnitus, musculoskeletal pain), mental health problems (insomnia, anxiety, worry, depression) and self-injurious behaviour. In comparison, Fridh et al. (2015) addressed a group of general symptoms called "subjective health complaints". Participants' health status was determined through responses to questions on headache, feeling low, irritability, nervousness, sleep disturbances and dizziness. The impact of cyberstalking was covered by Sheridan and Grant (2007), who concluded that well-being and economic consequences were comparable to the effects of traditional stalking. Further, significant differences specific to cyberstalking included the emergence of international perpetrators, threats of physical assault on the victims or people close to them, the need to change email addresses and loss of social relations (Sheridan and Grant 2007).

A recent study was conducted to examine the impact of cyber-victimisation on people with long term conditions in the UK (Alhaboby et al. 2017b). The impact was examined using both quantitative and qualitative methods and was found to be multifaceted including biomedical, mental, social, and financial impact. More importantly, cyber-victimisation was found to negatively impact the chronic condition's self-management (Alhaboby 2018) which is necessary to cope with the health condition and prevent life-threatening complications.

3.2 Available Support for Victims of Cyber-Victimisation

Response and support available to victims of cyber offences could be divided into informal support and instrumental support. Informal support includes approaching friends and family, while instrumental help is the formal support through channels available to victims to help in coping with the experience of cyber-victimisation (Galeazzi et al. 2009; Reyns and Englebrecht 2014). Instrumental support includes health and psychological strategies such as mental health support, and problem-solving strategies such as employing lawyers and actions by the Police.

Within the UK, there are a number of legislative acts to respond to cyber-harassment such as the Protection from Harassment Act 1997, the Malicious Communications 1988, the Communications Act 2003, the Crime and Disorder Act 1998 and the Equality Act 2010 (CPS 2016). When the victim is labelled as disabled, the harassment could also be addressed under the Disability Discrimination Act 1995 (DDA 1995), the Equality Act 2010 or the Communications Act 2003, section 127 for disability hate crime (CPS 2016). Despite the availability of a number of legal remedies, victims with disabilities seem to be struggling to get support (Alhaboby 2016b). This could be either due to the relative ambiguity of cyber offences accompanied by the unclear thresholds in legal acts, where people working in instrumental support channels lack sufficient training.

Another issue with support are the cases of cyber-victimisation. In the UK, 50% of offline victims complained that family and friends did not take them seriously, 50% were told they were going mad, 42% reported to Police and 61% thought they were helpful (Sheridan 2005). This might not be very different from the professionals' responses, the majority of cyber-victims had little support and this was accompanied by blaming the victim, especially by the Police (Short et al. 2014). The combination of the lack of support for cyberstalking victims and the vulnerability of people with disabilities to cyberstalking (Sheridan and Grant 2007), victims are being disempowered with a potentially significant impact on them. Hence, in order to provide proper remedies to people with long term conditions, further training of supportive channels and increased public awareness are required.

General Practitioners (GPs) and the Police build on their roles as helping professions in offline victimisation cases. A European-based study examined the recognition of victimisation in a sample of 50 GPs and 50 Police officers (Fazio and Galeazzi 2004) shows that in Italy GPs gave higher recognition of abnormality than Police officers, probably due to their awareness with psychopathologies. The researchers concluded that recognition and response are influenced by profession and personal differences. They recommended increasing awareness via targeted information, training and multidisciplinary effort (Fazio and Galeazzi 2004). However, the findings of this study can not be generalised because it was conducted in one country and the study population included only female victims.

To extend these results, using case scenarios, the Modena Group on Stalking (MGS) conducted a study in three European countries to examine the awareness and recognition of stalking by Police and GPs as they represented the first line of professionals contacted by victims (Fazio and Galeazzi 2004). Researchers attempted to examine recognition and attitudes among GPs and Police officers in a cross-national study in the European Union (Kamphuis et al. 2005). The researchers used case scenarios and standardised questions, and found that differences in responses depended on the country, profession and personal subjectivity (Kamphuis et al. 2005). Abnormal behaviour could be identified by the GPs, and to less extent among Police officers, which is in line with the findings of Fazio and Galeazzi (2004). Subjective differences among GPs and Police officers were also observed, such as considering stalking as a flattering relatively harmless behaviour and blaming the victim, but GPs in the UK in comparison to Police officers and GPs in other EU countries showed less individual variations and blaming victims (Kamphuis et al. 2005).

Furthermore, the researchers assumed that exploring real stories told by victims give more useful information. Thus, the MGS explored the experiences of stalking victims in the EU, reporting results from Belgium, Italy and Slovenia (Galeazzi et al. 2009). Researchers from the UK, Belgium, Italy, Netherlands, Slovenia and Spain took part in this study and data was collected in the context of a research project sponsored by the European Commission Daphne Research Programme. The online survey was available at the website in five languages and advertised via the press, radio and in collaboration with agencies to support victims. Out of the 391 included participants, 80.9% were females and they were aged between 15 and 64, with a mean age of 29.2 years. The study revealed that 78.8% of the cases included phone calls, 57% texting SMS, 26.6% sending emails, with 13.8% contacting the person via the Internet. With regards to the impact, 48.6% of the victims reported extreme levels of fear, 39.4% of participants had a low WHO wellbeing index, and 70.1% had a high score of general health questionnaire indicating clinical health consequences. Most victims looked for support from family and friends (86.7%), followed by colleagues (42.5%) and the Police (42.5%). Of those who contacted healthcare professionals, 25.1% contacted GPs, 19.7% communicated with mental health professionals, and only 14.8% contacted victim support groups (Galeazzi et al. 2009). The perceived quality of help received from victims' perspectives varied, mental health professionals were on the top of the list followed by family and friends, lawyers, victim support groups, colleagues, GP, social support groups and lastly, the Police. With regard to the perception of being taken seriously, GPs were ranked fourth after mental health professionals, lawyers and family. The Police were in last position on the list, this was partly explained by reasons related to not being taken seriously, stalking had stopped or when victims felt it was not a Police issue or they could do nothing about it. Regarding the perceived effectiveness of intervention provided by these groups, GPs were ranked last, with the Police, victim support groups and mental health professionals also ranking low down on the list compared to family, friends and lawyers (Galeazzi et al. 2009).

The assumed role of healthcare professionals in the self-management of chronic diseases is to educate and explain to patients. This is challenging in the case of cyber-victimisation because GPs recognise the problem, but do not provide effective support (Galeazzi et al. 2009). This highlights the importance of exploring GPs' encounters with cyber-victimisation victims and providing health promotion tools to increase the awareness of this issue. This is supported by previous findings, where victims felt being taken seriously by agencies would help them, which could be though increasing awareness on stalking and getting practical advice (Sheridan 2005). A possible challenge to address this issue in the UK is the limited participation by GPs. In the MGS research, the response rate was lower among GPs compared to Police officers, and low in the UK compared to other EU countries. GPs in the UK stated that they were supportive of the research, but because the methodology was overextended, they did not complete it (Kamphuis et al. 2005). Accordingly, multidisciplinary work is needed to incorporate different professions' work to mitigate the impact on the victims who are already in the process of managing their chronic conditions. One of the successful approaches in managing chronic conditions is disease-management programmes.

4 Disease Management Programmes (DMP) and Online Health Support

4.1 The Historical Development of Disease Management Programmes

The term Disease Management (DM) was used in early 1980s in the United States (US) for public-health campaigns such as influenza vaccinations or physical activity promotion (Stefan Brandt and Hehner 2010). In the 1990s, disease management programmes were introduced as an attempt to improve the quality and to reduce the cost of caring for people with chronic diseases. During the 1990s, care transformed from acute to chronic, with people living longer with their chronic conditions. Treating such long term conditions has become the biggest burden on healthcare expenditures, especially those chronic conditions that were not adequately treated and/or prevented (Super 2004). In 1990s, the disease management market in the US was led by pharma companies due to the fear that any governmental initiatives for health management would reduce their profits from sales. Often, those programs focused on one disease without realising the impact of the co-existence of other morbidities (Todd 2009). In 1999, the Disease Management Association of America (DMAA), known nowadays as the Population Health Alliance (PHA), was formed and regulated the concept of disease management and its standards and evaluation guidelines. Between 1999 and 2000, more than 200 disease management companies were active in the US, most of which were not pharma companies, as disease management services and objectives were set by the DMAA. The disease management programs were of different sizes and covered the five big chronic diseases, with a DMP specifically for diabetes always available. The direction towards managing a whole case rather than a single chronic disease started in the late 1990s. For example, the American Diabetes Treatment Center became American Healthways, demonstrating the practice of dealing with a whole case rather than diabetes only. We can describe the US disease management experience as the commercial presentation of disease management administrated by private disease management organizations selling their services to payers (Todd 2009).

In 2006, there were over 169 disease management vendors in the US and almost all interventions were phone-based to improve health behaviours, self-management and medication adherence. However, due to the poor control of the outcomes many payers turned to in-house disease management services such as Aetna, which brought the number of vendors down to 80 in 2007 and brought about a market shift towards wellness programs since then (Todd 2009). An important checkpoint in US DM evolution was the introduction of Medicare Health Support Services in 2004 (Super 2004; Barr et al. 2010) in which DM vendors were paid to offer DM services to the government-insured elderly and were challenged to prove their cost-effectiveness and improvement of the quality of care. The US example showed improved quality of care but could not clearly demonstrate any cost reduction. However, the take-home message for international practices from the US experience in DMP was to ensure the money invested in DMPs was used properly for DM interventions within the healthcare organisation to improve the quality of care by utilising it in the best possible way to be able to prove a cost reduction and/or effectiveness.

The evolution of disease management in Europe was parallel to DM evolution in the US, although with differences. National programmes for chronic disease management existed in Austria, Denmark, England, Finland, France, Germany, Italy, the Netherlands, and Poland, while regional or private initiatives were also available in England, France, Italy, Spain, and Sweden (Gemmill 2008). In the 1990s in Sweden, trained nurses led clinics in primary care, managing diabetes and hypertension and cooperating with the treating physician, while the Netherlands introduced "transmural care", which involved specialised nurses trained in the care of patients with specific chronic conditions. This was then followed by a move towards disease management models with nurse-led clinics. In 2002, Denmark introduced a nationwide vision of chronic disease control, which was translated later into structured disease management (Gemmill 2008; Nolte et al. 2008).

Probably the most comprehensive example of DMP experience in Europe was in Germany in 2002 when the government formally introduced DMPs to improve the quality of care for patients with chronic illnesses (Stefan Brandt and Hehner 2010; Gemmill 2008; Nolte et al. 2008). The structure of implementation of statutory German disease management initially resulted in a tendency to enrol low-risk patients. Diabetes disease management was the first introduced program and was also the focus of DM in Europe across 11 countries (Gemmill 2008). Germany also implemented a number of regional and private initiatives; however, most of the effectiveness studies focused on the statutory DM experience, which focused on the physician to improve the quality of care. At a wider international level, DMP initiatives were also seen in Australia, New Zealand, Japan, China and Canada (Stefan Brandt and Hehner 2010; Nolte et al. 2008).

4.2 DMP Practices

Disease management practices in the US had a significant role in guiding international DMP practices. American Healthways is one of the big American health management organizations (Pope et al. 2005). Its disease management programmes have often been presented as good US DMPs. Healthways' programmes are accredited by the National Committee for Quality Assurance (NCQA), offering DM services for all the five big chronic conditions, enrolling members according to risk stratification algorithms on an opt-out basis. Their interventions are conducted basically through the phone and web-based applications to change behaviours and improve self-management, with very strong cooperation with the treating physician. In 2004, Healthways services were offered to the governmental Medicare plan Medicare Health Support Pilot (Rula et al. 2011). This experience is quite different from that of Aetna, which is another American disease management organisation and one of the big private health insurance companies. Aetna was one of the first players to start offering in-house disease management programs, benefiting from having all the claims and the control to communicate with the patients and physicians. Aetna's DMPs are also phone-based and tailored to the participants' needs (Aetna 2016). A unique example of DMPs in the US is Kaiser Permanente (Sekhri 2000), which is an integrated managed care organisation offering health insurance and health management services throughout its integrated healthcare providers. Kaiser Permanente programs have also been evaluated in many studies (Wallace 2005).

In Europe, DMP practices were mixed, as some were phone-based to improve self-management and change behaviours for example physical activity; such DMPs could be found in Sweden and Finland and were extensively evaluated. Another form of DMP was found in the Netherlands and to some extent in Finland as well. This form was a clinic-based nurse-led DMP based on educating patients and closely cooperating with the treating physician to improve self-management and clinical outcomes (Gemmill 2008; Nolte et al. 2008).

In Germany, specifically integrated care was enabled in the Health Care Reform Act 2000, while DMPs started in 2002 (Nolte et al. 2008). The integrated care models were practically in implementation by 2004 (Stefan Brandt and Hehner 2010). The main characteristics of the German DMPs include the coordination of care provided by general practitioners, ensuring continuous care for patients, adherence to evidence-based guidelines, patient education and active involvement of patients, documentation (lab readings, diagnostic and therapeutic interventions, participation in patient education), quality assurance of process and outcome, incentives for participation for physicians and patients, voluntary participation (physicians and patients), and an obligatory structure of quality standards for participating physicians and hospitals (van Lente 2012).

Australia has also had some experience in DMPs, one example being the COACH initiative, which was a phone-intervention disease management program for post-cardiac episodes. The COACH DMP was effective and was followed by the PEACH initiative, which supported patients with type 2 diabetes using the same phone intervention used in COACH.

Disease management interventions also took a number of other forms, especially in Europe and Australia, including peer coaching (Johansson et al. 2015; Joseph et al. 2001; Moskowitz et al. 2013) and group education-based in the healthcare provider's setting to facilitate behaviour change.

4.3 DMP Evaluation

The Population Health Alliance (PHA) evaluation framework proposed that regardless of the structure of the DM program or the process measures, whether patient- or provider-related, all DMP outcomes and impacts should be evaluated within the following areas: psychosocial outcomes, clinical/health status, behaviour change, patient/provider productivity/satisfaction, quality of life (QOL) and financial outcomes (Alliance 2012). Over the last two decades most of the DMPs have been evaluated using either pre-post evaluation (Annalijn Conklin 2010), matched-pair evaluation or randomized/non-randomized controlled trials. However, so far prospective randomized controlled trial RCTs have been considered a gold standard in evaluating DMPs. Many of the examples mentioned above were evaluated accordingly despite the challenges and limitations associated with this evaluation design, such as randomization and participants' engagement.

Weingarten et al. (2002) reviewed 112 disease management interventions for chronic conditions to compare the different approaches. Patient education was the most popular intervention in 92 out of the 112 programs, followed by physician education in 47 interventions. However, while the review concluded that all studied interventions were associated with some kind of improvement in provider adherence to practice guidelines and disease control, future studies should compare different types of intervention to find the most effective (Weingarten et al. 2002). Mattke et al. (2007) looked at the evidence of disease management effectiveness across four meta-analyses, five reviews and three population-based interventions focusing on diabetes. It was concluded that although disease management seemed to improve quality of care, its effect on cost was uncertain (Mattke et al. 2007).

More recently, Kivelä et al. (2014) reviewed the effect of health coaching in managing chronic conditions. Thirteen studies were reviewed in which the majority of the studies were carried out in the US for phone-based and web-based health coaching interventions. RCT was the main design used, with samples ranging between 56 and 318 participants (n = 56–318). However, all those interventions where short, ranging from 2 to 6 months except one (n = 1755) that looked at weight management and was for 15–18 months. These studies evaluated clinical outcomes, QOL, and patients' behaviors and activation. The authors concluded that health coaching does improve the management of chronic diseases. However, further research into the cost-effectiveness of health coaching and its long term effectiveness for chronic diseases was recommended (Kivelä et al. 2014).

In a systematic review aimed at evaluating the effectiveness of self-management education in type 2 diabetes, Norris et al. (2002) reviewed 31 RCTs carried out during the 1990s (n = 20–532), with the length of the intervention ranging from 1.5 to 19 months. The review concluded that self-management education improves HbA1c levels at immediate follow up, and increased contact time increased the effect (Norris et al. 2002). The benefit declined 1–3 months after the intervention stops and further research was recommended to develop interventions effective in maintaining long term glycemic control. In another review looking at the effect of self-management training in type 2 diabetes, Norris et al. (2001) reviewed 72 RCTs. The positive effects of self-management training on knowledge, frequency and accuracy of self-monitoring, self-reported dietary habits, and glycemic control were demonstrated in studies with short follow-ups (<6 months). With longer follow-ups, interventions that used regular reinforcement throughout the follow-up were sometimes effective in improving glycemic control (Norris et al. 2002). This highlights that existing evidence supported the effectiveness of self-management training in the short term management.

In summary, over the last two decades several disease management programs have been implemented worldwide in response to the escalating burden of chronic conditions. The implementation of DMPs was an attempt to improve the quality of care for the five big chronic conditions that are influenced by lifestyle and behaviour change. Such programs came in all sizes, different interventions, different designs and even different evaluation designs. These programmes included an online disease management approach.

5 Towards Forensics-Enabled DMP to Support People with Long Term Conditions

Online coaching as part of DMPs utilises web-based applications as an intervention method aiming to encourage behavioural change and enhance self-management (Tang et al. 2013). A clinical trial (Lorig et al. 2006) was conducted to test web-based chronic diseases self-management of 958 patients with cardiovascular diseases, respiratory diseases and diabetes. The outcome was evaluated based on health status, health behaviour, as well as emergency and doctor visits. This trial showed that these web-based interventions were comparable to 'offline' chronic diseases self-management (Lorig et al. 2006). Furthermore, a review (Merolli et al. 2013) of 19 included studies to evaluate the impact of utilising social media in chronic diseases management concluded improvements in psychosocial aspects of management. However, it was acknowledged that further research is needed due to the lack of reporting negative findings, these studies advocated the use of technology in health interventions with a tendency to report positive findings. Investigating negative aspects while accounting for diverse populations (age groups, chronic conditions etc) is vital to reduce potential harm in the use of technology towards self-management planning for people living with long term conditions who require support on regular basis.

While non-adherence to self-management planning is a major instability factor, DMPs and Online Coaching Programmes – as demonstrated in earlier sections of this study- are cornerstones to support the stability of long term self-management for people with long term conditions. Likewise, we argue the inverse correlation on stability between the impact of cyber-victimisation versus the ability to forensically document all submitted data for such incidents as demonstrated in Fig. 1. The benefit

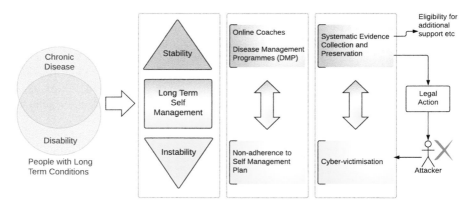

Fig. 1 A demonstration of factors contributing towards the stability of long term self-management for people with long term conditions versus instability factors. In the case of cyber-victimisation, a forensics-enabled system could eliminate the source of the problem, facilitate risk assessment, or support the victim's eligibility for additional support

of extending DMPs to be forensics-enabled by design has many advantages; First, it allows the victims, or a third-party acting on their behalf, to trigger a legal action against the source of the problem. For instance, a recent survey (Al-Khateeb et al. 2017) showed that victims of cyberstalking support third-party intervention to mitigate risk, and that victims would seek help from independent anti-cyberstalking organisation and the Police. In an ideal case, documented incidents should be preserved as evidence, contain incriminating material (e.g. breaking the Protection from Harassment Act 1997 in the UK), be admissible to a court of law, and can be attributed to the attacker. Second, the evidence could be utilised to support the victim's eligibility for extended instrumental support from national health services. Finally, this level of documentation offers an opportunity to implement more accurate methods to assess risk associated with victimisation.

Various architectures can be planned and implemented for such systems. To share an exemplar demonstrating a widely adopted client-server model, we consider a mobile application as a client-side software aiming to empower end users with the right tool to collect and report user-informed (victim-informed) 'sound' digital evidence (McKemmish 2008) of online harassment and stalking. Since both internet and cellular communications can be recorded at the victim's side, malicious data (potential evidence e.g. audio message) could be logged by the user accompanied by both: manual data written by the user to add further context, and metadata auto-generated by the client-side system such as GPS coordinates and timestamps. The server-side of the system will link submitted files to registered individuals. Useful statistics can be generated based on this data with various risk assessment threshold to trigger suitable response.

Working towards forensics readiness incorporates several security requirements including compliance with local laws and regulations. Starting with the Confidentiality, Integrity, Availability (CIA) Triad, a security model demonstrating key principles to maintain. Non-repudiation is another required assurance to the integrity and origin of communicated data. This can be achieved with a solid Public Key Infrastructure (PKI) implementation, a set of roles, policies and procedures to manage encryption and digital certificates (Salomaa 2013). While symmetric cryptography provides fast and efficient encryption solution especially for large volumes of data. Transport Layer Security (TLS) is the de facto and widely utilised protocol combining PKI and symmetric cryptography to secure web application. Hence the communication link between the client (mobile application) and the server should be secured by TLS.

Further to securing communication, the main new functionality of the system is to preserve submitted files as digital evidence. In the cyber, digital evidence can be any data that is stored or transmitted in a digital form (binary numeral system) to provide value to support a claim within a legal context. For instance, an offensive text sent via a Short Message Service (SMS) could constitute evidence and be forensically retrieved from a mobile phone. Examples of other invaluable data include:

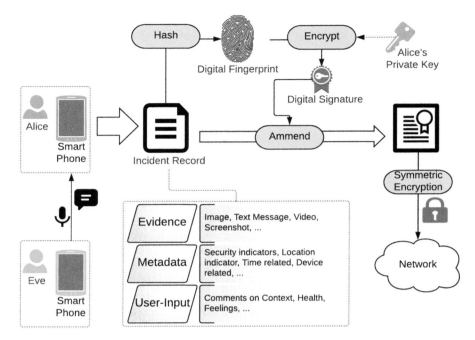

Fig. 2 Harassing content sent by Eve can be captured, signed and sent to a server as part of an online coaching programme. This process is automated but the victim (Alice) should choose to select the content and optionally amend notes to add further context

- Incoming mobile calls from pre-defined number and/or unknown numbers
- Incoming VoIP calls
- Apps Instant Messages (e.g. WhatsApp)
- Social media messages/stream (e.g. Facebook, Twitter)
- Local files (e.g. existing records, notes, video)

Additionally, many other artefacts can be automatically logged to contribute to the value or the admissibility of the digital evidence as shown in Fig. 2 and explained below.

Device related identifiers. Capturing values such as Device ID, Build No and Kernel version helps identifying the device from which the evidence was captured.

Location indicators. GPS coordinates, connected Wi-Fi and Network Operator data can be invaluable to recover the location of an incident captured by the mobile e.g. when using the mobile's camera to record an incident.

Time related. File system timestamps shows the time when each file was created, accessed and modified.

Security indicators. A 'rooted' device during the time of evidence capture could indicate a trust problem to the integrity of the reported data, either intentionally or due to Malware infection (Irshad et al. 2018).

Integrity checks. Captured data must be hashed to maintain the integrity of the file at the time of acquisition or submission. Multiple hashes are recommended to avoid errors within this process. Examples of hash functions currently used include: MD5, SHA-1, SHA-512 and SHA-256.

Digital Evidence in traditional cases are documented by a qualified Digital Investigator. This is a typical admissibility requirement included within guidelines such as the principles published by the Association of Chief Police Officers (ACPO) in the UK, knows as the ACPO Principles (Sutherland et al. 2015). Therefore, for forensic readiness to be maintained, the software should automate the process of data acquisition with reference to these principles, and the software code should go through a review process to meet the reliability requirement with reference to standards (e.g. the Daubert standard).

Nonetheless, it must be acknowledged that the software aims to preserve and transfer a certified copy of the original evidence to a remote server together with Digital Signatures for verification purposes. Based on the individual circumstances of the case, and local laws, users might be advised to keep the evidence available in their device from which it was initially captured/submitted. This could be an inevitably requirement to fully satisfy a court of law.

6 Conclusions

The cyber-victimisation of vulnerable groups is prevalent and causes multifaceted impact upon the victims. People with long term conditions and disabilities are frequently labelled as vulnerable, and commonly victimised online. Different definitions were given to such experiences, including online harassment, stalking, bullying, trolling, or disability hate. These variations were mostly dependant on elements such as age, power relations, duration, context, discipline or the concept of vulnerability itself. Despite these differences, the impact upon the victims is chronic, such as physical wellbeing, long term mental health impact, economic and social consequences. Additionally, the inconsistency in training of support channels and responding to these cases result in further distress and impact on the victims. Hence, cyber-victimisation of people with chronic conditions and disabilities is a complex issue that requires multi-disciplinary long term action and follow up.

Online Coaches within Disease Management Programmes (DMP) are one approach to facilitate effective intervention. These systems can be extended to incorporate third-party interventions as well (e.g. a legal action) for cases were the cyber-victimisation of a vulnerable victim escalates beyond control. This will provide risk assessors and digital investigators with reliable information to cross-check and determine the integrity and value of the reported incident(s), and the identity of the perpetrator when possible. Designing forensics-enabled Electronic Coaches is technically possible given that compliance with the legal admissibility of evidence is planned for as a core part of the system, and that applicable data protection laws such as the EU's General Data Protection Regulation (GDPR) are also met. This study demonstrated the need and explained the means towards covering the collection and preservation of incidents reported by individuals. However, more research is needed understand associated resources (cost, training etc) and quantify the impact of such implementation.

References

Aetna. (2016). *Disease management*. Available from: https://www.aetnabetterhealth.com/pennsylvania/health-wellness/special/disease-management.

Alexy, E. M., Burgess, A. W., Baker, T., & Smoyak, S. A. (2005). Perceptions of cyberstalking among college students. *Brief Treatment and Crisis Intervention, 5*(3), 279.

Algtewi, E. E., Owens, J., & Baker, S. R. (2015). Analysing people with head and neck cancers' use of online support groups. *Cyberpsychology: Journal of Psychosocial Research on Cyberspace, 9*(4).

Alhaboby, Z. A. (2018). *Written evidence on the impact of cyber-victimisation on people with long term conditions*. Online abuse and the experience of disabled people. Available from: http://data.parliament.uk/writtenevidence/committeeevidence.svc/evidencedocument/petitions-committee/online-abuse-and-the-experience-of-disabled-people/written/77961.pdf.

Alhaboby, Z. A., Barnes, J., Evans, H., & Short, E. (2016a). Cyber-victimisation of people with disabilities: Challenges facing online research. *Cyberpsychology: Journal of Psychosocial Research on Cyberspace*.

Alhaboby, Z. A., Al-Khateeb, H. M., Barnes, J., & Short, E. (2016b). 'The language is disgusting and they refer to my disability': The cyberharassment of disabled people. *Disability & Society, 31*(8), 1138–1143.

Alhaboby, Z. A., Barnes, J., Evans, H., & Short, E. (2017a). Cyber victimisation of people with chronic conditions and disabilities: A systematic review of scope and impact. *Trauma, Violence & Abuse: A Review Journal*. 1524838017717743.

Alhaboby, Z. A., Barnes, J., Evans, H., & Short, E. (2017b). Cyber-victimisation of people with disabilities: Challenges facing online research. *Cyberpsychology: Journal of Psychosocial Research on Cyberspace, 11*(1).

Al-Khateeb, H. M., Epiphaniou, G., Alhaboby, Z. A., Barnes, J., & Short, E. (2017). Cyberstalking: Investigating formal intervention and the role of corporate social responsibility. *Telematics and Informatics, 34*(4), 339–349.

Alliance, C. C. (2012). *Implementation and evaluation: A population health guide for primary care models*. Washington, DC: Care Continuum Alliance.

Anastasiou, D., & Kauffman, J. M. (2013). The social model of disability: Dichotomy between impairment and disability. *Journal of Medicine and Philosophy, 38*(4), 441–459.

Anderson, J., Bresnahan, M., & Musatics, C. (2014). Combating weight-based cyberbullying on Facebook with the dissenter effect. *Cyberpsychology, Behavior and Social Networking, 17*(5), 281–286.

Annalijn Conklin, E. N. (2010). *Disease management evaluation. A comprehensive review of current state of the art*. Available from: http://www.rand.org/pubs/technical_reports/TR894.html.

Annerbäck, E.-M., Sahlqvist, L., & Wingren, G. (2014). A cross-sectional study of victimisation of bullying among schoolchildren in Sweden: Background factors and self-reported health complaints. *Scandinavian Journal of Public Health, 42*(3), 270–277.

Barr, M. S., Foote, S. M., Krakauer, R., & Mattingly, P. H. (2010). Lessons for the new CMS innovation center from the Medicare health support program. *Health Affairs, 29*(7), 1305–1309.

Bocij, P., & McFarlane, L. (2003). Cyberstalking: The technology of hate. *The Police Journal, 76*(3), 204–221.

Boynton, P. M., & Greenhalgh, T. (2004). Selecting, designing, and developing your questionnaire. *BMJ, 328*(7451), 1312–1315.

Briant, E., Watson, N., & Philo, G. (2013). Reporting disability in the age of austerity: The changing face of media representation of disability and disabled people in the United Kingdom and the creation of new 'folk devils'. *Disability & Society, 28*(6), 874–889.

CDC. (2003). *Costs of intimate partner violence against women in the US*. National Center for Injury Prevention and Control.

Chen, P.-Y., & Schwartz, I. S. (2012). Bullying and victimization experiences of students with autism spectrum disorders in elementary schools. *Focus on Autism and Other Developmental Disabilities, 27*(4), 200–212.

CPS. (2016). *Impact and dynamics of stalking and harassment*. 2-2-2016. Available from: http://www.cps.gov.uk/legal/s_to_u/stalking_and_harassment/#a05a.

Davis, K. E., Coker, A. L., & Sanderson, M. (2002). Physical and mental health effects of being stalked for men and women. *Violence and Victims, 17*(4), 429–443.

DDA. (1995). *Disability discrimination act 1995*.

DH. (2012). *Long term conditions compendium of information*. London: Department of Health.

Didden, R., Scholte, R. H. J., Korzilius, H., de Moor, J. M., Vermeulen, A., O'Reilly, M., Lang, R., & Lancioni, G. E. (2009). Cyberbullying among students with intellectual and developmental disability in special education settings. *Developmental Neurorehabilitation, 12*(3), 146–151 6p.

Dreßing, H., Bailer, J., Anders, A., Wagner, H., & Gallas, C. (2014). Cyberstalking in a large sample of social network users: Prevalence, characteristics, and impact upon victims. *Cyberpsychology, Behavior and Social Networking, 17*(2), 61–67.

EA. (2010). Guidance on matters to be taken into account in determining questions relating to the definition of disability. In *Equality act 2010*.

Emerson, E., & Roulstone, A. (2014). Developing an evidence base for violent and disablist hate crime in Britain: Findings from the life opportunities survey. *Journal of Interpersonal Violence, 29*, 3086–3104.

Fazio, L. D., & Galeazzi, G. M. (2004). *Women victims of stalking and helping professions: Recognition and intervention in the Italian context*. Slovenia: Faculty of Criminal Justice, Univeristy of Maribor.

Fisher, B. S., Cullen, F. T., & Turner, M. G. (2002). Being pursued: Stalking victimization in a national study of college women. *Criminology & Public Policy, 1*(2), 257–308.

Forhan, M. (2009). An analysis of disability models and the application of the ICF to obesity. *Disability and Rehabilitation, 31*(16), 1382–1388.

Fridh, M., Lindström, M., & Rosvall, M. (2015). Subjective health complaints in adolescent victims of cyber harassment: Moderation through support from parents/friends – a Swedish population-based study. *BMC Public Health, 15*(1), 949–949.

Galeazzi, G. M., Bučar-Ručman, A., DeFazio, L., & Groenen, A. (2009). Experiences of stalking victims and requests for help in three European countries. A survey. *European Journal on Criminal Policy and Research, 15*(3), 243–260.

Gemmill, M. (2008). *Research note: Chronic disease management in Europe, European Commission Directorate-General "Employment, Social Affairs and Equal Opportunities" Unit E1-Social and Demographic Analysis*. London: London School of Economics and Political Science.

Gibson-Young, L., Martinasek, M. P., Clutter, M., & Forrest, J. (2014). Are students with asthma at increased risk for being a victim of bullying in school or cyberspace? Findings from the 2011 Florida youth risk behavior survey. *Journal of School Health, 84*(7), 429–434.

Greenhalgh, T. (2009). Chronic illness: Beyond the expert patient. *BMJ: British Medical Journal, 338*, 629–631.

Gulley, S. P., Rasch, E. K., & Chan, L. (2011). The complex web of health: Relationships among chronic conditions, disability, and health services. *Public Health Reports, 126*(4), 495–507.

Haegele, J. A., & Hodge, S. (2016). Disability discourse: Overview and critiques of the medical and social models. *Quest, 68*(2), 193–206.

Hamiwka, L. D., Cara, G. Y., Hamiwka, L. A., Sherman, E. M., Anderson, B., & Wirrell, E. (2009). Are children with epilepsy at greater risk for bullying than their peers? *Epilepsy & Behavior, 15*(4), 500–505.

Hugh-Jones, S., & Smith, P. K. (1999). Self-reports of short- and long-term effects of bullying on children who stammer. *British Journal of Educational Psychology, 69*(2), 141–158.

Humpage, L. (2007). Models of disability, work and welfare in Australia. *Social Policy and Administration, 41*(3), 215–231.

Irshad, M., Al-Khateeb, H. M., Mansour, A., Ashawa, M., & Hamisu, M. (2018). Effective methods to detect metamorphic malware: A systematic review. *International Journal of Electronic Security and Digital Forensics, 10*(2), 138–154.

Johansson, T., Keller, S., Winkler, H., Ostermann, T., Weitgasser, R., Sönnichsen, A. C. (2015). Effectiveness of a peer support programme versus usual care in disease management of diabetes mellitus type 2 regarding improvement of metabolic control: A cluster-randomised controlled trial. *Journal of Diabetes Research, 2016*.

Joseph, D. H., Griffin, M., Hall, R. F., & Sullivan, E. D. (2001). Peer coaching: An intervention for individuals struggling with diabetes. *The Diabetes Educator, 27*(5), 703–710.

Kamphuis, J. H., Galeazzi, G. M., De Fazio, L., Emmelkamp, P. M., Farnham, F., Groenen, A., James, D., & Vervaeke, G. (2005). Stalking—perceptions and attitudes amongst helping professions. An EU cross-national comparison. *Clinical Psychology & Psychotherapy, 12*(3), 215–225.

King-Ries, A. (2010). Teens, technology, and cyberstalking: The domestic violence wave of the future. *Texas Journal of Women and the Law, 20*, 131.

Kivelä, K., Elo, S., Kyngäs, H., & Kääriäinen, M. (2014). The effects of health coaching on adult patients with chronic diseases: A systematic review. *Patient Education and Counseling, 97*(2), 147–157.

Kouwenberg, M., Rieffe, C., Theunissen, S. C. P. M., & de Rooij, M. (2012). Peer victimization experienced by children and adolescents who are deaf or hard of hearing. *PLoS One, 7*(12), e52174–e52174.

Kowalski, R. M., & Fedina, C. (2011). Cyber bullying in ADHD and Asperger syndrome populations. *Research in Autism Spectrum Disorders, 5*(3), 1201–1208.

Krahn, G. L., Reyes, M., & Fox, M. (2014). Toward a conceptual model for national policy and practice considerations. *Disability and Health Journal, 7*(1), 13–18.

Kropp, P. R., Hart, S. D., Lyon, D. R., & Storey, J. E. (2011). The development and validation of the guidelines for stalking assessment and management. *Behavioral Sciences & the Law, 29*(2), 302–316.

Levine, C., Faden, R., Grady, C., Hammerschmidt, D., Eckenwiler, L., & Sugarman, J. (2004). The limitations of "vulnerability" as a protection for human research participants. *The American Journal of Bioethics, 4*(3), 44–49.

Lorig, K. R., Ritter, P. L., Laurent, D. D., & Plant, K. (2006). Internet-based chronic disease self-management: A randomized trial. *Medical Care, 44*(11), 964–971.

Maple, C., Short, E., & Brown, A. (2011). *Cyberstalking in the United Kingdom: An analysis of the ECHO pilot survey*. Bedfordshire: University of Bedfordshire: National Centre for Cyberstalking Research.

Mattke, S., Seid, M., & Ma, S. (2007). Evidence for the effect of disease management: Is $1 billion a year a good investment? *American Journal of Managed Care, 13*(12), 670.

McEwan, T. E., MacKenzie, R. D., Mullen, P. E., & James, D. V. (2012). Approach and escalation in stalking. *Journal of Forensic Psychiatry & Psychology, 23*(3), 392–409.

McKemmish, R. (2008). When is digital evidence forensically sound? In *IFIP international conference on digital forensics*. Springer.

Merolli, M., Gray, K., & Martin-Sanchez, F. (2013). Health outcomes and related effects of using social media in chronic disease management: A literature review and analysis of affordances. *Journal of Biomedical Informatics, 46*(6), 957–969.

Moskowitz, D., Thom, D. H., Hessler, D., Ghorob, A., & Bodenheimer, T. (2013). Peer coaching to improve diabetes self-management: Which patients benefit most? *Journal of General Internal Medicine, 28*(7), 938–942.

Mueller-Johnson, K., Eisner, M. P., & Obsuth, I. (2014). Sexual victimization of youth with a physical disability: An examination of prevalence rates, and risk and protective factors. *Journal of Interpersonal Violence, 29*(17), 3180–3206.

Nolte, E., Knai, C., & McKee, M. (2008). *Managing chronic conditions: Experience in eight countries*. Copenhagen: WHO Regional Office Europe.

Norris, S. L., Engelgau, M. M., & Narayan, K. V. (2001). Effectiveness of self-management training in type 2 diabetes a systematic review of randomized controlled trials. *Diabetes Care, 24*(3), 561–587.

Norris, S. L., Lau, J., Smith, S. J., Schmid, C. H., & Engelgau, M. M. (2002). Self-management education for adults with type 2 diabetes a meta-analysis of the effect on glycemic control. *Diabetes Care, 25*(7), 1159–1171.

ONS. (2014). *Internet access – households and individuals 2014*. 2–12–2014. Available from: http://www.ons.gov.uk/peoplepopulationandcommunity/householdcharacteristics/homeinter netandsocialmediausage/bulletins/internetaccesshouseholdsandindividuals/2014-08-07.

Oxford. (2015). *Definition of chronic*. 20–5–2015. Available from: http://www.oxforddictionaries.com/definition/english/chronic.

Pinel, J. P. (2009). Biopsychology of emotion, stress and health. *Pearson Education*, 468–475.

Pope, J. E., Hudson, L. R., & Orr, P. M. (2005). Case study of American Healthways' diabetes disease management program. *Health Care Financing Review, 27*(1), 47.

Reyns, B. W., & Englebrecht, C. M. (2014). Informal and formal help-seeking decisions of stalking victims in the United States. *Criminal Justice and Behavior, 41*(10), 1178–1194.

Reyns, B. W., Henson, B., & Fisher, B. S. (2012). Stalking in the twilight zone: Extent of cyberstalking victimization and offending among college students. *Deviant Behavior, 33*(1), 1–25.

Rula, E. Y., Pope, J. E., & Stone, R. E. (2011). A review of healthways' medicare health support program and final results for two cohorts. *Population Health Management, 14*(S1), S-3–S-10.

Salomaa, A. (2013). *Public-key cryptography*. Springer.

Seale, J., & Chadwick, D. (2017). How does risk mediate the ability of adolescents and adults with intellectual and developmental disabilities to live a normal life by using the Internet? *Cyberpsychology: Journal of Psychosocial Research on Cyberspace, 11*(1).

Sekhri, N. K. (2000). Managed care: The US experience. *Bulletin of the World Health Organization, 78*(6), 830–844.

Sentenac, M., Gavin, A., Arnaud, C., Molcho, M., Godeau, E., & Gabhainn, S. N. (2011a). Victims of bullying among students with a disability or chronic illness and their peers: A cross-national study between Ireland and France. *Journal of Adolescent Health, 48*(5), 461–466.

Sentenac, M., Arnaud, C., Gavin, A., Molcho, M., Gabhainn, S. N., & Godeau, E. (2011b). Peer victimization among school-aged children with chronic conditions. *Epidemiologic Reviews*, mxr024.

Sentenac, M., Gavin, A., Gabhainn, S. N., Molcho, M., Due, P., Ravens-Sieberer, U., de Matos, M. G., Malkowska-Szkutnik, A., Gobina, I., Vollebergh, W., Arnaud, C., & Godeau, E. (2013). Peer victimization and subjective health among students reporting disability or chronic illness in 11 Western countries. *European Journal of Public Health, 23*(3), 421–426.

Sheridan, L. (2005). *University of Leicester supported by network for surviving stalking: Stalking survey*. 13–5–2015. Available from: http://www.le.ac.uk/press/stalkingsurvey.htm.

Sheridan, L. P., & Grant, T. (2007). Is cyberstalking different? *Psychology, Crime & Law, 13*(6), 627–640.

Sheridan, L., Blaauw, E., & Davies, G. (2003). Stalking knowns and unknowns. *Trauma, Violence & Abuse, 4*(2), 148–162.

Short, E., Linford, S., Wheatcroft, J. M., & Maple, C. (2014). The impact of cyberstalking: The lived experience – a thematic analysis. *Studies in Health Technology and Informatics, 199*, 133–137.

Stefan Brandt, P. J. H., & Hehner, S. (2010). *How to design a successful disease-management program*.

Super, N. (2004). *Medicare's chronic care improvement pilot program: What is its potential?* Washington, DC: National Health Policy Forum.

Sutherland, I., Spyridopoulos, T., Read, H., Jones, A., Sutherland, G., & Burgess, M. (2015). Applying the ACPO guidelines to building automation systems. In *International conference on human aspects of information security, privacy, and trust*. Springer.

Tang, P. C., Overhage, J. M., Chan, A. S., Brown, N. L., Aghighi, B., Entwistle, M. P., Hui, S. L., Hyde, S. M., Klieman, L. H., & Mitchell, C. J. (2013). Online disease management of diabetes: Engaging and motivating patients online with enhanced resources-diabetes (EMPOWER-D), a randomized controlled trial. *Journal of the American Medical Informatics Association, 20*(3), 526–534.

Taylor, L. A., Saylor, C., Twyman, K., & Macias, M. (2010). Adding insult to injury: Bullying experiences of youth with attention deficit hyperactivity disorder. *Children's Health Care, 39*(1), 59–72.

Todd, W. E. (2009). *Disease management: A look back & ahead*. International Disease Management Alliance.

van Lente, E. J. (2012). In E. J. van Lente (Ed.), *Analysis of documentation data by Infas, Bonn on diabetes type 2 of all AOK – no control group*. Brussels, May 23rd., 2012. Berlin: AOK.

Wallace, P. J. (2005). Physician involvement in disease management as part of the CCM. *Health Care Financing Review, 27*(1), 19.

Webster. (2015). *Medical definition of CHRONIC*. 20-5-2015. Available from: http://www.merriam-webster.com/medical/chronic.

Weingarten, S. R., Henning, J. M., Badamgarav, E., Knight, K., Hasselblad, V., Gano, A., Jr., & Ofman, J. J. (2002). Interventions used in disease management programmes for patients with chronic illnesswhich ones work? Meta-analysis of published reports. *BMJ, 325*(7370), 925.

Wells, M., & Mitchell, K. J. (2014). Patterns of internet use and risk of online victimization for youth with and without disabilities. *The Journal of Special Education, 48*(3), 204–213.

WHO. (2015). *The impact of chronic diseases in the United Kingdom*.

Yen, C.-F., Chou, W.-J., Liu, T.-L., Ko, C.-H., Yang, P., & Hu, H.-F. (2014). Cyberbullying among male adolescents with attention-deficit/hyperactivity disorder: Prevalence, correlates, and association with poor mental health status. *Research in Developmental Disabilities, 35*(12), 3543–3553.

Zinner, S. H., Conelea, C. A., Glew, G. M., Woods, D. W., & Budman, C. L. (2012). Peer victimization in youth with Tourette syndrome and other chronic tic disorders. *Child Psychiatry and Human Development, 43*(1), 124–136.

An Investigator's Christmas Carol: Past, Present, and Future Law Enforcement Agency Data Mining Practices

James A. Sherer, Nichole L. Sterling, Laszlo Burger, Meribeth Banaschik, and Amie Taal

1 Introduction and Framework

Data mining, or probabilistic machine learning, is the process of finding interesting patterns or "nontrivial and potentially useful information"[1] in a data set, whether a "set of rules, a graph or network, a tree, one or several equations," or more.[2] This process may utilize heuristics based on mining event-data logs to produce models,[3] and most often prediction models built from prior data. Essentially, "machine learning takes raw data...and tries to predict something" from it.[4]

---

[1] E. V. Ravve, "MOMEMI: Modern Methods of Data Mining," ICCGI2017, November 2016.

[2] R. J. Roiger, "Data Mining: A Tutorial-Based Primer," CRC Press, 2017.

[3] A. K. A. de Medeiros, W. M. P van der Aalst, and A. J. M. M. Weijters, "Workflow Mining: Current Status and Future Directions," OTM Confederated International Conferences – On the Move to Meaningful Internet Systems, Springer Berlin Heidelberg, 2003.

[4] M. Moore, "The Realities of Machine Learning Systems," Software Dev. Times, April 2017.

J. A. Sherer · N. L. Sterling
Information Governance Team, Baker & Hostetler LLP, New York, NY, USA
e-mail: jas217@columbia.edu; nsterlin@umich.edu

L. Burger
Attorney, Munich, Germany
e-mail: Laszlo.Burger@alumni.uni-heidelberg.de

M. Banaschik
Ernst & Young – Forensics & Integrity Services, Cologne, Germany
e-mail: Meribeth.Banaschik@de.ey.com

A. Taal (✉)
Stratagem Tech Solutions Ltd, London, UK
e-mail: amie@stratagemtechsolutions.com

© Springer Nature Switzerland AG 2018
H. Jahankhani (ed.), *Cyber Criminology*, Advanced Sciences and Technologies for Security Applications, https://doi.org/10.1007/978-3-319-97181-0_12

These predictions can vary and may be sorted into two general categories: one, a probabilistic measure that says, based on an emerging data trend, further similar activity is likely to occur in some proportion of future instances; and two, a demonstration to LEA of how "people and things work," a window into the real practices of individuals and societal structures.

The first category, probabilistic measures,[5] utilized for a variety of different LEA activities (such as lineup fairness,[6] police patrol areas,[7] or even internal investigations[8]) is (more-or-less) supported by the data and the algorithms involved, although we discuss a number of challenges that erode a fundamentalist-type approach to those conclusions.

The second category is an extension of what humans have done since the advent of communication: determining the narrative of events—that is, making sense of the facts as presented and writing a cohesive history (e.g., constructing explanations) to support our observations and extending them in this case, to observations never before available.[9] The ability of LEA to tell a story is so important to the outcome of a matter that the process of developing a narrative itself has been heavily scrutinized and in some instances, demonized as wielding too much power and providing little probative value and instead prejudicing the accused.[10] This process is perhaps unsurprisingly, specifically directed by LEA agents who understand its importance and state unequivocally that LEA reports are, "by far, the most important part of the job."[11] As understood by at least one LEA representative, before "events are recorded, written down for others to read, understand, and comprehend technically, nothing has transpired. Events only become events when they are recorded for posterity . . . ."[12]

One of the fundamental strengths of data mining is therefore its ability to "turn low-level data, usually too voluminous to understand, into higher forms (information or knowledge) that might be more compact (for example, a summary),

---

[5]P. De Hert and S. Gutwirth, "Privacy, Data Protection and Law Enforcement. Opacity of the Individual and Transparency of Power," Privacy and the Crim. L., 61–104, 2006.

[6]C. G. Tredoux, "Statistical Inference on Measures of Lineup Fairness," Law and Human Behav. 22.2: 217, 1988.

[7]K. M. Curtin, K. Hayslett-McCall, and F. Qiu, "Determining Optimal Police Patrol Areas with Maximal Covering and Backup Covering Location Models," Networks and Spatial Econ. 10.1: 125–145, 2010.

[8]R. Innes, "Remediation and Self-Reporting in Optimal Law Enforcement," J. of Public Econ. 72.3: 379–393, 1999.

[9]C. Fray, "Narrative in Police Communication: The Art of Influence and Communication For The Modern Police Organization," Illinois State University ReD: Res. and eData, Theses and Dissertations 753, 2017.

[10]A. B. Poulin, "The Investigation Narrative: An Argument for Limiting Prosecution Evidence," 101 Iowa L. Rev. 683, 2016.

[11]A. Hoots, "The Importance of Quality Report Writing in Law Enforcement," Sch. of L. Enforcement Supervision, Undated.

[12]*Id.*

more abstract (for example, a descriptive model), or more useful (for example, a predictive model)."[13] That is, skillful use of data mining can take noise and be molded by LEA into a narrative that answers the six underlying questions (the who, what, where, when, why, and how),[14] or at least creates the underpinnings of an explanation of what happened.

These insights can be used by LEA at the global, national, regional, and community levels to predict and prevent crime. Predictive policing works to "harness the power of information, geospatial technologies, and evidence-based intervention models to reduce crime and improve public safety."[15] As a part of predictive policing, data mining has helped to move law enforcement activities from the **reactive**—responding to crimes committed—to the **proactive**—understanding the nature of the problem and developing strategies to prevent or mitigate future harm.[16] In addition, the use of data mining techniques in predictive policing can assist with solving past crimes, and the use of such techniques may also craft the narratives so necessary to bring perpetrators to justice for those crimes. For scholars examining how data mining and narrative development is utilized by LEA, even further data mining (in one particular instance, mining of social media) can be used to develop more granular descriptions of exactly how many of these new techniques, data sources, and methods of interaction are utilized in the service of LEA aims.[17]

Although LEA have mapped crime hotspots for decades,[18] recent trends in data and analytics have drastically improved LEA ability to use real-time forecasting and direct LEA resources effectively. In addition, the acceptance level of the use of such techniques in a variety of professions, as well as the credibility associated with their determinations, has grown.[19] Specifically, as data sets have grown, data mining in conjunction with Artificial Intelligence (AI) and machine learning have allowed LEA to combine and process huge amounts of data. The techniques for processing, understanding, and using this data effectively continue to improve and expand in tandem,[20] but this unidirectional trend might be changing. The data that LEA have

---

[13] K. A. Taipale, "Data Mining and Domestic Security: Connecting the Dots to Make Sense of Data," Colum. Sci. and Tech. L. Rev., vol. 5, p. 22, 2003.

[14] A. Hoots, "The Importance of Quality Report Writing in Law Enforcement," Sch. of L. Enforcement Supervision, Undated.

[15] National Institute of Justice, "Predictive Policing," June 9, 2014.

[16] Id.

[17] C. Fray, "Narrative in Police Communication: The Art of Influence and Communication for the Modern Police Organization," Illinois State University ReD: Research and eData, Theses and Dissertations 753, 2017.

[18] K. M. Curtin, K. Hayslett-McCall, and F. Qiu, "Determining Optimal Police Patrol Areas with Maximal Covering and Backup Covering Location Models," Networks and Spatial Econ. 10.1: 125–145, 2010.

[19] E. B. Larson, "Building Trust in the Power of 'Big Data' Research to Serve the Public Good," Viewpoint, JAMA 309(23):2443–2444, 2013.

[20] H. Chen, W. Chung, J. J. Xu, G. Wang, Y. Qin, and M. Chau, "Crime Data Mining: A General Framework and Some Examples," Computer 37.4: 50–56, 2004.

access to, at least in some parts of the world, may start shrinking as individuals demand and gain more rights over their own personal identifiable data (PII), which means that LEA may face additional challenges not only in access to data but in its evaluation as evidence in their pursuit of justice. Furthermore, challenges associated with the use of such data and the potential influence such data usage wields are magnified in the face of increased scrutiny and understanding of how data-driven decisions are made and implemented.[21]

In Europe, for example, Europol (the law enforcement agency for the European Union) is already addressing how it may deal with these new requirements.[22] Europol has indicated that it may seek approval from the European Council for collection and processing of personal data in the framework of the European Information System in order, to support case-specific aims. Such requests seem to focus on specific use cases that would necessitate such use, rather than blanket or categorical requests.

Europol has already moved forward with several databases that will be used to predict certain events based on social media behaviour or money transfer tracking, not only to predict criminal behaviors but also to identify potential witnesses, victims, contacts, associates, and other persons who could provide information about the suspect or criminal offences under consideration.[23] These Europol databases do not include categories of sensitive personal data, but they do link to various databases that provide identifiable DNA data supplied by LEA in EU Member States,[24] giving rise to the differential privacy concerns discussed elsewhere in this paper.

Although certain general limitations on data processing are set by law, Europol is also considering additional approval on a case-by-case basis from the European Council to avoid the more general rules,[25] focusing on LEA-specific platforms, such as the Europol Information System (EIS) or the Europol Computer System (TECS). In addition, Europol has created a European Cybercrime-Platform (ECCP), which is intended to exchange analytics and information with European national LEA.[26] These exchanges are meant to further amplify the reach, ambit, and predictive power

---

[21] M. Hu, "Big Data Blacklisting," Fla. L. Rev. 67:5, 2016.

[22] Europol, "Data Protection at Europol," March 21, 2018.

[23] Article 14 of Council Decision of 6 April 2009 establishing the European Police Office (Europol) (2009/371/JHA).

[24] D. Meinicke, "Big Data und Data Mining: Automatisierte Strafverfolgung als neue Wunderwaffe der Verbrechensbekämpfung," Kommunikation und Recht, p. 377, 2015.

[25] *See* Council Decision of 6 April 2009 establishing the European Police Office (Europol) (2009/371/JHA), in particular Article 10.

[26] For a structured introduction of the European Criminal Authorities' cooperation see (Sieber, Satzger, and von Heintschel – Heinegg, Europäisches Strafrecht, [European Criminal Law], Teil 4: Zusammenarbeit in Strafsachen [Part 4. Cooperation in Criminal Cases] and 11. Kapitel: Datenverkehr und Datenschutz im Rahmen der polizeilichen und justiziellen Zusammenarbeit [Data exchange and data protection in the framework of the cooperation of police and judiciary] Rn. 5–19.

associated with big data, and reflect the current thought that bigger does equal better when it comes to the value and accuracy of such predictions.[27]

While exceptions such as the EIS, TECS, and ECCP exist[28] to allow LEA and other public authorities to continue to access personal data in the interest of public safety, increased concerns about limiting the rights of LEA and other public agencies to this data may demand more transparency and ultimately mean less overall data and shorter data retention periods.[29] When implemented in conjunction with regulations such as the General Data Protection Regulation (GDPR) coming into force on 25th May 2018, these limitations may also be subject to review by Data Privacy Officers (DPOs) acting on behalf of external organizations, or external public entities themselves in the form of Data Privacy Authorities (DPAs). These additional forms of scrutiny could possibly operate to restrict what might otherwise be unfettered collection and limitless use of a variety of data and gathering methods.

This state of affairs may hinder some otherwise promising methods of predictive policing that rely on large data sets for analysis, such as advanced hotspot identification; risk terrain analysis; regression, classification, and clustering models; near-repeat modeling; spatiotemporal analysis methods; computer-assisted queries and analysis of intelligence and other databases; statistical modeling to perform crime linking; geographic profiling tools; and computer-assisted queries and analysis of sensor databases.[30] These types of applications are not based in sci-fi speculation; the amount of Internet of Things (IoT) devices coming online alone (numbered conservatively in the billions) will provide amplifying data points that can and likely will add to these types of analysis.[31]

Focusing solely on one of the methods listed above, near-repeat modeling, demonstrates just how connectivity may dramatically impact policing. In this example, analysts closely track behaviours of interest (often specific crimes) and link data associated with those crimes or specific individuals.[32] These methods

---

[27] E. B. Larson, "Building Trust in the Power of 'Big Data' Research to Serve the Public Good," Viewpoint, JAMA 309(23): 2443–44, 2013.

[28] F. Bignami commented on proportionality as a privacy safeguard in her assessment of the European Union's now defunct Data Retention Directive; a provision she saw as "designed to prohibit data mining—hi-tech fishing expeditions." F. Bignami, "Privacy and Law Enforcement in the European Union: The Data Retention Directive," Chi. J. Int'l L., vol. 8, pp. 233–55, 252, 2011.

[29] For instance, the right not to be subject to automated decision making, including profiling, is part of the European Union's General Data Protection Regulation (GDPR) (Regulation (EU) 2016/679), which takes effect on May 25, 2018. The limits of this right have been discussed by S. Wachter, B. Mittelstadt, and L. Floridi, "Why a Right to Explanation of Automated Decision-Making Does Not Exist in the General Data Protection Regulation," Int'l Data Privacy L., June 2017.

[30] W. L. Perry, B. McInnis, C. C. Price, S. C. Smith, and J. S. Hollywood, "Predictive Policing: The Role of Crime Forecasting in Law Enforcement Operations," Santa Monica, CA: RAND Corp., 2013, pp. 15–16.

[31] A. Nordrum, "Popular Internet of Things Forecast of 50 Billion Devices by 2020 Is Outdated," IEEE Spectrum, Aug. 18, 2016.

[32] M. B. Short, M. R. D'Orsogna, P. J. Brantingham, and G. E. Tita, "Measuring and Modeling Repeat and Near-Repeat Burglary Effects," J. of Quantitative Criminology, 2009.

are effective; behaviors of multiple people are tied together to create a framework similar to that found in the traditional "known criminal associates" method. But utilizing a variety of data sources, especially those tied to IoT devices, does two things that conflict with one another: it increases the signal while also increasing the noise.

This dual increase is troubling because it negates principles of traditional policing when evaluating "known criminal associates." Data mining in a much richer data environment and will likely at least at first, be overbroad and will supplant the advice normally given to investigators—that "a common starting point is to identify the criminal's associates however, the objective should always be to identify relationships between individuals and their roles in the criminal activities, rather than identifying associates for their own sake."[33] The use of a myriad of data points, easily seen but not exclusively associated with IoT issues will cast a much wider net of suspicion and ensnare if only momentarily, many more people.

In addition to the concerns associated directly with the veracity of predictive policing and just how intrusive it may be or become, other concerns regarding how judges, jurors, and fact finders may react to the pictures and narratives that data may present are coming under more scrutiny. Narrative development has already been subject to some criticism, specifically in those instances where a prosecutor "presents testimony instructing the jurors how to view the evidence, sharing the law enforcement perspective on what might otherwise seem to be inconsequential or innocent action."[34] This is true power asserted towards an outcome and *not* as an absolute if indeed there is not a true objective good where specific individuals are punished; how much more then will this narrative structure be reinforced, even if it is part of an educated guessing framework, when it is presented as the opposite? While there are active efforts to build further public trust and confidence in the use of big data mining and those outputs that *may* be used to improve the public good,[35] there are scholars and institutions who are firmly in the camp of positive application, and who have moved past the inherent fuzziness in big data analytics, instead averring that "[d]ecision-making is no longer an educated guess, but a scientific approach on which improvements can continually be made."[36]

---

[33] United Nations Office on Drugs and Crime, Criminal Intelligence Manual for Analysts, 2011.

[34] A. B. Poulin, "The Investigation Narrative: An Argument for Limiting Prosecution Evidence," 101 Iowa L. Rev. 683, 2016.

[35] E. B. Larson, "Building Trust in the Power of 'Big Data' Research to Serve the Public Good," Viewpoint, JAMA 309(23): 2443–2444, 2013.

[36] S. Smith, "Big Data. Predictive Analytics. Forecasting." Regis U. Criminology Resource Center, Undated.

2 Past Law Enforcement Agency Data Mining Practices and Concerns

LEA have used statistical and geospatial analyses to map out crime hotspots and forecast crime levels for decades,[37] but in the "last decade or so, new technologies have been brought to bear upon the information management challenge posed by this deluge of data."[38] LEA have focused on the potential in analytical tools applied to enormous data sets to make predictions that can assist in crime prevention.[39] These new techniques are aimed at three areas:

> First, they have enabled the cataloging of human behaviors that were previously ephemeral.... Second, semantic query systems and 'big data' analytical engines have introduced an approach to discerning patterns in data that prior systems lacked.... Third, these new techniques of surveillance gathering and data analysis have begun to transition into their next phase, prediction and scoring of individuals' risk of criminal behavior...trigger[ing] individualized suspicion.[40]

Detailed further, the first area of cataloging previously ephemeral behaviour carries with it a variety of sub-issues. This cataloging is itself advanced recordkeeping, not only increasing the number of items that can be recorded and recalled, but amplifying the power by which such items may be referenced, cross-referenced, and reviewed in the context of multiple items. Investigators can share such detail with other similarly situated professionals much more easily, without the cost traditionally associated with copying and organizing records. This lack of "friction" may amplify the investigatory power to a startling degree, and methods by which such sharing may be automated as a matter of course further increase its potential use.

The second area, semantic query systems and "big data" analytical engines writ large, take these amplified catalogs of easily sorted and sifted data and allow human beings to augment their own powers of reasoning and pattern recognition,

---

[37] *See* C. R. Shaw and H. D. McKay, "Juvenile Delinquency in Urban Areas," 1931 (correlating physical status, economic status, and population composition with delinquency rates); *see also* K. Miller, "Total Surveillance, Big Data, and Predictive Crime Technology: Privacy's Perfect Storm," J. Tech. of L. and Pol'y, vol. 19, p. 106, 2014; and M. Curtin, K. Hayslett-McCall, and F. Qiu, "Determining Optimal Police Patrol Areas with Maximal Covering and Backup Covering Location Models," Networks and Spatial Econ. 10.1: 125–145, 2010.

[38] K. Miller, "Total Surveillance, Big Data, and Predictive Crime Technology: Privacy's Perfect Storm," J. Tech. of L. and Pol'y, vol. 19, p. 106, 2014.

[39] W. L. Perry, B. McInnis, C. C. Price, S. C. Smith, and J. S. Hollywood, "Predictive Policing: The Role of Crime Forecasting in Law Enforcement Operations," Santa Monica, CA: RAND Corp., 2013, p. 2.

[40] Miller also notes that "these enhanced cataloging powers have coincided with an increasing willingness by law enforcement agencies to conduct – and courts to condone – widespread, total surveillance of citizens in the name of national security." K. Miller, "Total Surveillance, Big Data, and Predictive Crime Technology: Privacy's Perfect Storm," J. Tech. of L. and Pol'y, vol. 19, p. 106, 2014.

sometimes using the investigators' own words as the strongest starting points. Investigators can query such indexed data sets at will, rather than poring over records manually and allow the systems to self-correct and suggest certain words, phrases, documents or collections of information that are anomalous within the set itself or in comparison with other connected or indexed sets.[41]

The third area, that of prediction, ties closely into data set size and indexing, searching, highlighting, and automation. It is the combination of these powerful processes that, when focused on a subject, a person, an issue or otherwise draws conclusions about what is likely to happen. Notably, this is not just whether an event will likely happen, but also who is likely to perpetrate it.

Beginning in the 1990s,[42] the National Institute of Justice (NIJ) began using geographic information system tools to map crime data, and researchers applied regression analysis and mathematical models to attempt to forecast crime. In 2006, Police Chief (ret.) William J. Bratton and the Los Angeles Police Department (LAPD) along with researchers at the University of California, Los Angeles and University of California, Irvine led the development and expansion of one of the nation's first Computer Statistics programs (COMPSTAT), championing predictive policing and using predictive analytics to monitor and anticipate gang violence in the early years of this century.[43] A key COMPSTAT operational goal was to make the LAPD's hotspot maps predictive instead of descriptive,[44] but these programs are meant to perform more generally and flexibly in the face of evolving data and criminal behavior.[45]

The LAPD was among seven police agencies that received NIJ planning grants to develop predictive policing projects in 2009. Moving from annual to real-time crime mapping analysis over the course of the previous decade, the LAPD began crime forecasting in 2010. The LAPD also developed three projects: a debriefing project aimed at collecting information unrelated to the crime for which a suspect was arrested; social networking analyses specific to gang investigations; and a project to map gang homicides in order to predict future murders.[46]

---

[41] A. Taal, J. Le, and J. Sherer, "A Consideration of eDiscovery Technologies for Internal Investigations," IGS3, CCIS 534:59–73, 2015.

[42] By the early 1990s, data mining was seen as a sub-process within Knowledge Discovery in Databases (KDD), and the 1990s saw a significant increase in the interest in data mining generally with the establishment of a number of regular conferences. See F. Coenen, "Data Mining: Past, Present, and Future," The Knowledge Engineering Rev., vol. 26(1), 2011.

[43] See W. J. Bratton, S. W. Malinowski, "Police Performance Management in Practice: Taking COMPSTAT to the Next Level," Policing, vol. 2(3), pp. 259–265, 2008.

[44] M. Hvistendahl, "Can 'Predictive Policing' Prevent Crime Before It Happens?" Science, September 28, 2016.

[45] Bureau of Justice Assistance, "COMPSTAT: Its Origin, Evolution, and Future in Law Enforcement Agencies," Police Executive Res. Forum, 2013.

[46] National Institute of Justice, C. Beck, "The LAPD Experiment," January 6, 2012.

The same year, the Boston Police Department used an NIJ grant to "develop, test, implement, and evaluate a predictive policing model for property crime."[47] The New York Police Department (NYPD) applied NIJ grant funds to review analytic options to apply to its updated records management system, new data warehouse, and upgraded tracking systems.[48] The NYPD had been running statistical analyses on the city's crime reports and arrests since the mid-1990s, and the upgrades allowed improved analytics. The Maryland State Police Department similarly focused on "analysis tools and technology infrastructure."[49] Shreveport, a smaller police department, applied its grant to tactical crime prevention using "out-of-the-box software."[50] In 2011, Chicago and Shreveport were awarded competitive NIJ grants to continue into their second phases of implementation, and the Chicago Police Department created an in-house predictive analytics unit.

A second predictive policing symposium hosted by NIJ in 2010 highlighted "privacy and civil liberty issues... critically interrelated with predictive policing," emphasizing the need to engage privacy advocates and community leaders and to ensure that predictive policing was constitutional from the beginning.[51] But the public concern over widespread data collection and covert surveillance has only increased in the wake of details about the data gathering practices of the United States National Security Agency (NSA) leaked by Edward Snowden in 2013. This, "coupled with the cavalier attitude of current and former NSA directors and charges by security experts that the NSA has for several years attempted to introduce subtle flaws into cryptographic encryption standards in order to make communications easier to analyze," should serve to put both Americans and foreign citizens on notice of their lack of personal privacy.[52]

Gary T. Marx, Professor Emeritus of Sociology at Massachusetts Institute of Technology, said in a recent interview that "technology such as predictive policing creates 'categorical suspicion' of people in predicted crime areas, which can lead to unnecessary questioning or excessive stopping-and-searching."[53] Marx additionally noted that he was worried that machine analysis and decision making could lead to "the tyranny of the algorithm."[54] Privacy and racial justice groups doubt the technologies as well, questioning the secrecy surrounding the formulas they use

---

[47] National Institute of Justice, H. Gunaratne, "Discussion on the Predictive Policing Demonstration Projects and Evaluation," January 6, 2012.

[48] Id.

[49] Id.

[50] Id.

[51] National Institute of Justice, Predictive Policing Symposiums, "Privacy and Legal Issues," January 6, 2012.

[52] K. Miller, "Total Surveillance, Big Data, and Predictive Crime Technology: Privacy's Perfect Storm," J. Tech. of L. and Pol'y, vol. 19, p. 107, 2014; *see also* K. Zetter, "How a Crypto 'Backdoor' Pitted the Tech World Against the NSA," WIRED, September 24, 2013.

[53] N. Berg, "Predicting Crime, LAPD-Style," The Guardian, June 25, 2014.

[54] Id.

and expressing concern that the "practice could unfairly concentrate enforcement in communities of color by relying on racially skewed policing data."[55]

The Obama White House too noted this tension, stating that "[t]he technical capabilities of big data have reached a level of sophistication and pervasiveness that demands consideration of how best to balance the opportunities afforded by big data against the social and ethical questions these technologies raise."[56] These concerns were echoed by the European Data Protection Supervisor, who opined that "[t]here are serious concerns with the actual and potential impact of processing of huge amount of data on the rights and freedoms of individuals, including their right to privacy. The challenges and risks of big data therefore call for more effective data protection."[57]

In the face of these challenges, and despite underlying privacy concerns, the use of predictive policing methods has grown exponentially in the last decade. For instance, big data analytics as methods of predictive law enforcement have been used by the European Border and Coast Guard Agency (Frontex) in the preparation of its pre-frontier (or border) intelligence picture by text mining algorithms that predict migration.[58] Although the method is the same, no personal data is used in Frontex's analytic approach, unless there is personal information that has been made publicly available. Many LEA predictive policing tools do, however, rely on non-public data.

Consultants and private companies quickly began providing professional services and software to utilize the ever-growing pool of data, and media interest in what LEA were doing with this set of data increased. PredPol,[59] a popular predictive policing tool, in particular, received a good deal of early coverage by the media, which claimed PredPol could actually predict when and where crime would occur.[60] PredPol itself distributed news articles about the success of its predictive policing, suggesting that its use in Los Angeles and Santa Cruz saw "reductions in

---

[55] J. Jouvenal, "Police are Using Software to Predict Crime. Is It a 'Holy Grail' or Biased Against Minorities?," The Washington Post, November 17, 2016; *see also* K. Miller, "Total Surveillance, Big Data, and Predictive Crime Technology: Privacy's Perfect Storm," J. Tech. of L. and Pol'y, 19:106, 2014 (arguing that "the move toward predictive policing using automated surveillance, semantic processing, and analytics tools magnifies each technology's harms to privacy and due process, while further obfuscating the systems' technological and methodological limitations").

[56] Executive Office of the President (Barack H. Obama), "Big Data: Seizing Opportunities, Preserving Values," May 2014.

[57] European Data Protection Supervisor, "Meeting the Challenges of Big Data: A Call for Transparency, User Control, Data Protection by Design and Accountability," European Data Protection Supervisor (EDPS) Opinion 7/2015.

[58] J. Piskorski, M. Atkinson, J. Belyaeva, V. Zavarella, S. Huttunen, and R. Yangarber, "Real-Time Text Mining in Multilingual News for the Creation of a Pre-Frontier Intelligence Picture," ACM SIGKDD Workshop on Intelligence and Security Informatics, July 2010.

[59] PredPol, http://www.predpol.com/.

[60] E. Goode, "Sending the Police Before There's a Crime," N.Y. Times, August 15, 2011; N. Berg, "Predicting Crime, LAPD-Style," The Guardian, June 25, 2014; J. Jouvenal, "Police are Using Software to Predict Crime. Is it a 'Holy Grail' or Biased Against Minorities?" The Washington

crime of 12 percent and 27 percent respectively."[61] While the media and PredPol itself have at times been accused of exaggerating its capabilities,[62] a "21-month single-blind randomized control trial in three LAPD divisions found PredPol to accurately predict twice as much crime as existing best practices."[63] Not all uses of predictive policing uses have been met with similar media approval. Certain police departments have met criticism due to such technology uses, and in particular, a Manhattan judge recently ordered the New York City Police Department to release documentation about its own predictive policing partnership following a lawsuit filed by the Brennan Center for Justice.[64]

PredPol's technology uses three data points: crime type, crime location, and crime date/time.[65] Other types of predictive policing rely on data gathered in other potentially linkable ways: for example, biometric surveillance through linking video surveillance cameras with facial recognition software and facial image databases makes it possible for LEA to find individuals in almost any public space.[66] LEA can now use mobile phones to take pictures that can be identified through facial recognition technology.[67] One out of every two Americans is likely already in an LEA-accessible facial recognition database.[68] Chip-enhanced identification is safer in terms of authentication but simultaneously feeds information directly back to LEA. Data from various types of GPS and automobile tracking devices as well as cell phone towers give LEA the ability to track the movements of individuals.

A next logical step is the unification of all of this data into one enormous centralized database that would assist a variety of LEA interests. A project, unsurprisingly, that the Federal Bureau of Investigation already has underway with its Next Generation Identification (NGI) addresses exactly that issue.[69] The capabilities of NGI include advanced fingerprint identification technology; a repository for data

---

Post, November 17, 2016; J. Smith, "Crime-Prediction Tool PredPol Amplified Racially Biased Policing, Study Shows," Mic., October 9, 2016.

[61] D. Bond-Graham, "All Tomorrow's Crimes: The Future of Policing Looks a Lot Like Good Branding," SFWeekly, October 30, 2013.

[62] J. Jouvenal, "Police Are Using Software to Predict Crime. Is It a 'Holy Grail' or Biased Against Minorities?" The Washington Post, November 17, 2016.

[63] N. Berg, "Predicting Crime, LAPD-Style," The Guardian, June 25, 2014.

[64] A. Winston, "Transparency Advocates Win Release of NYPD 'Predictive Policing' Documents," The Intercept, January 27, 2018.

[65] PredPol, "How PredPol Works," http://www.predpol.com/how-predictive-policing-works.

[66] M. Hu, "Biometric ID Cybersurveillance," Ind. L.J., vol. 88, pp. 1475–81, 2013.

[67] J. Lynch, "Face Off: Law Enforcement Use of Face Recognition Technology," Electronic Frontier Foundation, February 12, 2018.

[68] C. Garvie, et al., "The Perpetual Line-Up," Geo. L. Center on Privacy and Tech., October 18, 2016.

[69] Federal Bureau of Investigation, "Next Generation Identification,"; *see also* M. Hu, "Biometric ID Cybersurveillance," Ind. L.J., vol. 88, pp. 1152–53, 2013. Hu calls this "bureaucratized surveillance," which amounts to the state automating a screen of all interactions with citizens in her opinion.

associated with individuals of special or LEA-noted concern; latent and palm prints; facial recognition; "Rap Back," a program which provides on-going criminal history status updates to authorized agencies; Cold Case/Unknown Deceased, which uses "advanced search algorithms within NGI, and the ability to cascade NGI searches against the criminal and civil files, as well as event-based searches" to identify individuals; and Iris Pilot, a program launched in 2013 to evaluate the technology for iris image recognition and build a criminal iris repository.[70]

Both the Department of Homeland Security's Future Attribute Screening Technology and the Transportation Security Administration pre-flight screening systems similarly combine and cross-reference ever-growing data sets.[71] A new system called LineSight developed by Unisys, a company that already provides screening systems to American, European, and Australian border patrol agencies, processes data from airline tickets and travel history, cargo manifests, and various organizations, including Interpol, to harness machine learning technology.[72] LineSight analyzes data to flag suspicious individuals and items at border crossings in "near real-time."[73] The aim is to help the agencies that patrol these borders make better decisions about admitting, denying, or further scrutinizing both individuals and cargo.

In Europe, a predictive system for air passenger data is already in place, initiated by Directive (EU) 2016/681 of the European Parliament and of the Council of 27 April 2016. This system uses passenger name record (PNR) data for the prevention, detection, investigation, and prosecution of terrorist offences and serious crime. The relevant German Act[74] implementing the Directive permits sample matching of personal data *without* any suspicion of wrongdoing.[75] This law prohibits the use of any sensitive personal data, such as, race or ethnic origin, religion or belief, political opinions, trade union membership, health data, or sexual orientation. Nor are the prospective dangers of a particular journey relayed to the passenger.[76]

---

[70] Federal Bureau of Investigation, "Next Generation Identification."

[71] *See* U.S. Department of Homeland Security, Privacy Impact Assessment for the Future Attribute Screening Technology (FAST) Project 3, 2008; in the United States, the Federal Agency Data Mining Reporting Act of 2007, 42 U.S.C. § 2000ee-3, requires federal agencies to report on data mining activities.

[72] S. Melendez, "A New Border Security App Uses AI to Flag Suspicious People in Seconds," Fast Company, March 6, 2018.

[73] Unisys, "LineSight," http://www.unisys.com/offerings/industry-solutions/public-sector-industry-solutions/justice-law-enforcement-and-border-security-solutions/linesight/Story/linesight-id-3610.

[74] Fluggastdatengesetz vom 6. Juni 2017 (BGBl. I S. 1484), das durch Artikel 2 des Gesetzes vom 6. Juni 2017 (BGBl. I S. 1484) geändert worden ist [German Act on Flight Passenger Data of 6 June 2017].

[75] Section 4, para 1–2 of the German Act on Flight Passenger Data.

[76] T. Rademacher, "Predictive Policing im deutschen Polizeirecht, [Predictive Policing in German Police Laws]," Archiv des öffentlichen Rechts 142 3:366–416; pp. 412–414, 2017.

The Australian government introduced the Identity-Matching Services Bill in February 2018. The bill, currently under review, would establish the Australian government's biometric identity system, authorizing the Department of Homeland Affairs to "collect, use and disclose identification information in order to operate" these newly created systems.[77]

Critics of these developments point out that there has been little meaningful oversight by legislators or the public when these new technologies are adopted, and few "legal protections to prevent the internal and external misuse."[78] They also question whether these systems have been adequately tested for accuracy, claiming this "has led to the development of unproven, inaccurate systems that will impinge on ... rights and disproportionately impact people of color."[79] Such developments have been confirmed by criminologists, who identify a concentration of the efforts of criminal prosecution based on predictive policing supported by statistical probabilities or methods of data analysis. The reason for this is that institutionalized predictive policing methods focus on the most accurate prediction. The predictive policing "hits" therefore are likely to concentrate on those specific crimes, places, and potential wrongdoers that have been determined to have the most significant probability. In addition, one must not forget that traditional criminologists supply the basic methods for training the systems, which leads to an emphasis on traditional crime scenes and crimes.[80]

Some European police agencies, as an example, approach new techniques relating to predictive policing cautiously. There are general concerns related to actual police measures as a deterrent to criminal behavior before any (predicted) wrongdoing has been commenced. One of the partially adopted new techniques is body cameras ("body cams"). The ostensible reason for the use of such body cams is to decrease the number of attacks on police officers; however, body cams may also serve as direct evidence regarding such attacks thereby reducing the likelihood of the escalation of a dispute. But, to use Germany as an example, there is no unified legal framework for the use of body cams on a German federal level; thus, there is no unified database of the recordings.[81] The Dutch have similarly experimented with body cams beginning the late twentieth century while other countries; among them France, Italy, and Sweden, have only recently adopted the technology.

---

[77] Parliament of Australia, "Review of the Identity-Matching Services Bill 2018 and the Australian Passports Amendment (Identity-Matching Services) Bill 2018," 2018.

[78] J. Lynch, "Face Off: Law Enforcement Use of Face Recognition Technology," Electronic Frontier Foundation, February 12, 2018.

[79] Id.

[80] T. Singelnstein, "Predictive Policing: Algorithmenbasierte Straftatprognosen zur vorausschauenden Kriminalintervention [Prognosis of Criminal Actions Based on Algorithms for Preventive Criminal Intervention]," NStZ, 1, p. 4, 2018.

[81] F. Ebert, "Entwicklungen und Tendenzen im Recht der Gefahrenabwehr [Developments and Tendencies in the Law of Mitigating Danger]" LKV – Landes- und Kommunalverwaltung, 10 p. 16, 2017.

3 Other Industry Data Mining Practices

Many of the applications that LEA have used in the past and continue to use today are, at least in part, developed for some commercial purpose that is not specifically LEA focused. Some were developed for and are used by the private sector to forecast consumer behavior or determine sales strategies through tracking consumer behavior. For example, major retailers might use data mining and analytics to determine how to stock stores.[82] This data can then be sold to other private companies or even provided to LEA. In recent years, following its acquisition of SPSS, IBM now offers its SPSS Crime Prediction Analytics Solution (CPAS) Service to LEA as a more or less standard solution at an arguably affordable price point.[83] At least one other program, successfully used by the Santa Cruz police to "generate projections about which areas and windows of time [were] at highest risk for future crimes by analyzing and detecting patterns in years of past crime data," uses modeling originally developed to predict earthquake aftershocks.[84] In fact, an approach used by lenders to pre-qualify mortgage applicants can also be used to assess the risk for escalation in a series of burglaries.[85]

LEA also makes use of the developments of private enterprises on a more localized level. For instance, the German Institute for Pattern-Based Prediction Technique has developed a system of predictive policing for burglaries that is based solely on the statistical data of the local police entities.[86] Because of this, predictions can only be made in relation to place and time, but not in relation to an actual person. As a result, the PRECOBS (PRE-Crime Observation System) forecasting software to predict burglary crime can be used relatively freely by the police without infringing on the rights of the individual. Essentially, the only consequence of a PRECOBS crime prediction is increased police presence and caution in near repeat areas. Recorded burglaries have decreased significantly in cities through the use of this predictive policing tool, leading to the conclusion that the predictions have been relatively accurate.[87]

---

[82] W. L. Perry, B. McInnis, C. C. Price, S. C. Smith, and J. S. Hollywood, "Predictive Policing: The Role of Crime Forecasting in Law Enforcement Operations," Santa Monica, CA: RAND Corp., p. 2, 2013.

[83] "IBM SPSS Crime Prediction Analytics," IBM Corp., July 2012.

[84] E. Goode, "Sending the Police Before There's a Crime," N.Y. Times, August 15, 2011.

[85] C. McCue, Data Mining and Predictive Analytics: Intelligence Gathering and Crime Analysis, 2007.

[86] Institut Für Musterbasierte Prognosetcknik, "Near Repeat Prediction Method – Predictive Policing made in Germany," Undated.

[87] D. Gerstner, "Predictive Policing als Instrument zur Prävention von Wohnungseinbruchdiebstahl [Predictive Policing as an instrument for the prevention of burglaries] in: Forschung Aktuell/research in Brief/50, Max Planck-Institut für ausländisches und internationales Strafrecht Freiburg im Breisgau, p. 37, 2017.

4 Present-Day Data Mining: Form, Participant, and (Another) Participant

> A HOMELESS GUY IN BURNSVILLE CALLED 911 IN FEBRUARY BUT DIDN'T SAY WHERE HE WAS. OFFICERS FOUND HIM BY CHECKING THE MOST RECENT VIDEO IN HIS YOUTUBE CHANNEL.[88]

Modern-day data mining is science fiction in practice, insofar as it would have been unheard of 10 years ago to say that police could locate someone by pinpointing the geographic location where a homeless person had taken a video, uploaded it, and shared it for all to see. This underscores both the existing forms of data that can be "mined," as well as the new—and sometimes unexpected—participants in the process. These concepts of form and participant underpin the available data, knitting both together into a narrative that law enforcement agencies (and others) can follow. However, additional considerations about the validity of the data, both inherently and in connection with still other data sources, soon follow. As discussed above, the increased number of online IoT devices, soon to be in the billions, will dramatically increase the connections LEA may be able to use to determine locations of individuals, crimes, and other behaviors that may not *be* crimes, but that may be suggestive.

But increased data, however large the volumes are, does not advance policing or predictive power without more. This process of developing data mining learning and extending it to sets of new observations, is instead predicated on a number of concurrent advances in how data is recognized, including "advancements in voice recognition, image recognition, statistical translation, and semantic indexing of knowledge."[89] It is also supported by vast new sources of data, including the transactional wireless surveillance data contained within the United States Stingray program.[90] In the United States, federal agencies, such as Immigration and Customs Enforcement, are also using regional database information, which provides "phone numbers, addresses, and comments about individuals' scars, marks and tattoos that may have not made it into federal records."[91] Further, LEA are collecting information from other sources, including "[b]ody cameras, [c]ellphone hacking devices, license plate scanners, and [s]oftware that can identify faces in surveillance video."[92]

But without an overlay, for all their detail, these data points are just that—numbers in an array. For an LEA to develop a narrative, an additional step joins

---

[88] J. E. Shiffer, "Police's Growing Arsenal of Technology Watches Criminals and Citizens," Star Tribune, May 1, 2017.
[89] M. Moore, "The Realities of Machine Learning Systems," Software Dev. Times, April 25, 2017.
[90] J. Kelly, "Cellphone Data Spying: It's Not Just the NSA," USA TODAY, December 8, 2013.
[91] G. Joseph, "Where ICE Already Has Direct Lines to Law-Enforcement Databases with Immigrant Data," NPR – Code Switch, May 12, 2017.
[92] J. E. Shiffer, "Police's Growing Arsenal of Technology Watches Criminals and Citizens," Star Tribune, May 1, 2017.

the process. In one heavily reviewed instance, data or opinion mining determines specific "sentiment" points contained in data, which are then reviewed further by humans, using the AI strata to first process and structure the data before the human team verifies the data for nuance, sentiment, and overall topics.[93] The techniques are not yet automated and require some kind of human intervention currently; however, AI may hold the promise for taking such human behavior, tracking and modeling it, and then subsequently formalizing it such that it can be repeated *without* such human intervention.[94]

Moreover, it is essential to implement a human decision regarding the police measures as a consequence of the prediction that result from the data mining activities. This is even more important for assessing if and to what level in-depth analysis covering more sensitive data of persons may be continued. Currently, automated background searches and/or analysis of an individual may be justified by indication from various pieces of evidence relating to (potential) wrongdoings or wrongdoers. Also, there is a notable difference between the permissible use of personal data of individuals with and without prior convictions.[95]

Likewise, social media is used to track diseases and outbreaks despite privacy and security concerns associated with sensitive data.[96] This mirrors requirements by health organizations that "require accurate and timely disease surveillance techniques in order to respond to emerging epidemics."[97] These requirements are otherwise difficult to meet when relying on sick patients to respond accurately and in a timely fashion to in real-time requests from physicians, who in turn must update hospital systems.

Despite some articles to the contrary,[98] these processes are not all automatic, and most continue to require yet another participant, a human, to review the process and make key decisions. Indeed, human interaction is still part, if not the overall equation, of the end result and evolving strategy.[99] This same function is present in financial analysis—human interaction after data is modeled, cleaned, and presented,

---

[93] J. Kloppers, "Data Mining for Social Intelligence – Opinion Data as a Monetizable Resource," Dataconomy, May 12, 2017.

[94] M. Lewis, D. Yarats, Y. N. Dauphin, D. Parikh, and Dhruv Batra, "Deal or No Deal? End-to-End Learning for Negotiation Dialogues," arXiv: 1706.05125, Jun. 16, 2017.

[95] T. Singelnstein, Predictive Policing: Algorithmenbasierte Straftatprognosen zur vorausschauenden Kriminalintervention [Prognosis of criminal actions based on algorhythms for preventive criminal intervention], NStZ, 2018 1, p. 7

[96] L. Bandoim, "Surprising Ways Researchers Use Social Media to Track Disease," EmaxHealth – Family Health, May 21, 2017.

[97] D. A. Broniatowski, M. J. Paul, and M. Dredze, "National and Local Influenza Surveillance through Twitter: An Analysis of the 2012–2013 Influenza Epidemic," PloSOne, vol. 8.12, 2013.

[98] S. Yan, "Artificial Intelligence Will Replace Half of All Jobs in the Next Decade, Says Widely Followed Technologist," CNBC – Tech, April 27, 2017.

[99] M. Moore, "The Realities of Machine Learning Systems," Software Dev. Times, April 25, 2015.

still carries opportunities for analysis that is qualitative or discretionary, rather than entirely quantitative, in nature.[100]

As discussed above, opinion mining also highlights the necessity for this type of approach. Prior practices and existing mechanisms are not geared towards new and massive data sources that do not allow experts to "vet" them,[101] and scaling existing practices to encompass these new amounts of data would be too expensive,[102] especially in recovering economies. AI and data mining activities are "making possible things in business which no human could realistically achieve—at least not while maintaining profitability."[103] And the possible is not just the possible—it is essential, as some consider it to be "a foregone conclusion that a better understanding and application of Big Data will be key to long-term success in a variety of industries."[104]

In the LEA context, these "complex computer algorithms... try to pinpoint the people most likely to be involved in future violent crimes—as either predator or prey."[105] This strategy of "predictive policing" combines those same types of data that ICE is interested in, as well as "information about friendships, social media activity and drug use to identify 'hot people' and aid the authorities in forecasting crime."[106] But note that the data itself does not have a "hot person" data point; that determination ultimately still resides with an individual LEA representative. And, for the time being, likely should; early attempts at replicating human behaviors, even when utilizing other humans as benchmarks for behavior, have presented significant ethical and practical challenges.[107]

---

[100] G. Action, "We're Seeing How Far We Can Push Artificial Intelligence in Asset Management: Man Group's Lagrange," CNBC Tech Transformers, May 17, 2017.

[101] L. Chambers, "How Artificial Intelligence Can Break a Business in Two Minutes," Rude Baguette, May 4, 2017.

[102] J. Kloppers, "Data Mining for Social Intelligence – Opinion Data as a Monetizable Resource," Dataconomy, May 12, 2017.

[103] L. Chambers, "How Artificial Intelligence Can Break a Business in Two Minutes," Rude Baguette, May 4, 2017.

[104] D. Hendrick, "Study Lists 5 Big Data Obstacles and 5 Firms Embracing Analytics," Claims Journal, April 26, 2017.

[105] J. Eligon and T. Williams, "Police Program Aims to Pinpoint Those Most Likely to Commit Crimes," N.Y. Times, September 24, 2015.

[106] Id.

[107] O. Tene and J. Polonetsky, "Taming the Golem: Challenges of Ethical Algorithmic Decision Making," N.C. J. of L. and Tech., June 6, 2017.

5 Current Concerns

THE PRODUCTION OF BUTTER IN BANGLADESH HAD A REASONABLE COR-RELATION WITH THE S&P 500 FROM 1981 TO 1993, BUT THAT'S PURE DATA MINING.[108]

THE TROUBLE WITH THE INTERNET... IS THAT IT REWARDS EXTREMES. SAY YOU'RE DRIVING DOWN THE ROAD AND SEE A CAR CRASH. OF COURSE YOU LOOK. EVERYONE LOOKS. THE INTERNET INTERPRETS BEHAVIOR LIKE THIS TO MEAN EVERYONE IS ASKING FOR CAR CRASHES, SO IT TRIES TO SUPPLY THEM.[109]

Certainly such prediction models contain issues of fundamental fairness in initial application, where the models return representations of past behaviors regardless of whether all of the participants in the process were behaving appropriately.[110] That is, many of the models incorporate past questionable behaviors, including discrimination and the exclusion of "generations of minorities."[111] There are concerns within many of these data sets that past data practices will be geared incorrectly towards disparate impact, especially where judges note that studies addressing these issues "raise concerns regarding how [this type of] assessment's risk factors correlate with race."[112]

The inclusion of minorities, especially to the exclusion of others, should be front-of-mind in those instances, for example, where individuals are added to a list of "future" criminals; critics who wonder about the predictive value of such an addition have raised exactly these concerns.[113] A recent study found that while there is a "reasonable concern that predictive algorithms encourage directed police patrols to target minority communities with discriminatory consequences... no significant differences in the proportion of arrests by racial-ethnic group" existed between the Los Angeles predictive policing experiments and the regular analyst-driven police practices.[114] The authors of the study ultimately determined that predictive policing does not seem to increase bias, instead augmenting existing patterns and biases. They conclude that "future research could seek to test whether

---

[108] S. Moore, "The Surprisingly Strong Data Behind 'Sell In May,'" Forbes, April 30, 2017.

[109] D. Streitfeld, "The Internet Is Broken": @ev Is Trying to Salvage It," N.Y. Times, May 20, 2017.

[110] J. Sherer, "When is a Chair Not a Chair? Big Data Algorithms, Disparate Impact, and Considerations of Modular Programming," DESI VII Workshop on Using Advanced Data Analysis in eDiscovery and Related Disciplines, 2017.

[111] Id.

[112] A. Liptak, "Sent to Prison by a Software Program's Secret Algorithms," N.Y. Times, May 1, 2017.

[113] M. Davey, "Chicago Police Try to Predict Who May Shoot or Be Shot," N.Y. Times, May 23, 2016.

[114] P.J. Brantingham, M. Valasik, and G.O. Mohler, "Does Predictive Policing Lead to Biased Arrests? Results from a Randomized Controlled Trial," DOI, 2018, https://doi.org/10.1080/2330443X.2018.1438940.

the situational conditions surrounding arrests and final dispositions differ in the presence of predictive policing."[115] Other critics have cautioned that the predictive policing focuses on the "punitive element" of the justice system to the exclusion of reform, and by targeting "high-risk individuals," predictive policing precludes "reasonable chance[s] to improve their behavior or learn the lessons from their past," potentially encouraging "an endless cycle of recidivism."[116]

In addition to concerns regarding digital redlining and past practices permeating data sets and models drawn from data mining, volatility is an increasing concern, where a dynamic environment can present "ever changing patterns" leading to three data mining challenges: "change of the target variable, change in the available feature information, and drift."[117] This change in available feature information can be further affected by a changing approach to information.

In one powerful example that combined two different processes by AI data mining, an Associated Press twitter account was hacked in 2013 and (incorrectly) tweeted that then President of the United States Barack Obama was injured in a White House explosion.[118] This news feed was plugged into a number of proprietary data monitoring systems, which in turn sent direction to trading algorithms that executed flash trades and crashed the market.[119] This concern is paramount for the use of these platforms, as developers of the systems need "to see where [the] data might come from, [and] see when it is corrupted or valueless."[120]

Finally, there are concerns raised about the general lack of awareness and oversight of the use of predictive policing technologies. Critics note that there are "plenty of ways that police attention is undesirable even if it does not lead to a warrant, an arrest or criminal charges."[121] Public oversight of and transparency in the use of these new technologies may be critical moving forward.

---

[115] Id.

[116] A. Johansson, "5 Lessons Learned from the Predictive Policing Failure in New Orleans," Venture Beat, March 19, 2018.

[117] G. Kremplet, I. Zilobaite, D. Brezezinski, E. Hullermeier, M. Last, V. Lemaire, T. Noack, A. Shaker, S. Sievi, M. Spiliopoulou, and J. Stefanowski, "Open Challenges for Data Stream Mining Research," ACM SIGKDD Explorations Newsletter, vol. 16(1), pp. 1–10, September 25, 2014.

[118] H. Moore and D. Rober, "AP Twitter Hack Causes Panic on Wall Street and Sends Dow Plunging," The Guardian, April 23, 2013.

[119] L. Chambers, "How Artificial Intelligence Can Break a Business in Two Minutes," Rude Baguette, May 4, 2017.

[120] M. Moore, "The Realities of Machine Learning Systems," Software Dev. Times, April 25, 2015.

[121] N. Feldman, "The Future of Policing is Being Hashed Out in Secret," Bloomberg, February 28, 2018.

6 Future Possible Practices

Data mining will continue aided by new computing advances that generate, both by volume and by increases in the amounts of transactions, "logs or user-generated content."[122] Firms fully engaged in this space have recognized this point-of-no-return, and instead of narrowing their focus, big data analytics firms have turned their practices towards "feel-good projects such as ending homelessness in Santa Clara County, distributing aid to Syrian refugees, fighting human trafficking and rebuilding areas devastated by natural disasters."[123] Likewise, new applications in medical techniques and advancements in epidemiological research shine as beacons of hope for big data use.[124] These demonstrate scientists' trust in the use of big data, which translates into public acceptance of its usage—sometimes without question and with poor results.[125]

Governments, other state bodies, and investigators generally are "likely to turn to social media analysis in the search for greater clarity"[126] and rely on novel data generation approaches just as they will need to incorporate new algorithmic structures to deal with the additional data generated. Of course, new approaches and techniques "will undoubtedly... require the use of computers and advanced algorithmic techniques[127] (incorporating machine learning),[128] due to the big data size, complexity, and nature of the task."[129] This additional complexity on the analytic side may lead to increased calls for auditability of those algorithms to determine, even when they seem to work, whether they are working appropriately.[130]

---

[122] G. Kremplet, I. Zilobaite, D. Brezezinski, E. Hullermeier, M. Last, V. Lemaire, T. Noack, A. Shaker, S. Sievi, M. Spiliopoulou, and J. Stefanowski, "Open Challenges for Data Stream Mining Research," ACM SIGKDD Explorations Newsletter, vol. 16(1), pp. 1–10, September 25, 2014.

[123] M. Kendall, "Palantir Using Big Data to Solve Big Humanitarian Crises," The Mercury News, October 4, 2016.

[124] S. J. Mooney, D. J. Westreich, and A. M. El-Sayed, "Epidemiology in the Era of Big Data," Epidemiology 26(3):390–395, May 2015.

[125] S. Shah, A. Horne, and J. Capellá, "Good Data Won't Guarantee Good Decisions," Harv. Bus. Rev. – Decision Making, April 2012.

[126] J. Kloppers, "Data Mining for Social Intelligence – Opinion Data as a Monetizable Resource," Dataconomy, May 12, 2017.

[127] M. Feldman, S. A. Friedler, J. Moeller, C. Scheideggerand, and S. Venkatasubramanian, "Certifying and Removing Disparate Impact," BIGDATA Program, July 16, 2015.

[128] K. Guruswamy, "Data Science – Data Cleansing and Curation," Teradata - Aster Community, July 15, 2016.

[129] J. Sherer, "When is a Chair Not a Chair? Big Data Algorithms, Disparate Impact, and Considerations of Modular Programming," DESI VII Workshop on Using Advanced Data Analysis in eDiscovery and Related Disciplines, 2017.

[130] C. S. Penn, "Marketers: Master Algorithms Before Diving into Machine Learning," February 1, 2017.

An additional danger, especially for law enforcement, occurs where there is no signal in the noise—but a mirage within the data that leads to action.[131] This is where it is "prudent to be skeptical of relationships that could merely be the result of running large batches of tests and only reporting the few examples that look impressive."[132] In concert, adaptive privacy mechanisms focus on the challenge presented where fixed privacy preservation rules may no longer hold: women do not remain pregnant or bicycling patterns change with the seasons (in climates with variable weather).[133] Even if a viable pattern exists within the data mined and analyzed, LEA practitioners need to examine whether the situation to which the pattern may be applied still exists to avoid the "hammer looking for a nail" approach.[134]

In practice, U.S. practices can range from, as discussed above, tracking diseases,[135] detecting plagiarism,[136] trading stocks, and managing assets[137] to sending people to prison[138] in those instances where an algorithm "calculates the likelihood of someone committing another crime."[139] Where trades are made on the basis of faulty information—or a faulty application—the market can and does correct without moral blame. In the area of law enforcement, the sentiment may be quite different.

These applications will (very likely) only continue to grow. We have noted that, "[s]ince global organizations are retaining larger and larger volumes of structured and unstructured data due to legislative, regulatory, and procedural requirements, investigators face increasingly complex challenges in how to analyze and answer" the narrative model and the six underlying questions addressed above.[140] This also likely means that the applications associated with this data will grow, and in the face of attempts to encourage public trust in big data use,[141] the data itself may outstrip

---

[131] M. Hu, "Big Data Blacklisting," Fla. L. Rev. 67:5, 2016.

[132] S. Moore, "The Surprisingly Strong Data Behind 'Sell in May,'" Forbes, April 30, 2017.

[133] G. Krempl et al., "Open Challenges for Data Stream Mining Research," ACM SIGKDD Explorations Newsletter, vol. 16(1), pp. 1–10, June 2014.

[134] A. Maslow, "The Psychology of Science," Harper & Row, p. 15, 1966.

[135] D. A. Broniatowski, M. J. Paul, and M. Dredze, "National and Local Influenza Surveillance through Twitter: An Analysis of the 2012–2013 Influenza Epidemic," PloSOne, vol. 8.12, 2013.

[136] E. V. Ravve, "MOMEMI: Modern Methods of Data Mining," ICCGI2017, November 2016.

[137] G. Action, "We're Seeing How Far We Can Push Artificial Intelligence in Asset Management: Man Group's Lagrange," CNBC Tech Transformers, May 17, 2017.

[138] A. Liptak, "Sent to Prison by a Software Program's Secret Algorithms," N.Y. Times, May 1, 2017.

[139] M. Smith, "In Wisconsin, a Backlash Against Using Data to Foretell Defendants' Futures," N.Y. Times, June 22, 2016.

[140] A. Taal, J. Le, and J. Sherer, "A Consideration of eDiscovery Technologies for Internal Investigations," IGS3, CCIS 534:59–73, 2015.

[141] E. B. Larson, "Building Trust in the Power of 'Big Data' Research to Serve the Public Good," Viewpoint, JAMA 309(23): 2443–2444, 2013.

our attempts to explain it, and instead merely confirm whatever the viewer or report analyst thought to begin with.

7 Mitigating Issues

Even so-called industry "watchdogs," such as the Electronic Frontier Foundation, acknowledge that "[n]ot all uses of big data implicate dangers to privacy or rights, such as datasets that are not about people or what they do."[142] Big data contains promise for a wide variety of people, with medicine and epidemiological applications at the forefront of many of these hopes.[143] But concerns may apply— and certainly draw additional scrutiny—when "big data is used to individually target people in a certain group found within a dataset."[144]

One concern is that a new phase of predictive policing "will use existing predictive analytics to target suspects without any firsthand observation of criminal activity, relying instead on the accumulation of various data points. Unknown suspects will become known to police because of the data left behind."[145] This use of big data is at odds with, and may ultimately undermine, the "small data" on which reasonable suspicion has traditionally relied.[146] Reasonable suspicion relies on the observable actions of suspects, and the reasonable suspicion test requires an "articulate, individualized, particularized suspicion" about an action, not an individual.[147] With predictive policing able to harness big data in order to target the individuals likely to commit or to have committed crimes, what becomes of the need for reasonable suspicion?

Other concerns may go deeper than intentional dataset collection and utilization in the first instance, as subsequent use of datasets in combination can also give rise to an individual application or approach triangulated from disparate data sources.[148] In particular, the "versatility and power of [some types of] re-identification algorithms imply that terms such as 'personally identifiable' and 'quasi-identifier' simply have

---

[142] Electronic Frontier Foundation, "Big Data in Private Sector and Public Sector Surveillance," April 8, 2014.

[143] E. B. Larson, "Building Trust in the Power of 'Big Data' Research to Serve the Public Good," Viewpoint, JAMA 309(23): 2443–2444, 2013.

[144] Electronic Frontier Foundation, "Big Data in Private Sector and Public Sector Surveillance," April 8, 2014.

[145] A. Guthrie Ferguson, "Big Data and Predictive Reasonable Suspicion," U. Pa. L. Rev., vol. 163, p. 331, 2015.

[146] Id. at 331–32.

[147] Id. at 332.

[148] J. Sherer, J. Le, and A. Taal, "Big Data Discovery, Privacy, and the Application of Differential Privacy Mechanisms," The Computer and Internet L., 32:7, July 2015.

no technical meaning" and, while "some attributes may be uniquely identifying on their own, any attribute can be identifying in combination with others."[149]

The data, once collected, is by its very nature additive if not in isolation, then certainly in conjunction or collaboration with other data sources. That is both the promise and threat provided by so-called differential privacy mechanisms. One data set, intentionally created, may demonstrate a particular instance or theme, whether the behavior of people on a street corner over a 24-h period or the use of an ATM. But when taxi cab data and credit card receipts are combined with the location of an individual on a street corner, or when ATM use is correlated with phone GPS signaling, individual identification becomes an issue. Further, when such identification is then combined with the scrutiny of LEA and other "trusted" actors, a proposition of "guilty until proven innocent" may emerge.[150]

Acknowledgment The authors would like to thank Brittany Yantis and Michael Del Priore for their assistance with this article.

---

[149] A. Narayanan and V. Shmatikov, "Myths and Fallacies of Personally Identifiable Information," Viewpoints, ACM, 53: 6, 2010.

[150] M. Hu, "Big Data Blacklisting," Fla. L. Rev. 67:5, 2016.

DaP∀: Deconstruct and Preserve for All: A Procedure for the Preservation of Digital Evidence on Solid State Drives and Traditional Storage Media

Ian Mitchell, Josué Ferriera, Tharmila Anandaraja, and Sukhvinder Hara

1 Introduction

Preserving evidence is important. Without preservation measures, all cases are jeopardised. Digital Forensics has no special dispensation, preservation matters. To understand how preservation has affected the Standard Operating Procedures (SOP) of digital forensics a brief background on the data acquisition of a HDD is given. McKemmish (1999) encourages the minimal handling of evidence and in Rule 1 states, "this is to be achieved by duplicating the original and examining the duplicate data". This SOP, whereby an exact duplicate is created, is known as imaging. There are some basic guidelines given to imaging a device, see Williams (2018) for more details:

1. Avoid mounting the device and use a write prevention device, e.g., Tableau Write-Blocker (Tableau 1996);
2. Use NIST approved software to complete the image, e.g., dcfldd (2013), Harbour (2002); and,
3. Verify and store the digital fingerprint of the image, e.g., use hash algorithm SHA256 (180-1, F.I.P.S.F. 1996).

Essentially, this can be broken down to three steps: (i) protect the device from contamination; (ii) data acquisition; and, (iii) verification for reproducibility. The last step confirms that all future data acquisition of that device should match a unique digital fingerprint, known as a hash algorithm. If the digital fingerprint does not match then either the device has been incorrectly imaged or contaminated.

I. Mitchell (✉) · J. Ferriera · T. Anandaraja · S. Hara
Middlesex University, London, UK
e-mail: I.Mitchell@mdx.ac.uk; s.hara@mdx.ac.uk

© Springer Nature Switzerland AG 2018
H. Jahankhani (ed.), *Cyber Criminology*, Advanced Sciences and Technologies for Security Applications, https://doi.org/10.1007/978-3-319-97181-0_13

McKemmish's (1999) Rule 2 states, "Where changes occur during a forensic examination, the nature, extent and reason for such changes should be properly documented". This is where our journey begins, the understanding of why and how the original state of a digital device requires deconstruction for future data acquisition in order to preserve its evidential integrity.

1.1 Background

Solid State Drives, SSD, are different to traditional Hard Disk Drives, HDD. Let us be clear from the start both are non-volatile storage devices, in other words, data saved on these devices can be recalled without error. It is the software supporting these mechanisms that are different.

HDDs have an overwrite facility, thus if a sector requires changing then an overwrite is executed. The co-evolution of the development of file systems and HDDs has seen many innovations, however, all require 'dead data' (Krishna Mylavarapu et al. 2009). The definition of "dead data" is data that is no longer relevant, however, in the context of storage devices, it refers to deleted data. Modern file systems (Carrier 2005) when deleting files only make changes to the meta-information about that file, e.g., a file allocated to contiguous blocks 4000–4009 after being deleted would make change these blocks to unallocated and make the record of that file deleted. The actual data in blocks 4000–4009 would remain until another file overwrites them, this data is no longer relevant and is known as 'dead data'. This was efficient and exploited the overwrite facility.

SSDs do not have an overwrite facility, thus if a sector requires changing then a combination of reset and write are executed. In addition, SSD components have a limited number of writes, say 100,000, and thus limiting the endurance of SSDs. An array of Wear-Levelling (WL) and Garbage Collection (GC) algorithms have been developed (Subramani et al. 2013) and deployed by manufacturers to reduce the number of writes, and increase the endurance of the SSD. Briefly, WL algorithms ensure that writes to components are equally distributed, whilst GC ensures that components containing 'dead data' are reset. WL and GC programs are stored in the control units of SSDs, and thus, *write prevention devices cannot stop WL and GC algorithms from making any changes to the device*. Returning to the file allocated to contiguous blocks 4000–4009. On deletion of the file, two things occur: the file system will update appropriate entries and the blocks 4000–4009 will become unallocated; and, a TRIM command is sent to the SSD control unit (Shu and Obr 2007) that initiates the GC. The GC algorithms are deployed and reset the 'dead data' to zeroes.

1.2 Motivation

Unlike HDDs, SSDs have a dynamic state due to GC or WL algorithms. The self-corrosion on SSDs is automated and is a challenge for Digital Forensics (Bell and Boddington 2010). The challenge is getting reproducible results for the imaging of SSDs using SOPs. Such inconsistencies in results could lead to: legal representatives questioning the competence of DFAs; increased administrative burden on DFA, to document the differences that have occurred; potential loss of digital evidence, due to GC and WL algorithms activated in Digital Forensic Lab; and, cognitive burden on DFA, to explain why the differences have occurred.

With the advent of SSDs the SOPs are being challenged (Bell and Boddington 2010; King and Vidas 2011; Nisbet et al. 2013). Just to re-iterate, using write-blockers does not stop automatic self-corrosion of SSDs, and current SOPs fail to give reproducible results. Change is required, the proposed SOP works by the deconstruction of a drive, putting it in a state that renders WL and GC algorithms futile, making it difficult for non-malicious contamination, and giving reproducible results for imaging of SSDs. Essentially, the deconstruction removes important blocks that refer to the partition structure. These blocks are stored, and then later used in the reconstruction of the image of the SSD. The device can be imaged using traditional standard techniques, but cannot be analysed until reconstructed. This simple method is explained in Fig. 1 combined with the results in Mitchell et al. (2017) proved stable for DOS/MBR partitions. The tests here extend to GPT and add a further database to ensure and verify the reconstruction of the device.

The motivation for this research is three-fold: (i) technological advances (SSD and GPT); (ii) challenge to develop a reproducible SOP for SSDs; and, (iii) quality assurance for all imaging of all devices (ISO17025). The aims are to develop a new SOP that will enable preservation of data for current and subsequent data acquisition on GPT formatted SSDs and HDDs, and DOS/MBR SSDs and HDDs, and is named Deconstruct and Preserve for all, or $DaP\forall$.

$$H(I_0^{'}) = H(I_i), \forall i \geq 1 \qquad (1)$$

$$H(I_0) = H(I_i^{'}), \forall i \geq 1 \qquad (2)$$

2 DaP

McKemmish (1999) mentions that change can be justified when acquiring evidence, especially if without change it is virtually impossible to acquire any digital evidence, e.g., see Sylve et al. (2012). SSDs are storage devices, and therefore have to be imaged by all parties (defence, prosecution and subsequent appeals that may require

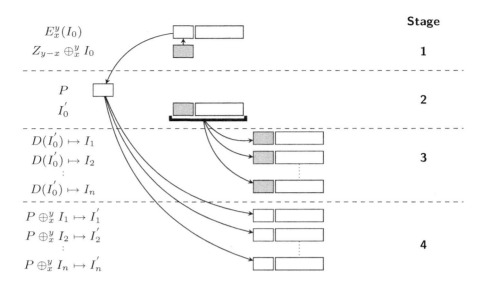

Fig. 1 Three stages of DaP: (1) Deconstruction: extraction of critical GPT component; (2) Preservation: Critical GPT component, P, and Deconstructed partition, I_0'; (3) Acquisition and imaging of deconstructed partition, $I_1, I_2, \cdots, I_n$; and (4) Reconstruction by repatriation of P with images in stage 3, $I_1, I_2, \cdots, I_n$, yielding new complete images, $I_1', I_2', \cdots, I_n'$, for further analysis and investigation

an independent review) with the same results. Essentially, DaP (Mitchell et al. 2017) is explained in four simple stages in Fig. 1 as follows:

Deconstruction extract and record identified blocks from the device, e.g., an HDD or SSD, to render WL and GC algorithms ineffective. These algorithms will still try to run, however, the extraction of the identified blocks means they are unable to find and reset 'dead data'.

Preservation replace identified blocks with zeroes on the device, i.e., wipe identified blocks.

Acquisition Image partition, make byte-for-byte duplicate copy of the device.

Reconstruction Move the extracted and recorded blocks in the Deconstruction stage to the same location in the image, obtained from the Acquisition stage. The image is then ready for analysis.

The results in Mitchell et al. (2017) exhibited stability and provided reproducibility on a variety of devices, including SSDs. The Deconstruction stage is only executed once on the storage medium. The blocks extracted are stored for the later reconstruction stage, also stored are the location the blocks were extracted from. The Acquisition stage can be executed many times, with verification and reproducibility. The Reconstruction stage can also be executed many times and returns an exact copy of the original storage medium. This duplicate image can be analysed and has the following advantages (McKemmish 1999): allows the DFA to change content and reconstruct events, without damaging the original device; ensures the protection of the original device; allows several DFAs to work on the image simultaneously. Such advantages can only be capitalised on if the reproducibility of digital evidence is reliable and accurate.

3 Method

DaP∀ shares all the aims with DaP, but tries to make improvements by becoming storage device independent and proceduralise DaP (Mitchell et al. 2017).

In Mitchell et al. (2017) it is shown how to stabilise a seized SSD. However, this did not include GPT formatted SSDs. DOS/MBR formats have only 4 partitions, which can be increased by sub-partitions. GPT format allows 128 partitions and thus increased storage capabilities. So, a set of experiments was designed to identify the best GPT component to extract. These experiments comprised of extracting GPT components, which are: Protective MBR (MBR); GPT Header (GPT1); GPT Header and GPT Header Copy (GPT2); Partition Table (PT1); and, Partition Table and Partition table copy (PT2). The experiments then tested the stability of the SSD by reproducing and verifying the hash values.

The location of the GPT components can be found using TSK's (Carrier 2011) `mmls` command, other techniques are covered in Nikkel (2009). Figure 2 gives an overview of the structure of a GPT formatted device, showing individual components.

A database is used to store the component extracted and its associated hash values, e.g., P, $H(P)$, $H(I_0)$ and $H(I_0^{'})$. Authorised access is allowed to Digital Forensic Analysts (DFA) completing subsequent images on the same device, and thus requiring the extracted component for the reconstruction stage. For authorised access to P, the passphrase is set to $H(I_0^{'})$ and provided by hashing of subsequent images, I_i. For example, $H(I_i)$ is sent to the database and if Eq. 1 is satisfied then

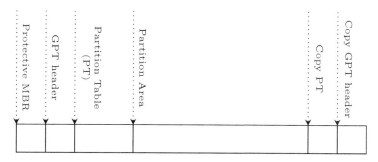

Fig. 2 GUID partition table (GPT) structure. There are four parts, (i) the protective MBR; (ii) GPT header; (iii) partition table; and (iv) partition or content. Copies of the partition table and GPT header are held at the end

access to associated P is permitted; else denied. On reconstruction the verification involves matching hashes from the original reconstructed image, $H(I_0)$, with the subsequent reconstructed images, $H(I'_i)$, left and right parts of Eq. 2, respectively. If Eq. 2 is satisfied, then the reconstruction is a success and analysis of the resulting image, I'_i, can commence; else the reconstruction stage has failed and the error is attributed to incorrect insertion of component, P, in image, I_i.

The procedures explained above are represented in Fig. 1 and as a SOP flowchart in Fig. 3.

4 Results

The experiments were completed on 4 different SSDs as shown in the list below. Each SSD underwent steps explained in the experimental framework described by Bell and Boddington (2010). Each experiment followed the stages explained in the SOP in Fig. 3. The first part was to complete the deconstruction of an identified component for a GPT formatted SSD. The latter part of the experiment was to leave the device for a duration of 7 days and duplicate results. If the hashes of the images matched then the experiment was a success and the deconstruction of GPT formatted SSDs preserved the digital evidence. The 4 SSDs used are listed below:

1. Kingston V300
2. Transcend TS64GSSD370
3. Zheino Q1 30GB mSATA
4. OCZ Agility 3

Control experiments have been completed in Mitchell et al. (2017), from these set of experiments it is known that a DOS/MBR formatted device can be stabilised by extraction of identified blocks. For DOS/MBR formatted SSDs the removal of the MBR for SSDs, and other storage devices, maintained the evidential integrity for multiple data acquisitions of the same device. In other words, the hashes matched and preservation was maintained. The objective of each of these experiments is to achieve the same outcome and discover which component in GPT formatted devices can be removed and preserve the evidence on the SSD.

The set up for all the experiments is detailed in Mitchell et al. (2017) and based on Bell and Boddington (2010). Briefly, the SSD is populated with files containing the repetitive string '01234567'. The files are deleted and then the extraction stage of DaP is completed. This will test if the identified blocks in the experiments preserved and maintain state on the GPT formatted SSD. So, the purpose of all these experiments is to analyse whether the TRIM commands, WL and GC algorithms are rendered ineffective when the identified blocks are removed and wiped.

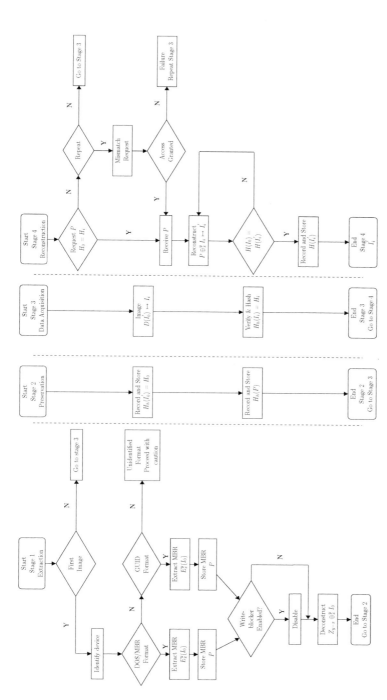

Fig. 3 DaP∀ SoP for stage 1: extraction; there are many assumptions here that include the set up of the Info. Mgt. system, and use of write-blocker. Stage 2: preservation: ensuring that the data on device is preserved by extracting component, P, which needs to be stored and accessed correctly. Stages 1 & 2 are only implemented on the inaugural image and is skipped on subsequent images. Stage 3: Data acquisition. Imaging device with write-blocker and producing output H_t. Stage 4: Reconstruction: password protected at digital forensic analyst level. Access to extract, P, is permitted provided correct $H_h(I_t')$. Case manager can approve access

4.1 Extract Protective MBR (MBR)

The protective MBR is for legacy Operating Systems (OS) that do not support GPT formats and prevents the partition from being reformatted.

In the absence of the protective MBR, it is expected the SSD will be automatically unmounted by the operating system,[1] preventing the TRIM commands, WL and GC algorithms from removing any traces of potential digital evidence stored on the device. The hashes generated from the SSDs, after the deconstruction of the GPT protective MBR, matched the original hash, confirming that the integrity of evidence stored on the SSD had not been compromised. The SSD without the Protective MBR was preserved and results are shown in column MBR in Table 1.

4.2 Extract GPT Header (GPT1)

The GPT header contains the pointers to the GPT partition table and backup copies of GPT and PT.

It is expected that, once the GPT header is removed, the TRIM commands, WL and GC algorithms become ineffective. The SSD without the GPT header became unmountable and unreadable. For all GPT1 extractions the hashes match, confirming that the integrity of evidence stored on the SSD has not been compromised and the results are shown in column GPT1 in Table 1.

Table 1 1–4 are four different TRIM enabled SSDs. A match indicates that $H_1 = H_2$, else a mismatch. The image for H_2, was completed 1 week after the image for H_1. Keys to experiments represent the identified blocks that are deconstructed and are as follows: *MBR* protective MBR, *GPT1* GPT header, *GPT2* GPT header and copy of GPT header, *PT1* partition table, and, *PT2* partition table and copy of partition table

SSD	Experiments				
	MBR	GPT1	GPT2	PT1	PT2
1	Match	Match	Match	Mismatch	Mismatch
2	Match	Match	Match	Mismatch	Mismatch
3	Match	Match	Match	Mismatch	Mismatch
4	Match	Match	Match	Mismatch	Mismatch

---

[1]Depending on the operating system, the drive should be unmounted, e.g. Kali in Forensic mode.

4.3 Extract GPT Header and Copy of the GPT Header (GPT2)

Logically, if GPT1 works then GPT2 should give the same results. It is possible to reconstruct the GPT header using the copy of the GPT header located at the end of the device. With this in mind, this experiment is similar to the GPT1 experiment above. It is expected, once the GPT header and GPT header copy are removed, that the same results are achieved as in GPT1. For all GPT2 extractions, the hashes match, confirming that the integrity of evidence stored on the SSD has not been compromised and the results are shown in column GPT2 in Table 1.

4.4 Extract Partition Table (PT1)

The GPT Partition Table (PT) contains the pointers to the starting and ending LBAs of each partition entry.

The PT was removed and wiped from the SSD. For all PT1 extractions, the hashes mismatch, confirming that the integrity of the evidence stored on the SSD was compromised and the results are shown in column PT1 in Table 1.

4.5 Extract PT and Copy of PT (PT2)

There is a copy of the PT. Both the PT and the PT copy was removed and wiped from the SSD. For all PT2 extractions, the hashes mismatch, confirming that the integrity of the evidence stored on the SSD was compromised and the results are shown in column PT2 in Table 1.

4.6 Summary

The results were surprising, particularly the removal of the PT components resulting in a mismatch, see columns PT1 and PT2 in Table 1. Also, a surprise was the removal of the protective MBR stabilised the SSD. A summary of the results are in Table 2, and the components of GPT formatted devices under consideration for extraction are: protective MBR (MBR); GPT Header (GTP1); and GPT Headers (GPT2).

Finding locations of GPT Headers, both primary and secondary, can take time and is prone to mistakes. It can also add complexities to the investigation when dealing with traditional DOS/MBR and GPT formats. For this reason, it is going to be the recommendation to extract the protective MBR for GPT formatted drives and MBR for DOS/MBR formatted drives, both located at LBA_0. This technique

Table 2 Summary of experiments

Experiment	$H_1 = H_2$
MBR	Match
GPT1	Match
GPT2	Match
PT1	Mismatch
PT2	Mismatch

is format and device neutral. The DFA can image different devices using the same method, DaP∀DaP∀ preserves digital evidence held on portable storage devices[2] and the results in Table 2 shows it preserves the devices for the future and ensures reliable and accurate reproducibility.

4.7 Recommendations and Guidelines

From these experiments a SOP has been developed for data acquisition of all storage devices and is shown as a flowchart in Fig. 3. The key points are described below:

Access Standard precautionary measures should be taken to access device, e.g, follow SOP from U.S. Department of Justice (2009) and Scientific Working Group on Digital Evidence (SWDGE) (2013) until data acquisition.

First Data Acquisition

- store item and case numbers
- Deconstruct
- Record and store $H(P)$ & P
- Image
- Record and store H_0
- Reconstruct
- Analysis
- Report

Subsequent Data Acquisitions

- Image
- Send Hash, H_i
- Match, $H_0 = H_i$
- Receive associated P
- Reconstruct
- verification
- Analysis
- Report

---

[2]Include HDD, Flash drives, SSD and similar storage devices.

5 Conclusions

Bell and Boddington (2010) suggest that continuing current practices for data acquisition of potential digital evidence from storage devices would be, "potentially reckless" and "imprudent". This is due to automatic self-corrosion and through no fault of the DFA. However, knowingly continuing such practices with full knowledge of such errors being produced, is not keeping abreast with new technologies and could be in breach of Professional Practice and borders on incompetence, e.g., see item 6 in Forensic Science Regulator (FSR) (2017).

The challenge set in Bell and Boddington (2010) was to make changes to the evidence acquisition process and this has, in part, been solved by DaP (Mitchell et al. 2017). DaP showed that SSDs were preserved when removing important partition information. This stability allowed further data acquisition to be completed at different times without self-corrosion. However, these trials were limited to a DOS/MBR formatted SSDs and hence the improvements in DaP∀. DaP∀ introduces two important themes: (i) ability to preserve SSD and other storage devices independent of the formatting or file system, and; (ii) a fully developed SOP with the ability to ensure that correct steps are taken in all images of SSDs and other storage devices.

5.1 Contention

Deconstruction stage of DaP∀ could be viewed as contamination, particularly since you are overwriting data on areas of a storage device. In the unlikely event that any additional information has been stored in these areas, it can be retrieved and during reconstruction, it will be repatriated with the original device's image and preserved for future analysis. Altering devices to complete information is not new to Digital Forensics. For example, Mobile Forensics has required the use of uploading software to the device in order to acquire data from the mobile phone (MSAB 2015). Also, memory forensics requires similar techniques to complete memory acquisition (Ligh et al. 2014, ch.19). Both techniques could overwrite important user-generated information and are considered standard. DaP∀ does not lose any information, user-generated or computer-generated, it simply deconstructs, stores and reconstructs.

Further assurance is in place ensure that upon receiving of P, it should be re-hashed to confirm it is correct. The locations (x, y) are also given during this exchange to ensure the correct reconstruction of the image. The overall reconstructed image is hashed and confirmed with the stored hash of the original image. The authors' advice is to complete stage 1 as early as possibly allowed in the investigation, and thus preserve any data that may be lost due to self-corrosion, King and Vidas (2011) shows that an SSD can be wiped efficiently and permanently under certain conditions.

5.2 Discussion

Employing DaP∀ will allow all future images to be bit-for-bit identical. New training for employees is expected, but this is consistent with any new SOP. DaP∀ rises to the challenge originally detailed in Bell and Boddington (2010) and results show that this is met. DaP∀ is proposed here as a new SOP to forensically preserve storage devices and reduce the risk of contamination, both due to human error and self-corrosion. The additional database storage of verified hash values allows an audit trail for showing any mistakes and elucidates at what stage those mistakes occurred. Each stage records one or two hash values and verifies that these are correct when future duplicate acquisitions are required.

The main contribution of this research is to develop a new SOP for the data acquisition stage for a wide range of storage devices, which will optimise the preservation of evidence. The results of this research show that the simple deconstruction of a single block from the device, LBA_0, in either the MBR or the protective MBR results in the stabilisation of the device. If all other procedures are followed, e.g., use of write prevention devices, then the subsequent images of the device yielded from data acquisition will be identical.

Finally, with the advent of sales of SSDs (Statista.com 2016) there is a need for a solution to the preservation of evidence and it is recommended that DaP∀ resolves this issue.

References

180-1, F.I.P.S.F. (1996). Secure hash standard.

Bell, G. B., & Boddington, R. (2010). Solid state drives: The beginning of the end for current practice in digital forensic recovery? *Journal of Digital Forensics, Security and Law, 5*(3), 1–20.

Carrier, B. (2005). *File system: Forensic analysis*. Boston: Addison-Wesley.

Carrier, B. (2011). The sleuth kit. TSK – sleuthkit.org.

DCFLDD 1.3.4-1. (2013). *Test results for digital data aquisition tool* (Technical report), Homeland Security.

Forensic Science Regulator (FSR). (2017). *Codes of practice and conduct for forensic science providers and practitioners in the criminal justice system* (Technical report), UK Govt, Birmingham.

Harbour, N. (2002). dcfldd. Defense Computer Forensics Lab. http:/dcfldd.sourceforge.net 5(5.2), 1.

King, C., & Vidas, T. (2011). Empirical analysis of solid state disk data retention when used with contemporary operating systems. *Journal of Digital Investigation, 8*, S111–S117.

Krishna Mylavarapu, S., Choudhuri, S., Shrivastava, A., Lee, J., Givargis, T. (2009). Fsaf: File system aware flash translation layer for nand flash memories. In: *Design, Automation & Test in Europe Conference & Exhibition, 2009. DATE'09* (pp. 399–404). IEEE.

Ligh, M. H., Case, A., Levy, J., & Walters, A. (2014). *The art of memory forensics*. Indianapolis: Wiley.

McKemmish, R. (1999). *What is forensic computing?* (Trends and issues in crime and criminal justice, Vol. 118). Canberra: Australian Institute of Criminology.

Mitchell, I., Anandaraja, T., Hadzhinenov, G., Hara, S., & Neilson, D. (2017). Deconstruct and preserve (DaP): A method for the preservation of digital evidence on solid state drives (SSD). In *Global Security, Safety and Sustainability – The Security Challenges of the Connected World*

MSAB. (2015). XRY – Android basics: Debugging and extractions, available on XRY certification course.

Nikkel, B. (2009). Forensic analysis of GPT disks and guid partition tables. *Digital Investigation, 6*, 39–47.

Nisbet, A., Lawrence, S., & Ruff, M. (2013). A forensic analysis and comparison of solid state drive data retention with trim enabled file systems. In: *Australian Digital Forensics Conference* (pp. 103–11).

Scientific Working Group on Digital Evidence (SWDGE) (2013). *Model standard operation procedures for computer forensics (ver. 3)*. https://www.swgde.org/.

Shu, F., & Obr, N. (2007). Data set management commands proposal for ata8-acs2. *Management, 2*, 1.

Statista.com. (2016). *Global shipments of HDDs and SSDs in PCs from 2012 to 2017*. http://www.statista.com/statistics/285474/hdds-and-ssds-in-pcs-global-shipments-2012-2017/. Accessed June 2016.

Subramani, R., Swapnil, H., Thakur, N., Radhakrishnan, B., & Puttaiah, K. (2013). Garbage collection algorithms for nand flash memory devices–An overview. In *2013 European Modelling Symposium (EMS)* (pp. 81–86). IEEE.

Sylve, J., Case, A., Marziale, L., Richard, G. G. (2012). Acquisition and analysis of volatile memory from android devices. *Digital Investigations, 8*, 1–10.

Tableau sata/ide bridge (March 2018). https://www.guidancesoftware.com/tableau/hardware//t35u.

U.S. Department of Justice. (2009). *Electronic crime scene investigation: An on-the-scene reference for first responders*. National Institute of Justice, November 2009.

Williams, J. (2012). *Good practice guide for digital evidence* (Technical report), Association of Chief Police Officers (ACPO). http://library.college.police.uk/docs/acpo/digital-evidence-2012.pdf. Accessed March 2018.

Part IV
Education, Training and Awareness in Cybercrime Prevention

An Examination into the Effect of Early Education on Cyber Security Awareness Within the U.K.

Timothy Brittan, Hamid Jahankhani, and John McCarthy

1 Introduction

Cyber security awareness and general cyber skills are becoming a necessity to being able to obtain a career in almost every industry within the UK. The House of Commons Science and Technology Committee, (2016), Digital skills crisis Second Report of Session 2016–2017, has reported that, The UK will need 745,000 additional workers with digital skills to meet rising demand from employers between 2013 and 2017, and almost 90% of new jobs require digital skills to some degree. However, the skills gap from what is required within businesses and the graduates entering the workplace are increasing year on year. "... but opportunities are often constrained by a lack of relevant digital skills within the labour force. As demand for digital skills outstrips supply, employers across a wider range of sectors are experiencing digital skill gaps within their workforce, and encountering difficulties in filling advertised vacancies...", 'DIGITAL SKILLS for the UK ECONOMY', ECORYS UK (on behalf of the department for business innovation & skills, and department for Culture media & sport), (2016).

The typical practice for a new employee when they start work, after leaving school/ college/ university, they may get an initial information security briefing and they may get a refresher each year thereafter. "*46% of Organizations that provide ongoing information security awareness training beyond new starter Induction...*", 'Cyber Resilience: Are your people your most effective defence?', IPSOS MORI/Axelos, (2016). Some companies carry out regular IT awareness

---

T. Brittan · J. McCarthy
Northumbria University London and QAHE, London, UK

H. Jahankhani (✉)
QAHE and Northumbria University, Northumbria University London, London, UK
e-mail: hamid.jahankhani@northumbria.ac.uk

© Springer Nature Switzerland AG 2018
H. Jahankhani (ed.), *Cyber Criminology*, Advanced Sciences and Technologies for Security Applications, https://doi.org/10.1007/978-3-319-97181-0_14

campaigns; however, these are few and far between. According to the Institute of Directors, (2016), *"49% said they provided cyber awareness training for staff"*, 'Cyber Security Underpinning the digital economy'. However *"75% of employer's state that they are unwilling to interview candidates who do not have basic IT skills"*, 'A Leading Digital Nation by 2020: Calculating the cost of delivering online skills for all', McDonald (2014), for Tinder Foundation (now known as Good Things Foundation).

There is numerous research reporting that a number of companies carry out a one-off or annual cyber awareness program. In general, they appear to make a difference in the immediate aftermath, but rapidly trail off as the days, months/years go by. If the subject matter was ingrained into users whilst they are immersed in education, then it would become second nature to them when they enter the workforce. Coventry et al. (2014), using behavioural insights to improve the public's use of cyber security best practices has highlighted that "There is a need to move from awareness to tangible behaviours. Governments and Organizations need to be secure by default".

The government has responded by laying out a very well thought through programme of education, through all the key stages. This was created with advice from industry experts across education and technology as well as what the current business requirements are as far as skilled workers. This programme encompasses all age groups and goes into great detail over the General Certificate of Secondary Education (GCSE) which is an academic qualification, generally taken in a number of subjects by pupils aged 14–16 in secondary education in England, Wales and Northern Ireland and also at Advanced level (A-Level) in a number of subjects by pupils aged 16–18.

With the new era of Internet of Things, (IoT)), which is essentially *"...an ecosystem of discrete computing devices with sensors connected through the infrastructure of the internet."*, 'A brief history of IOT and computing', Information-age.com, (2017), with everything being connected, there is now a necessity to move ICT education into a more professional approach by making it one of the core subjects that should be taught to all students by specialists in the field.

2 Curriculum and Resources

The U.K. Government have published guides, publications as well as research in order to formulate and produce the new Pre-GCSE, GCSE & A-Level Computer Science curriculums. These were created after taking advice from numerous educators and business leaders, for their requirements and from across the country. The 'Computer Science GCSE subject content, lays out the aims and learning outcomes, however, the objectives provided by the government is sparse as shown in the Assessment Objectives shown in Fig. 1.

Although it has been formed into curriculums by the various UK examination boards, their curriculums are all based on the following Modules: System Architecture; Memory; Storage; Wired & Wireless Networks; Topologies, Protocols &

Computer science GCSE subject content		
Objectives	Requirements	Weighting
AO1	Demonstrate knowledge and understanding of the key concepts and principles of Computer Science	30%
AO2	Apply knowledge and understanding of key concepts and principles of computer science	40%
AO3	Analyse problems in computational terms: • To make reasoned judgement • To design, program, evaluate and refine solutions	30%

Fig. 1 Computer science GCSE subject content, Department for Education (2015)

Layers; System Security System Software; Ethical, Legal, Cultural & Environmental Concerns; Algorithms; Programming Techniques; Robust Programming; Computational Logic; Translators and Facilities of Language; Data Representation.

In order to deliver the curriculum UK examination board and other agencies including private training organisations have been providing materials for teachers/trainers to deliver the required objectives. These materials delivered through online and offline resources, under licence to act as a pool for teachers to educate in all matters related either to computer science or freely through individuals or educational organisations in order to support teachers and some cases pupils.

The aim of this chapter is to highlight the following;

- What is the effect the new syllabus has on improving the cyber awareness, within Key Stage 3 (KS3) & KS4 education?
- Does the KS3 curriculum improve the cyber awareness of those not taking GCSE computer Science.
- Can we provide a better framework for ensuring that those teaching ICT have the knowledge and skills required to deliver?
- Is there a need to seek external help from cyber security experts, to get the scheme up and running to a sufficient standard?

It is important to highlight that someone who happened to have a vague interest with IT has commonly taught ICT or was a maths or physics teacher that thought of as the best person available to teach the subject. Current statistics by the Department for Education ('School workforce in England', 9 November 2016.https://www.gov.uk/government/statistics/school-workforce-in-england-november-2016, backs this up by showing that only 30% of ICT teachers hold a relevant ICT degree or higher, and almost half (49.6%) hold no relevant post A level qualification.

3 Academic Studies Journals

In terms of academic studies, there are a few from different parts of the globe which have been used to help form the background literature for this research. There have been a number of studies into forming a cyber-security awareness programme, such as Eshet-Alkalai, (2004), 'Digital Literacy: A Conceptual Framework for Survival Skills in the digital Era'; and Basham and Rosado, (2005), 'A Qualitative Analysis of Computer Security Education and Training in the United States'. Which cover a lot of cyber security literature from the early noughties, as well as forming a framework for creating a programme of study for graduates and post graduates. These types of studies have helped form the cyber security training industry, but all are focused on the post-school education system.

There has recently been a follow-up study by the Royal Society, (2017), 'After the reboot: computing education in UK schools' into the implementation of the new computing curriculum through all age groups. This is an exceptionally detailed report on how well we believe the new curriculum is performing. The report includes several recommendations to ensure it will be a success, such as – *"Ofqual and the government should work urgently with the learned societies in computing, awarding bodies, and other stakeholder groups, to ensure that the range of qualifications includes pathways suitable for all pupils, with an immediate focus on information technology qualifications at Key Stage 4"*. However, the report doesn't directly inquire specifically on the issues with cyber security awareness, where this study attempts to address, although it does generally cover all aspects of the computing curriculum.

The study by the Royal Society (2017) also investigated the uptake of teachers into computer science and their background prior to teaching. The report also explains in depth the diversity within the uptake of GCSE Computer Science, or the lack thereof, from ethnicity, sex, social background and location (urban or rural). It discusses the potential disparity between living in an area which is more likely to have an increased number of technical jobs, therefore an increased number of children requiring technical skills. As well as the likelihood of having fast and reliable internet access and resources.

Another study 'A Recommended ICT Curriculum for K-12 education', Hu et al. (2014), carried out in Taiwan focuses on the whole school education system in Taiwan from primary up until school leaving age (typically 17–18 years old). It describes the implementation of their system *"... The curriculum contained a required course, called Introduction to Information Technology and three elective courses – Basic Programming, Advanced Programming, and Topics in Computer Science. The curriculum outline for Introduction to Information Technology revealed six themes including introduction, hardware, software, networks, problem solving, and computer and society..."*. This study was carried out after the Taiwanese 2006 version of Computer science was implemented and going through a review.

Brady (2010) research on 'Security Awareness for Children' gives a technical report on implementing a Cyber security awareness programme for children and demonstrates why it needs to be specifically targeted towards children. Within the report there is a framework for creating a cyber-awareness programme for children (based on 10 to 12-year olds) and the obstacles implementing such a scheme. The author is both a cyber-security professional and a primary school teacher so has insight and familiarity of both disciplines. Although the study was carried out in

Ireland from 2010 and on primary school children, it does bring an understanding of the opinions of parents, teachers and pupils on their understanding of being safe online.

The study by Brady, (2010), raises some questions on how cyber awareness is taught such as: Why are teachers unwilling, or not encouraged to carry out research into the topic they are teaching? *"Almost all teachers who participated in the survey (92.45% n = 49) replied that they did not research information on such initiatives"*. If they are not motivated to research the topic they are teaching, then they should not be teaching it. If it is a question of time to research the subject matter, then clearly there is a need to make it more prominent.

In reference to child e-safety initiatives; Why aren't parents informed of the initiatives? "A large majority of the sample population revealed that they never received information (81.08% n=90) or researched information (87.39% n = 97) on child e-safety initiatives. 80.18% of the parents responded that they would like more information on how to teach their child to surf the Internet safely", (Brady 2010). As well as how they possibly should be taught which in turn supports this research by showing where the level of cyber awareness in children should be at by the time they reach the beginnings of GCSE's i.e. secondary school age 12.

The UK Government have commissioned many studies into cyber security, digital skills, and future technology requirements (in relation to employers and employees). The majority of these reference education as a prime area in need of improvement although all of them are targeting the future technically able, rather than the future technical user, by concentrating on those who take up ICT as a qualification.

In 2012 the UK government published a paper along with the Royal Academy of Engineering (Royal Society) on the state of ICT within education called 'Shut down or restart? The way forward for Computing in UK schools'. The report examined the then ICT curriculum, which was a follow up to a government study into the 2011–2012, review of the National Curriculum in England. This study discovered major issues within the curriculum as it was and made several recommendations on how GCSE ICT should be improved in order to create a better educated pupil (in terms of computer science).

The report raised many issues and potential solutions but one of the findings of note was "... there needs to be recognition that Computer Science is a rigorous academic discipline of great importance to the future careers of many pupils. The status of Computing in schools needs to be recognised and raised by government and senior management in schools...".

Studies such as 'Wakeham Review of STEM Degree Provision and Graduate Employability', (2016), and 'Shadbolt Review of Computer Sciences Degree Accreditation and Graduate Employability', (2016), followed on from the earlier studies by both the government and Royal Society. These reviews focus on the need to have technically able graduates and employees entering the workforce, and show little about the technical abilities of those entering the workforce from a none technical discipline (at University level). From these reports, the government commissioned a further investigation into the 'Digital skills crisis', House of

Commons Science and Technology Committee, (2016), which examines the current lack of digital skills within the UK population. The report points at a lack of skills within the teaching of Computer science – *"many ICT teachers still do not have the qualifications or the knowledge to teach the computing curriculum"* as well as defining ways of counteracting this issue.

The UK Council for Child Internet Safety (UKCCIS) is a working group of more than 200 organisations whose goal is to help children stay safe online. They come from all sectors – government, industry, law, academia and charities. They carry out studies and research into all aspects of children's activity online. Primarily they cover children's activities online, and parental understanding of online security measures, through surveying both children and parents. In 2017, they published a paper on 'Children's online activities, risks and safety', (UKCCIS 2017). The study goes into detail about children's activities online, how their internet use is regulated and what they do online. One of the findings of this report demonstrates the importance of cyber security awareness *"Efforts to develop children's digital resilience (digital resilience – being able to 'negotiate online risk environments') should focus on critical ability and technical competency in order to support children in becoming active agents in their own protection and safety."*

4 A Psychological Viewpoint

As psychology is an important factor in how education is applied, to ensure a cyber-security awareness curriculum is implemented correctly, we should look to psychological studies, in particular the psychology of children, and to restrict the research to only articles which relates to this study. However, this proved to be an area with very few studies. Coventry et al. (2014), 'Using behavioural insights to improve the public's use of cyber security best practices', although written with the general public in mind as opposed to children is possibly the most relevant publication here. It covers the current basic best practices for users, why users do not comply with them, and what influencers (Social, Environmental, Personal) cause the non-compliance. It looks at current methods of improving cyber security behaviour and the use of the MINDSPACE framework as a way to influence user's behaviour.

The MINDSPACE framework is designed to alter user behaviour by making better public policies through scientific research to enforce improvement in various areas (predominantly economical). This could be applied to both teachers and students, in order to nudge them forwards by teachers to be persuaded into taking more of an interest in the topic of cyber security, which could influence the pupils they are teaching; Students could be encouraged to take the subject further than just KS3.

Coventry et al. (2014), applies this framework to create examples on how to improve cyber security awareness for the public through psychology. It goes on to summarise psychological influencers *"Good design is fundamental, and security must be designed in from the start. Security should not rely on the knowledge and behaviours of end-users and attempts should continue to be made to ensure people are secure by default."*, 'Using behavioural insights to improve the public's use of cyber security best practices'. Through this statement alone, this research believes that targeting cyber security awareness at an early age for all will aid in the improvement of the security of our society. As we are in an age of ever-increasing technology, the earlier children are taught about the risks and ethics, the easier the knowledge required will be ingrained into them by the time they are adults.

Despite gaps in literature for this research, however, there is more than enough materials to cover the majority of what is needed even though the content covered was indirectly related. Every report pointed to a need for better ICT education from various angles, however very little is written about Cyber Security Awareness – proving there is a need to produce research within this very select area.

A psychological viewpoint especially one carried out with a focus on Cyber Security awareness proves valuable insight into what influences people's behaviour, as well as ways to counteract bad practice. Although there are significant differences between the psychological behaviour of adults compared with children, or even children of differing age groups, it does give an overall perception on what influences the decisions people make.

After reviewing the literature for the curricula and the materials available to teach it, there appears to be a heavy emphasis on teaching coding over the other sub-topics. This is partly due to the complexity of teaching (and learning) this skill but also due to the limited time they have to teach the whole syllabus.

The current GCSE Syllabus, and most literature contains nine chapters teaching coding from the principals, to the fundamentals, to application of those skills. Which may aid the production of a number of future programmers, which is what the government aims to achieve, but only if there is enough uptake of the course. One which appears to rely almost completely on programming will discourage those who have an interest in I.T. but feel that learning a programming language may be either beyond them or be yet enough difficult subject on top of the pupils already challenging workload. There is typically only a single chapter on cyber awareness in GCSE Study books.

The Syllabus itself has a single paragraph determining what is required in terms of Cyber awareness – "use technology safely, respectfully and responsibly; recognise acceptable/unacceptable behaviour; identify a range of ways to report concerns about content and contact", 'Computing programmes of study: key stages 3 and 4.', Department for Education, (2013).

5 Research Design

This research is centred around collecting information from existing primary and secondary research. Firstly, from existing whitepapers and publications within the scope of 'cyber awareness education'. Secondly, from research on previous GCSE, pre-GCSE, GCSE and post-GCSE curriculums, along with the current best practices for cyber awareness training in the workplace.

The initial research is used to form the basis of an ordinal-polytomous questionnaire with an open answer category to provide those questioned an opportunity to provide their own opinion. This will assist the process in preventing bias created by the researcher. It will be targeted at those in education (teachers/headmasters), business education (companies that specialise in user awareness), and cyber training in schools (companies that specialise in training seminars for schools) within the UK.

Following the initial research, a questionnaire formed to garner the opinion of those in prime positions – I.E. Teachers, Head Teachers and School governors/executives – to gain primary research data. Some of the research will also focus on the best methods/ practices of delivering a cyber-awareness scheme and the psychology behind it.

By using both primary and secondary research data and using an ordinal-polytomous (Ordinal-polytomous – where the respondent has more than two ordered options) questionnaire, the researcher was able to utilise previous studies and operations and discover their effectiveness along with current opinion of those who are teaching the existing courses. Due to the recent upheaval of the curricula (GCSE Computer Science replaced the previous GCSE ICT in 2016, also, the whole ICT curriculum for all ages groups was revised at the same time), using this approach as opposed to only using secondary data would ensure the research is as up-to-date as possible. A quantitative style would provide pure statistical data that is easy for analysing and would be ideal if there had been plenty of earlier studies in this area, as it would be more likely to have covered the majority of possible questions.

This approach does have its drawbacks, for example, a lack of responses from the questionnaire would mean a lack of primary data. Also, due to the fact the curriculum has only recently been changed, a number of respondents may already have a bias against it, if they do not like change.

The use of a Likert-type scale (Likert-type scale – rating scale typically from 1 to 5, or 1 to 10), or a table style used in many questions was in order to gather the ordinal data for quantitative analysis. The open questions were formed to try and prevent bias on the part of the questioner, whilst at the same time be directive – They were phrased in a way to encourage the users to not just put one-word answers, as well as notice the slight differences between the questions and to what they pertain to. This was done with a target of ensuring they answer the specific part of the question, for example pre-GCSE, not GCSE, or how its Taught rather than how well its Grasped (by teacher, student, or general public).

6 Delivery Methods

The approach to gain responses to the questionnaire was through researching and collecting contact details for over 3000 schools within the U.K. This was done through the governments transparency data which contained a list of schools within the U.K. the Head teacher and their websites. Using Gmail as an email provider for convenience of use (it works seamlessly with Google Forms) and using a template created for the task, the researcher was able to privately email every school on the list.

Following on from this, a secondary approach was to sign up to Computing at School – a forum dedicated to teachers of ICT and ICT support personnel at schools – and post a request to complete the questionnaire. Finally, the researcher began calling schools directly asking for their participation.

7 Critical Analysis and Discussion

The responses to the questionnaire were broken up into two sections. The first consisting of the fifteen closed questions to be used for quantitative analysis, which enables the research to have a statistical backbone. The second section containing the six open-ended questions, is used to gain an overall opinion on missing elements and individual issues with the curriculum to form the qualitative analysis.

The overall lack of responses (n = 10, % of requests = 0.33%) to the questionnaire compared to the number of forms sent out possibly shows an apathy to either the subject matter, the format, the method it was sent or it was due to the time of year it was received (over the busiest term of the year for teachers). This was earmarked as the most prominent risk during the planning phase of the research. The majority of responses came from IT Teachers, with no head teachers participating, and one school governor. Despite this, researcher believes there is enough data to form an opinion on the current state of cyber security awareness of children, the system that educates them, and what can be done to improve it.

One of the first questions asked was to rate the respondent's own knowledge of cyber security, on a Likert scale of 1–5, to gauge the audiences level of understanding of a topic which they have to teach, or in the case of non-teaching participants, what they expect others to teach. The majority fell into the average/above average category, which shows they have at least a fundamental understanding of what is required.

When a similar question was asked about their students the overwhelming response was extremely poor or poor, with only one selecting ok/average. This shows that teachers who have an average understanding of cyber security believe their pupils knowledge of the subject to be grossly inferior to their own.

From this, we can deduce that pupil's overall knowledge of the subject is poor, but the reasons behind it may be something which they cannot control. In order to resolve the root cause of this, further questions were asked about the state of the current curriculum, at both pre-GCSE and GCSE levels. The pre-GCSE curriculum was rated at ok by a third of respondents, with the rest rating it at poor or extremely poor. However, when it came to the GCSE syllabus, the ratings were much higher – 66% saying it was average to above average standard. Thus, showing a disparity between what is taught to all students (pre-GCSE) and what is taught to those with a technological interest (GCSE).

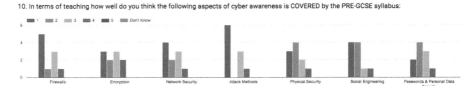

Fig. 2 'An examination into the effect of early education on cyber security awareness within the U.K., cyber security questionnaire', Brittan (2018)

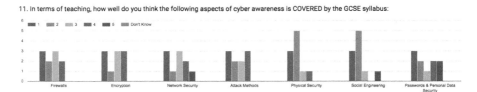

Fig. 3 'An examination into the effect of early education on cyber security awareness within the U.K., cyber security questionnaire', Brittan (2018)

This was then further broken down into subject matter areas (as directed by the current GCSE Computer Science syllabus): Firewalls; Encryption; Network Security; Attack methods; Physical Security; Social Engineering; Passwords and personal security. They were then asked how they rated how well each topic was covered in both pre-GCSE and GCSE curricula.

The responses as shown in Figs. 2 and 3, in the pre-GCSE question, nearly all of the responses for each subject were average/below average, with some scoring an abysmal 1 in the majority of responses. The GCSE response faired marginally better, although three of the most prominent topics, that almost every person will need as an adult in the modern world – Physical Security and Social Engineering were rated the lowest on average, however the diversity of opinion on the quality of the syllabus is worthy of further investigation. On closer examination, the teaching respondents scored it lower than the non-teachers.

This shows the opinion of the very people teaching the subject believe the coverage of cyber security material is decidedly poor. With some of the most basic skills that every adult has to learn, through company cyber awareness schemes, are not being covered to an adequate standard at an age when it can be more easily influenced. Jean Piaget (1964), who specialised in child development sums up what education should achieve: *"The principal goal of education is to create people who are capable of doing new things, not simply of repeating what other generations have done—people who are creative, inventive, and discoverer"*.

A follow-on question in the survey, was asked – "In terms of teaching, how well do you think the following aspects of cyber security awareness is being Taught" – although this can be very subjective as we're asking the very people teaching the subject, how well do they think they are doing!' A surprisingly candid response from the majority showing that most topics were not being taught to an 'ok' standard.

In order to discover the reasons behind such low ratings, a question on the teaching materials available was asked. The majority replied with a score of 1 (Likert 1–5 scale), with the others selecting either 2, or 3. This contradicts this researcher's opinions on the material which was uncovered during the literature review stage of the project, which was found to be both varied and in-depth. Therefore, prompting the need for further study into why the materials are not being made available, or being publicised enough, or why the students are not being engaged through the current material.

In the penultimate section of quantitative data, they were asked if they believed training of the teachers by cyber security experts and material that had been published by cyber security experts would be more or less effective than the current schemes/resources. Although these questions are slightly presumptuous – how would they know, whether it would be better without experiencing it? – they were designed to gauge their opinion on the quality of materials/training that was currently being offered. The overwhelming response was that it would be more effective in both cases (70%). This shows that there is a genuine need for dedicated expertise to be deployed in this area of computer science education.

The final section covered the relation of cyber awareness compared to life skills and other soft subjects (classified as those which are not core subjects), as well as general IT literacy. When it came to directly comparing it to other subjects, it was generally considered as either the same, or more important than subjects such as General IT awareness, Design Technology, the Arts, Humanities, Modern Foreign Languages, and other soft subjects.

When the question of how highly it was regarded as a required life skill, more than 80% regarded it as above average, or vital. Showing there is a desire for it to become a core part of the curriculum. As a comparison, they were asked a question to rate 'general IT ability' on a similar scale, which was also overwhelmingly regarded as an important or vital life skill. Cross referencing these responses to the ones on the pre-GCSE and GCSE curricula, this shows that there is a need for both specialist content as well as general IT literacy to be a part of what is taught to all students, not just those already with an ability or an interest within IT.

The qualitative section could easily be contained on a separate survey of its own and ideally would need far more responses to discover any realistic patterns in the answers. However, there is a need to ensure completeness of the survey and grant the opportunity to the respondents to give their genuine opinion without constraint. From the questionnaire, there were six open answer questions which can be used for the qualitative analysis.

- The first question asked respondents: What critical aspects of cyber security do you think is missing from either the pre-GCSE or GCSE syllabus?

The responses consisted of one common theme of spotting and avoiding cyber-attacks. There is a clear need to train people on what constitutes normal and abnormal behaviour of their devices. This would help them understand when they are potentially facing a cyber-attack or know when it's actually just an old piece of kit that is slowing down due to numerous other issues such as modern applications requiring more processing power.

- Another question asked: In terms of teaching cyber security, what do you think is the most difficult aspect to teach?

This resulted in some similarities in the answers, with issues on teaching 'malware' and the 'technical aspects' as well as the thrill (or lack thereof) of learning hacking. One of the most interesting responses was *"Most of it. Not everyone wants to/ is capable of learning Computer Science to a good enough standard to indulge in cyber security work or programming so it's hard to teach beyond the theoretical"*. Which raises several further questions about the standard of the GCSE course – should it be more difficult? Is it too hard to teach? Another response: *"Overall awareness. Students use tech daily and believe they are safe. Difficult to convince them of the risks."* Shows a need to ensure cyber security is given a more prominent position in the curriculum.

- The follow-on question asked: What is the most difficult aspect of cyber security for teachers to grasp?

They responded again with some similarity based around the number of ways a system can be compromised as well as putting it into a language the kids can understand. Once again showing the complexity of cyber security – If the teachers struggle with fully understanding the content that is in the curriculum, how are they expected to be able to adequately teach the same content to children? Again, one of the responses stands out, pointing to the technical complexity of the subject *"Teachers themselves don't necessarily understand the risks. Cyber security is a career in itself"*.

These raise further questions on whether GCSE computer science is too broad a subject to teach? For those of us who work within the IT industry we know there are overlaps over certain roles within the industry, and at certain levels there is no need to be an expert in a particular field. However, there are many different aspects of IT, all of which are as complex as each other and require specialist knowledge and training. Should the syllabus be broken up into several 'modules' which can be put together in different ways to achieve the required grade? Or would this lead to the more technically difficult aspects get dropped in order to just obtain the grade?

- In your opinion what would be the one thing the government could change that would improve teachers' ability to teach cyber awareness?

There were many commonalities based around resources and training with references to cost, time and material. There was again, one particular comment that stands out, pointing to the complexity of the subject *"Actually train teachers and Head teachers alongside each other. They have literally no concept of how hard this stuff is to teach yet it still only gets about an hour or a lesson per week at KS3 nationally"*. – This also shows of a lack of understanding of what is required to teach the technical aspects between those who determine the timetable/schedule, to those who teach the subject.

A further point, that was made more than once, was to make it compulsory. Showing that the teachers want to spend the time on training and teaching the subject, but because of its relative unimportance within the entire KS3 & KS4 curricula, there is not enough time/resources to dedicate to it.

- In your opinion what would be the one thing the government could change that would improve students' ability to retain cyber awareness knowledge?

There were a variety of answers, yet again, time and resources being mentioned as well as how relevant it was to the real world, but also some interesting points on how to do this:

> A cyber security game or play-pen where students could try and break into systems
>
> Access to real tools with risk precautions. Can't be taught via theory alone
>
> Exciting resources that bring the threat real to them.... Their data has been hacked!.

A similar question was asked – how to make a cyber-security awareness scheme work?

Many responded to similar answers as to the question above, with a high emphasis on making it fun, interactive and interesting. There were a couple of responses which both asked for gamifying the material as well. There were also more requests for time and resources, as well as *"A much bigger focus on this topic rather than as a minor topic."* Again, suggesting cyber security is too big a subject to be shoe-horned into the encompassing subject as computer science. One respondent asked for more cyber professionals to work with teachers to both the questions on the government's involvement.

Although from a limited breadth of data, there were many similarities in the responses given to both the quantitative and qualitative aspects of the survey. There appears to be a large emphasis on the lack of resources available to teachers as well as training and time. As far as resources are concerned this researcher initially believed there was both good quality, depth and variety of material that was available. However, on reflection the researcher is a cyber-security practitioner, with an obvious interest within the field, whereas the audience are far younger with little or no knowledge in the area. The materials need to be catered for the audience and designed to peek their interest without over-glossing the technical aspects.

There also appears to be a high regard for the importance of cyber security awareness, not only within GCSE Computer Science, but as part of the whole pre-GCSE and GCSE curriculum.

The teachers are aware of the importance of the subject within the modern world and want to invest more time into it but feel there is a lack of emphasis as directed by the current curriculum. There is also a sense that cyber security is a topic that needs more prominence throughout the school, not only for the children, but all teachers to create a *"...best practice by all teachers throughout school"* and *"A Culture of being cyber secure..."*, as methods to make a genuine awareness scheme work.

There is a possible bias in the sense that those who responded to the survey have an interest in this area and felt the need to reply, whereas some of those who didn't respond perhaps did so as they believe the current content suffices. A much larger number of responses would help minimise this bias.

8 Conclusion

From this research on cyber education, there does appear to be numerous companies, agencies and governmental guidelines into the training of children in cyber awareness. However, this researcher has found there to be very limited number of studies into cyber awareness in secondary education and almost none into cyber awareness within the U.K. There is a clear need for further study within Information technology education as a whole, as was pointed out by the Royal Society study 'After the reboot: computing education in UK schools', however this study proves there is a lot more work required within the security aspect alone before we reach the government's goal of closing the digital skills gap.

The Curriculum for The Pre-GCSE part (KS3) is clearly aimed at building knowledge for those going on to take the subject for GCSE, which should be a good way of building rapport with the students and the subject matter. However, from this study, there seems to be a disparity between the pre-GCSE content (or at least the available resources) and what is taught for GCSE, thus, losing the interest of many students before they even try it as one of their chosen subjects. According to the Royal Society, (2017) report Fig. 4, it can clearly be seen that the uptake for GCSE computing (both as GCSE IT and GCSE Computer Science) has dropped considerably since the change:

- In 2017 only 12% of those taking GCSE's took up Computer science,

ICT, Computing, History, Geography, Business Studies and Mathematics GCSE qualifications taken in England, Wales and Northern Ireland (2012 – 2017). All UK candidates aged 16						
	2012	2012	2014	2015	2016	2017
ICT	46,471	63,832	87,512	103,342	78,161	69,008
Computing	-	3,867	15,842	33,607	60,146	65,205
History	209,566	243,852	244,988	237,378	252,075	250,590
Geography	175,319	208,447	214,815	218,685	235,818	240,616
Business Studies	65,987	71,888	85,161	91,383	90,169	89,192
Mathematics	491,777	512,312	596,524	596,767	570,459	573,822
As Mathematics is mandatory, it has been included to provide a relative indicator of cohort size. Note: the Joint Council for Qualifications (JCQ) uses the category 'Computing' to include all GCSE qualifications in computing and computer science. Source JCQ						

Fig. 4 After the reboot: computing education in UK schools, Royal Society, (2017)

- Which is a marked improvement from the 9% who took it in 2012,
- In 2015 it was at high of 17.3%. – GCSE ICT and Computer Science available as options.

After dissecting the whole GCSE computer Science curriculum, and speaking to those who teach it, there is a heavy emphasis on programming both theoretically and practically, which creates an offset of having a lack of cyber awareness being taught. Due to the technicality of learning programming and the amount of time it takes to learn a language, this causes all of the other sub-topics to be marginalised. Understandably, from a teacher's point of view, they have a need to get their students to gain as high a grade as possible, and it is easy for anyone to see that by concentrating on programming, a pass mark is easier to achieve.

The number of responses urging for cyber security awareness to be given more of a priority (at all levels) shows that what is being taught currently is not enough. *"With a mismatch of knowledge, fears and expectations between parents and their children, and with technology developing at such a tremendous pace, children must be equipped from an early age to engage safely and resiliently with the internet."*, 'Growing up Digital', Children's commissioner, (2017).

A mandatory cyber security awareness program for all students would both empower those who intend on entering a non-technological field (which still requires some ICT skills) as well as give a better grounding for those wanting to find work within the tech sector. This may in turn help encourage more students to take up GCSE computer science by giving a better foundation of cyber skills and lessen the gap between the pre-GCSE and the GCSE syllabus. It would also help reduce the gap between those who are technically able and those who currently are not.

The prominence of cyber security awareness was clearly a big factor in the responses of the survey. A trial adding cyber security awareness as part of the core content (outside of computer science) to be learnt through both key stages 3 and 4, or perhaps, even the earlier (and later) key stages, would show what kind of improvement gains could be achieved.

References

Basham, M., & Rosado, A. (2005). A qualitative analysis of computer security education and training in the United States. *Journal of Security Education, 1*(2–3), 81–116.

Brady, C. (2010). *Security awareness for children*. Technical report RHUL–MA–2010–05 31st March 2010. Department of Mathematics Royal Holloway, University of London. https://www.ma.rhul.ac.uk/static/techrep/2010/RHUL-MA-2010-05.pdf. Accessed Jan 2018.

Brittan, T. (2018). *An examination into the effect of early education on cyber security awareness within the U.K., cyber security questionnaire*. MSc dissertation, Northumbria University.

Chaudron, S. (2015). *Young Children (0–8) and digital technology – A qualitative exploratory study across seven countries*. European Commission Joint Research Centre Institute for the Protection and Security of the Citizen, file:///C:/Users/wwjf6/Downloads/lbna27052enn.pdf. Accessed Feb 2018.

Coventry, L., Briggs, P., Jeske, D., & Van Moorsel, A. (2014). SCENE: A structured means for creating and evaluating behavioral nudges in a cyber security environment. In A. Marcus (Ed.),

Design, user experience, and usability. Theories, methods, and tools for designing the user experience, DUXU 2014. Lecture Notes in Computer Science (Vol. 8517). London: Springer.

Department for Education. (2013). *Computing programmes of study: Key stages 3 and 4*. https://www.gov.uk/government/publications/national-curriculum-in-england-computing-programmes-of-study. Accessed Jan 2018.

Department for Education. (2015). *Computer science GCSE subject content*. https://www.gov.uk/government/publications/national-curriculum-in-england-computing-programmes-of-study; https://www.gov.uk/government/publications/gcse-computer-science. Accessed Jan 2018.

ECORYS UK. (2016). Digital skills for the UK economy. Department for Business Innovation and Skills, Department for Culture Media and Sport. https://assets.publishing.service.gov.uk/government/uploads/system/uploads/attachment_data/file/492889/DCMSDigitalSkillsReportJan2016.pdf. Accessed Jan 2018.

Eshet-Alkalai, Y. (2004). Digital literacy: A conceptual framework for survival skills in the digital Era. *Journal of Educational Multimedia and Hypermedia*. (2004, *13*(1), 93–106.

House of Commons Science and Technology Committee. (2016). *Digital skills crisis*. https://publications.parliament.uk/pa/cm201617/cmselect/cmsctech/270/270.pdf. Accessed Nov 2017.

Hu, C., Lin, Y., Chuang, H., & Wu, C., (2014). *A recommended ICT curriculum for K-12 education*. In International conference on Teaching and Learning in Computing and Engineering, (LaTiCE) 2014, IEEE Xplore, Electronic ISBN:978-1-4799-3592-5.

Institute of Directors. (2016). *Cyber security, underpinning the digital economy*. Policy report. https://www.iod.com/Portals/0/PDFs/Campaigns%20and%20Reports/Digital%20and%20Technology/Cyber%20Security%20-Underpinning%20the%20digital%20economy.pdf?ver=2016-09-13-171033-407. Accessed Nov 2017.

McDonald, C. (2014). *Leading digital nation by 2020: Calculating the cost of delivering online skills for all*. Tinder Foundation. https://libraries.wales/wp-content/uploads/2016/06/a_leading_digital_nation_by_2020_0.pdf. Accessed Jan 2018.

RESILIA. (2016). *Cyber resilience: Are your people your most effective defence?*https://www.axelos.com/Corporate/media/Files/cyber-awareness.pdf. Accessed Jan 2018.

Royal Society. (2017). *After the reboot: Computing education in UK schools*. https://royalsociety.org/~/media/policy/projects/computing-education/computing-education-report.pdf. Accessed Mar 2018.

Shadbolt, N. (2016). *Shadbolt review of computer sciences degree accreditation and graduate employability*. http://dera.ioe.ac.uk/16232/2/ind-16-5-shadbolt-review-computer-science-graduate-employability_Redacted.pdf. Accessed May 2018.

UK Council for Child Internet Safety. (2017). *Children's online activities, risks and safety*, UKCCIS. https://www.gov.uk/government/groups/uk-council-for-child-internet-safety-ukccis. Accessed Jan 2018.

An Examination into the Level of Training, Education and Awareness Among Frontline Police Officers in Tackling Cybercrime Within the Metropolitan Police Service

Homan Forouzan, Hamid Jahankhani, and John McCarthy

1 Introduction

As our current society is becoming more dependent on technology there is a need for a better information security to ensure that such information assurance are more and more compatible with contemporary information systems and applications. Customarily, guarding digitally stored and accessed information against unauthorised access, hacking and breaches were not the priority of a UK Metropolitan Police Services (MPS 2017). Though, as computer devices, the internet, and smart phones began to form a fundamental part in all our lives, the use of technology has become the basis for all if not most establishments, and this is continuing to grow. Unfortunately, due to the rapid growth of technology a growing number of individuals use the cyberspace to destroy, steal and compromise critical data, information security has formed part of the core governance of the National Police Security Strategy (GOV UK 2015). Such breaches and loss of information has a detrimental impact on places of work, homes, key infrastructure and prosperity of any organisations.

As a consequence Cyber security in the recent years classified as a 'Tier 1' threat by The UK Government's 2010 National Security Strategy. This sets cybercrime as one of the country's highest priorities for action. The Government's 2010 National Security Strategy has characterised threats into Tier ranging from 1 to 3, where 'Tier 1' threat is the highest that needs to be prevented to protect government and industries functions within the UK. In addition, in the twenty-first century, there

---

H. Forouzan (✉) · J. McCarthy
Northumbria University London and QAHE, London, UK

H. Jahankhani
QAHE and Northumbria University, Northumbria University London, London, UK
e-mail: hamid.jahankhani@northumbria.ac.uk

are very few criminal offences committed where some digital evidence is not in existence. Therefore, a robust and adequate information assurance is compulsory for information security reliability and availability to address the needs of modern security systems within the Metropolitan Police service.

Before understanding the concept of cybercrime, it is vital to be able to differentiate between Cyber-Enabled crime and Cyber-Dependent crime. Cyber-Enabled crimes are crimes that have traditionally been in existence, however with the current level of technology advancement they are being benefited by the use of computer devices to enhance their crime methods in committing that crime. Whereas Cyber-Dependent Crime are new crime methods being introduced that would have never been in existence if computer where never invented. Such crimes consist of communication interceptions and system hackings. The home office describes such attacks as crimes where digital systems are subject to illegal attacks. Such attacks are intended to effect the functionality of the intended digital systems by either destroying, removing, altering or disrupting the data or the IT infrastructure of such a system (McGuire and Dowling 2013).

One of the authors of this chapter is a Police Officer who has been working in the Criminal Investigation Department (C.I.D) within the Metropolitan Police Service (MPS) since 2003 and has a first-hand knowledge around the function and current methodologies currently in practice. Over the past 15 years within the Metropolitan Police Service (MPS), the researcher has noticed how cybercrime has become a significant element of police investigations. This chapter aims to critically examine the current the level of training, education and awareness among officers in tackling cybercrime within the MPS. The chapter aims to start by comparing the physical crime to cybercrime and then determine the level of understanding, awareness of police officers who respond to cybercrimes. By understanding, the level of training police officers go through currently the chapter will critically analyse whether these officers are equipped with the specialist knowledge within the Police Service in order to effectively investigate and tackle cybercrime.

The term cybercrime is used to describe acts, which incorporates the unlawful usage of computer technology and the internet. The term can be split into two parts one being the use of computer technology and the other the usages of the internet to carry out criminal conduct (Bell 2004; Brown 2015). Internet related activities consist of using social media (Facebook, twitter, telegram etc.) with the intention to commit unlawful acts. Examples of such conducts consists of cyberbullying, harassment, child grooming and child sexual exploitation.

In March 2016 the Office of National Statistics (ONS) Survey of Crime in England and Wales report, highlighted a number of significant changes around cybercrime. Due to the high rise in cybercrime, for the very first time, the report had dedicated a section on fraud and cyber-crime, which discovered an estimate of 5.8 million fraud and cybercrime cases committed in England and Wales (NCA 2016). This would mean a total of 5.8 million victims of cybercrime raising the overall number of crime committed to over 12 million. The data illustrated that nearly half of all crime committed in England and Wales are fraud and cybercrime related. One might ask with such high rise in cyber-crime and increase in business demand are the police adequately equipped to combat cybercrime and serve justices for the victims. Whilst physical crime such as robbery, theft and burglary has seen a sharp decrease over the years, cybercrime through the cyberspace has seen a rapid increase year by year (NCA 2016).

Furthermore, a Cybercrime Assessment 2016 was compiled by the National Crime Agency (NCA 2016), which highlighted the fact that a large proportion of all the crimes committed in the UK are cyber related crimes. According to the NCA within the United Kingdom, cyber related crime has exceeded any other forms of crime. The report warned that this situation is accepted to become worse as cybercrime is increasing in numbers.

According to Ford 2016, the chances of one to falling short of being a victim of cybercrime in the UK then robbery is 20 times more and 10 times more likely to be victims of cybercrime then theft (Loveday 2017). Due to advancements in technology and increase in cybercrime the need of a police service to be fully equipped to identify and bring to justices perpetrators is more important than ever to provide a good service to the community and increase victim satisfaction.

On the other hand, the Information Commissioner's Office statistics divulge that the statistics obtained by the police or other national databases are not the accurate value of reported cyber-crimes as many cybercrimes go under reported (Kesar 2011). There are a number of factors as to why victims are not reporting cyber related crimes. They include:

- Victims are not being aware of such a breach or an attack due to lack of awareness or the attack has taken place inconspicuously.
- Some victims do not believe that the police will take cyber related crimes seriously as traditional crimes.
- Some business/company's believe this could have a reputational damage to their organisation.
- Some management teams may not inform their senior leadership team of fear of criticism.

The ONS (2016) also highlighted the fact that from the estimate number of crimes committed in England and Wales only about 30% of the victims made contact with the police. Such under reporting is preventing the law enforcement agencies is identifying, disturbing and prosecuting the perpetrators responsible (NCA 2016).

The Hiscox Cyber Readiness Report 2017 carried out a survey on companies across UK, USA and Germany to identify how readiness for possible cybercrime attack. The report initiated that only 30% of the companies where readily prepared for a possible cyber-attack. In 2016 the chief executive of Hiscox Insurance, Steve Langan stated, "cybercrime cost the global economy over $450 billion, over 2 billion personal records were stolen and in the U.S. alone over 100 million Americans had their medical records stolen,". As clearly illustrated within this report around 70% of the businesses in the U.S., U.K. and Germany are not sufficiently equipped to tackle potential cyber-attacks (Langan 2017).

The police services across the UK have benefited from the growth of technology on many fronts in particular in combating crime (Action Fraud 2017), but on the other hand, it has also brought into question the integrity, privacy and confidentiality, of their information systems due to the threats and vulnerabilities it poses on their information systems. The advancement of technology has its own implications for safety and security. Any possible hacking, breaches or misuse of information systems has a bearing on organisational reputation, customer satisfaction and public confidence. Having staff adequately trained and educated on the subject matter would prove pivotal to the organisation as this plays an essential role in preventing and early detecting of such breaches, allowing a thorough investigation and prosecution of the criminals.

The concept of information sharing has become more and more on the agenda in the world today as it has been proven that in collaborating with other agencies yield's better results. Sharing information is key as it will identify, prevent and protect vulnerabilities for organisations. Therefore, a robust successful information assurance for the Metropolitan Police Service is vital to protect the data held on its information systems.

Traditionally offenders who want to gain personal information to commit identify theft would have gone through the trash, office documents etc. ... of the intended victim to achieve their goal. With the scale of rapid advancement in technology and individuals relying on the internet to carry out their daily activities such as shopping, pay bills, purchase goods, using social media to upload personal information and many more actives, the access to such information becomes more readily accessible for criminals. The criminals will use the intended victim's information systems to access the systems trash, temporary, cache memory, cookies, file and much more. A more technical attacker will use malware to able to gain access to the intended information system (Clough 2015).

Once such information is gained, the individual would be able to carry out many different criminal actives. Such common approached are financial and telecommunication frauds were the individual would use the information gain for personal financial gain. Other purposes of Identity theft could be using the personal information gained and posing as that individual to commit criminal offences. Examples are prostitution, drugs trafficking, money laundering and many more other offences.

Physical crime and cybercrime both have the word 'crime', which incorporates unlawful act or omission regardless of one using a computer device or network as a tool to commit that crime or physically targeting that victim. The only difference between the two terms is that one uses a computer device or a network to orchestrate an illegal act of crime; this could mean that the person may or may not be present at that crime scene location and could be thousands of miles away from that particular crime (Wall 2007). However where a traditional crime has been committed, the perpetrator is or has been at the location where the crime is or has taken place. In both cases, a crime has been committed and needs to be further explored to identify and bring the perpetrators to justice.

During both incidents, the perpetrator will leave evidential traces behind either being physical at the scene (fingerprints, DNA, other physical evidences etc.) or using technology and or the internet to commit that crime. As conventionally, the legal authorities are duty bound to investigate and try to capture all materials committed in the crime regardless of that being committed physically or cyber related crimes (Swire 2009).

Furthermore, traditional crime will provide more physical evidences to the police, since the 1890's the police have mastered their evolving methods to trace and prosecute offenders. For example when a physical crime has been committed, there are evidence that can link the perpetrator to that scene or the victim. For instance CCTV footage, forensic evidence (fingerprints, DNA, scientific evidence etc.), videos, photos, witnesses etc.. in most cases this will provide a shorter time to investigate and more importantly for the courts and the judiciary to convict the criminals responsible.

However, the difficulty lies when the police have to go out of their normal practices and investigate more complex cyber related crimes, which have no jurisdiction and could incorporate multiple victims at any one time. In addition to that, the internet leaves a large number of footprints on the World Wide Web (WWW) and makes the investigation and the evidence gathering more difficult and time consuming for investigators that are not adequately trained or educated in the area. The perpetrator can use different identities and be physically in one country whilst using a proxy server of another country to target the victims in the third country, therefore, the investigation would require more resources, finance, and time and due to the nature of the work detailed evidential gathering in order to locate and prosecute offenders.

2 Police Response to Cybercrime

In recent years the police response to the cybercrime has seen a big improvement as now all Police Forces across the UK are training their staff and raising their awareness on the subject matter and also have taken the approach to create a Cyber Unit (Operation Falcon) to tackle cybercrime. Whereas a few years ago physical crime was seen as being a more of a real crime than cybercrime due to the victim impact it carried. In particular, in the Metropolitan Police Service, there are Electronic training packages for officers around the subject matter but the level of training and education is a bare minimum and only touches on certain aspects of cybercrime. The training package mainly focuses on acquisitive crimes but does not focus on relevant case studies, cyber bulling or harassment and legality.

The MPS Operation Falcon are responsible for tackling evolving and emerging crime types of Fraud and Cybercrime. Operation Falcon have a number of sub departments, where each department are aligned to detect, protect, and prevent a specific branch of cybercrime in order to provide the most effective response to both cyber enabled and dependent crime as well as traditional financially acquisitive offences. Fraud and cybercrime reported via Action Fraud and disseminated to forces by the National Fraud Intelligence Bureau. However, crime in action and vulnerable victims remain the responsibility of the local 32 London boroughs for initial engagement and investigation. The Her Majesty's Inspectorate of Constabulary (HMIC) report December 2015 has also touched upon this point by stating that:

> As such, it is no longer appropriate, even if it ever were, for the police service to consider the investigation of digital crime to be the preserve of those with specialist knowledge. The public has the right to demand swift action and good quality advice about how best to deal with those who commit digital crime from every officer with whom they come into contact – from the first point of contact to an experienced detective. It is for the police service at large to recognise that dealing with victims of digital crime is now commonplace. Treating such crime as 'specialist' or requiring expertise that is provided only by the few is outdated, inappropriate, and wrong. Every officer must be equipped to provide victims of digital crime with the help and support that they have a right to expect from those charged with the duty to protect them. (HMIC 2015)

2.1 Action Fraud

In 2009, Action Fraud was established by the City of London Police in conjunction with the National Fraud Intelligence Bureau and is the national fraud and cybercrime reporting centre in the UK. When a fraud or a cyber-related crime has taken place, the intended victim and or a third party shall make contact with Action Fraud and report the alleged offence. The report would then be assed and then allocated to the relevant police force or department to investigate. The system can be accessed by either calling the call centre via phone or using the online reporting service 24 h a day 7 days a week. The service also has a help and support via both the phone and online services where victims of crime can obtain useful crime prevention advice in safeguarding their devices and systems from possible future attacks. In addition, the site provides an up-to-date cyber or fraud crime related new crime methods to raise awareness and provide advice in how to detect, prevent and report such new cyber-crimes (Action Fraud 2017). When such crimes are reported the Action Fraud will then allocate the report to a dedicated department or unit to further investigate. Action Fraud is only a telephone and or online reporting system.

2.2 The Legal Framework

There are number of legislations and Laws that the Police Staff need to be aware of when dealing with information. These will either have a direct or indirect impact on the information security and safeguarding within all Police Forces in UK.

2.2.1 Computer Misuse Act 1990

There are three main offences under The Computer Misuse Act 1990: those are:

- Unauthorised access to computer material;
- Unauthorised access with intent to commit or facilitate the commission of further offences; and
- Unauthorised acts with intent to harm the process of computers.

The Computer Misuse Act 1990 went through to radical amendments by the Serious Crime Act 2015, and the Police and Justice Act 2006. As a result of the amendments The Computer Misuse Act 1990 now involves 'hacking' and any unauthorised access, by either using another's identity to log into another's system and make/obtain/supply articles in use of that offence.

One of the most significant changes that the Serious Crime Act 2015, brought to the Computer Misuse Act was the introduction of section 43, which made it an offence if breach was committed by the perpetrator outside the United Kingdom. In addition, the Police and Justice Act made a noteworthy contribution to the Computer Misuse Act by making the Denial-of-Service attacks as an offence.

2.2.2 Data Protection Act 1998 (DPA) and General Data Protection Regulation (GDPR) 2018

Data Protection Act 1998 is an important piece of legislation in the context of the security and safeguarding of information composed, stored and used by the Police Forces, as a high proportion of such information will be 'personal data' and 'sensitive personal data' relating to persons. Such personal data is not limited to that contained on information communications and technology (ICT) systems but can exist in any format including CCTV images, photographs, notes book entries etc. so long as they can be linked to individual living persons.

The General Data Protection Regulation (GDPR) effective from May 2018, is designed to enable individuals to better control their personal data. It is hoped that these modernised and unified rules will allow businesses to make the most of the opportunities of the Digital Single Market by reducing regulation and benefiting from reinforced consumer trust.

Areas to look for the implementation of the GDPR should be as follows;

- **Incidents**: The GDPR will now make it a legal duty to report any incidents to the ICO within 72 h. This legislates similarly to current working arrangements of contacting the ICO in a similar time period.

The potential financial penalties could be significantly higher for incidents where the regulations were not adhered to. Article 83 states the general conditions for imposing administrative fines. Paragraphs 4 and 5 state the limits of administrative fines that potentially can be issued (this is dependent on member state approval of what the maximum fine will be). The maximum fines for failing to comply with the regulations could potentially be 20 million Euros or 4% or of the total worldwide annual turnover of the preceding financial year.

- **Fair Processing and legal basis for processing data**: Data subjects must have fair processing information made available to them "at the time when personal data are obtained" (Article 13). Further to this, it must also include the following on top of what is currently provided:

    the contact details of the data protection officer
    the right to lodge a complaint with a supervisory authority
    the existence of automated decision-making, including profiling, referred to in Article 22(1) and (4) and, at least in those cases, meaningful information about the logic involved, as well as the significance and the envisaged consequences of such processing for the data subject

Data subjects have the right to know what safeguards are in place for any personal data that is transferred to a "third country or an international organisation" (i.e. outside the UK). The applicable safeguards that can be used in international transfers are stated within Article 46.

Recital 47 states; in reference to the legitimate interests of a controller as legal basis for processing a data subjects personal data "Given that it is for the legislator to provide by law for the legal basis for public authorities to process personal data, that legal basis should not apply to the processing by public authorities in the performance of their tasks". This in effect means that processing personal data using legitimate interests as a basis is no longer an option.

- **Data Protection Impact Assessments (Article 35)**: Privacy Impact Assessments (PIAs) within the Act are named 'data protection impact assessments' (DPIAs).

There are limited differences to the two, except; that the appointed Data Protection Officer's (DPO) advice must be sought (currently SIRO) and, 'where appropriate, the controller shall seek the views of data subjects or their representatives on the intended processing'. Currently patient reps are required sit on Procurement Panels and provide input, but this is not replicated widely across all new processes that currently would require a PIA to be undertaken, or in future a DPIA.

Recital 84 states; "Where a data-protection impact assessment indicates that processing operations involve a high risk which the controller cannot mitigate by appropriate measures in terms of available technology and costs of implementation, a consultation of the supervisory authority should take place prior to the processing".

Article 25 states that data protection must not only be by default but must be by design (Privacy by design).

- **Data Protection Officer Role**: According to Article 37 paragraph 1; the organisations must appoint a Data Protection Officer (DPO). The details of the appointed DPO must be published and submitted to the ICO.

In some regards the DPO role has general similarities to the SIRO role. For example, all PIA's will need to have consultation with the DPO (our current sign-off mechanism for PIA's covers this).

However, article 38 states 'Data subjects may contact the data protection officer with regard to all issues related to processing of their personal data'. It is unclear from this whether tasks such as subject access requests can be delegated, As the ICO would be the supervisory body in this respect, advice should be sought from them.

Similarly, guidance issued by the Working Party (EU joint committee of Member States' Supervisory Authorities [ICOs]) about the role of DPO are not clear. Guidance states that the "The personal availability of a DPO is essential to ensure that data subjects will be able to contact the DPO" in relation to Article 38 paragraph 4, in regards to "to all issues related to processing of their personal data and to the exercise of their rights under this Regulation". The level of expertise, skill, knowledge etc. stated in Articles 37–39 and Recital 97 that are relevant to the role, and given that the DPO must be available to communicate directly with data subjects and the Supervisory Board.

- **Subject Access**: Article 12 paragraph 3; states that SAR's (as stated under article 13) need to be complied with 'without undue delay' and within a month of receipt of the request.
- **Processing Activities**: Article 30 states that Data Controllers 'shall maintain a record of processing activities under its responsibility'.
- **Strategic work to undertake in partner with other organisations and potential future work**: Article 40 paragraph 1; states that the drawing up of codes of conduct in keeping with the regulation should be encouraged. Paragraph 2 states that 'bodies representing categories of controllers or processors' should help prepare the codes of conduct. The DSPU team are in discussions with the IGA to have relevant work on this undertaken.

In relation to the above, Article 41 states that compliance with the code of conduct must be undertaken by an appropriate body which has an appropriate level of expertise (expertise in this case is decided by the regulatory body – the ICO).

- **Other Considerations**: Current legislation maintains that only the Data Controller for information may be liable for non-compliance with the Data Protection Act. Article 82 paragraph 2 states; "A processor shall be liable for the damage caused by processing only where it has not complied with obligations of this Regulation specifically directed to processors or

2.2.3 Regulation of Investigatory Powers Act 2000 (RIPA)

RIPA relates to the interception of communications, which provides the police with the relevant power to intercept recording transmissions to combat crime and makes it an offence for unauthorised interception without warrant. Therefore, any form of intentional communication interception is illegal under RIPA.

2.2.4 The Human Rights Act 1998

The United Kingdom is duty bound to obey by the Human Rights Act 1998 when considering legally intercepting another's communication devices as this could be a breach of their Article 3 of the Human Right Act 1998 "The right to respect for private and family life".

However, the act does also include:

> ...There shall be no interference by a public authority with the exercise of this right except such as is in accordance with the law and is necessary in a democratic society in the interests of national security, public safety or the economic well-being of the country, for the prevention of disorder or crime, for the protection of health or morals, or for the protection of the rights and freedoms of others.

3 Metropolitan Police Services Cybercrime Training and Awareness

The Metropolitan Police Service (MPS) is good at understanding and tackling serious and organised crime however the questions that could be asked is can the same methodology be implemented in combating cybercrime. The basis of dealing with any crime is the same however; cybercrime needs adequate training and education due to its technicality and lack of awareness. This section of the chapter will be looking at the current training and education that front line police officers receive around cybercrime within the Metropolitan Police Service.

One of the authors of this chapter brings with him an in-depth knowledge and understanding of the subject matter, currently employed by the Metropolitan Police Service as a Detective Inspector and dealing with physical crime and cybercrime daily.

When a physical crime takes place, the crime reported to the police who takes immediate action, however, reporting is somehow differs when cybercrime occurs. When a cybercrime is reported it will be allocated to an officer who has already numerous other priority physical crimes to investigate. As a result, those time-consuming investigations (cybercrime being one of them) inevitably pushed down the list. Additionally, due to the technicality of cybercrime and the lack of knowledge and awareness within the police, a high number of perpetrators get away from prosecution and victims are not getting their justices (Bidgoli and Grossklags 2016).

Due to the high rise in cybercrime the Metropolitan Police Service have begun to raise awareness of the cybercrime across the service and have developed a handful of electronic training packages on the subject matter and since 2015 one particular training package has been made as mandatorily for all officers in the MPS to complete. This electronic training and learning package compiled by the college of policing in partnership with the National Centre for Applied Learning Technologies (Ncalt). The context and the level of education that the training package provides to its users and the areas of cybercrime that the package focuses on are subjects of discussion here.

Through a set questionnaire distributed to 250 MPS front line borough based police officers a bigger picture of the police's knowledge and understanding is ascertained. The questions sought to identify if officers were aware of cyber related crimes, the level of training, if they have completed the mandatory electronic Ncalt cybercrime learning and training package and if they deem the training package to be adequate to investigate cyber related crimes. All participants' anonymity were respected. The Operation Falcon officers did not receive the questionnaire, as they are cybercrime specialist who have undertaken a number of cybercrime courses and are in fact experts in the matter. The aim here was to understand and examine the learning and training of front line officers who would be the initial respondents to the incident.

By reviewing the MET London Crime Statistics 2015/16, four boroughs were selected out of the 32 London boroughs to take part in the pilot study. It is important to note that the MET London Crime Statistics 2015/16 did not have a cybercrime category so that a more specific pilot boroughs could have been selected.

In addition, in order to get a better understanding of the issue, it was vital to listen to both sides of the argument. Therefore, a number of interviews with victims of cybercrime were conducted in order to seek their feedback on the police's response to the matter. The impact experienced by the victims (financial, psychological, and emotional) and the aftercare they received from the police was also touched upon during the interviews. All participants' anonymity were respected.

Having gained the data from the questionnaires and interviews, the result were analysed and discussed through an interview with one of the managers at the Metropolitan Police Service training school and Collage of Policing (CoP). Result was to propose a bespoke cybercrime electronic training package to assist front line officers in investigating cyber related crimes. This would be beneficial to the officers, as it will provide a more detailed knowledge and understanding to the matter.

3.1 The Ncalt Training Package Analyses

Since 2015, the MPS have completed an online electronic Ncalt learning and training programme and have made that mandatory for all officers within the MPS to complete (MPS 2016). The Ncalt training package provides basic understanding of cybercrime and sets a scenario-based incident asking the user to incorporate the tools and skills within the Ncalt package to complete the scenario. The mandatory Ncalt learning and training package is more like a basic awareness package, which touches upon some of the cyber related crime incidents and does not detail cybercrime types and cybercrime prevention.

For example, the electronic Ncalt training package briefly looks at some of the reasons as why one would use cyberspace to commit crime. There is no training in how officers should respond to a victim of cybercrime and what actions to taken when dealing with a cyber-related crime. Furthermore, there are no sections that provide officers guidance and advice on cybercrime prevention to cybercrime victims. One of the MPS's pledges is to increase public confidence and customer satisfaction, therefore, the question is what MPS is doing in fulfilling that pledge in relation to cybercrime victims.

During the interview with the victims, it was clear that they were rather unhappy with the response police provided to their cybercrime investigation. The victims felt that there was little to none investigation carried out and felt that the officer's main concerns were around if the financial loss covered by the bank or not. Further, victims received very basic crime prevention advice. This clearly illustrates that the MPS needs to educate and train its staff in order for them to be able to investigate cybercrime and be able to provide a better service to its customers.

The 250 officers across MPS London boroughs were asked the following 10 questions:

Question 1: How knowledgeable are you with cybercrime?
Question 2: Are you aware that according to recent government statistics cybercrime has taken over physical crime in reporting numbers?
Question 3: What level of training have you received in order to deal with cybercrime?
Question 4: Do you believe that the current level of training and education you receive is adequate in tackling cyber related crimes?
Question 5: Do you believe the Metropolitan Police Service (MPS) are tackling cybercrime as seriously as acquisitive crime?
Question 6: How cyber security aware are you?
Question 7: What level of cyber security training and awareness have you received?
Question 8: How regularly do you change your work computer login passwords?
Question 9: Have you completed the current MPS Cybercrime Ncalt Training?
Question 10: Do you believe the current Ncalt training package provides you with the relevant skills and knowledge to combat Cybercrime?

The results as illustrated in Fig. 1, below have highlighted the fact that the majority of front line officers are of an opinion that they are not adequately trained and educated to be able to tackle cyber related crimes. A total of 95.8% of the participants were either unsure or did not agree that the current mandatory Ncalt training package provided front line officers with the skill set and knowledge to be able to tackle such a high rise and complex crime. Such high numbers hinders one of the main MPS pledges, which is to gain their customers satisfaction and confidence, which could lead to an organisational failure and have a detrimental reputational effect on the MPS.

The above results shared with one of the managers from the MPS Crime Academy during an interview. The response was; the College of Policing (CoP) has decided that the current electronic Ncalt training and learning package provide the right level of training for officers to effectively deal with Cybercrime. However, we are also aware that currently there are no set training inputs or seminars

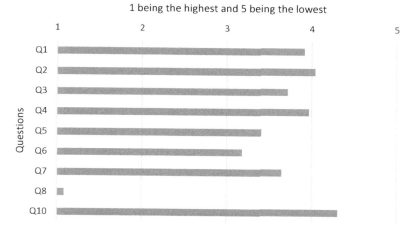

Fig. 1 Illustrates the average score for each question

around cybercrime that is embedded into the front line officer's mandatory training programme in order to provide officers with the relevant skill set and knowledge to empower them to tackle cyber related crime investigations.

Another conclusion that drawn from the questionnaire is that approximately 30% of the participants were not even aware that there was a mandatory online eLearning Ncalt training package that needed to be complete. One might ask, If this is the only learning and training package available to frontline officers why the United Kingdom's largest police service has no monitoring process in place to check if officers have completed the mandatory electronic Ncalt learning and training package. Consequently, the questionnaire result would indicate that about 30% of front line officers are not sufficiently trained and educated to be able to investigate such and up and rising crime.

It is understandable that the MPS has recognised cyber-crime as a fast emerging crime and as a result of that implemented Operation Falcon to tackle such a high rise in cyber-crime. The officers within that command are normally well trained and educated who would also have to undertake a number of mandatory financial investigation courses, cyber-crime course and also have to completed a four-day lecture by a cyber-security matter expert. However, as stated above not all fraud/cyber-crimes are subject to investigation by Operation Falcon.

Nevertheless, as highlighted in the victims interview they were unhappy with the level of service and handling of the officers who dealt with the crime. One thing that is for sure is that the officers who responded to victim of that crime were borough front line officers and not part of Operation Falcon. As HMIC, (2015) specified:

> ...Treating such crime as 'specialist' or requiring expertise that is provided only by the few is outdated, inappropriate, and wrong. Every officer must be equipped to provide victims of digital crime with the help and support that they have a right to expect from those charged with the duty to protect them. (HMIC 2015)

Furthermore, as a result of the questionnaire 73.4% of officers were of the opinion that the MPS was not taking cyber related crimes as serious as acquisitive crime. Thus indicating that the MPS are still focusing on investigating traditionally crime despite the fact, that cyber related crime has taken over physical crime.

4 Proposed Improvement to the Cybercrime Training Package

After carefully analysing the interviews, questionnaire feedbacks, the current MPS Ncalt training package and liaising with the MPS training school and Collage of Policing (CoP) a number of recommendations are identified that that would seek to enhance cybercrime awareness among frontline police officers, raise their knowledge and understanding and empowering officers to tackle and investigate cyber related crimes.

Having a more user-friendly interactive electronic cybercrime package with mandatory class based training days for all the officers will yield better understanding and raise awareness among police officers. The electronic package needs to be more relevant and up to date covering all the potential cyber related crimes such as in details cyber bullying, cyber stalking, revenge porn etc. in more details. As illustrated in the questionnaire question ten 81.7% of the officers were of an opinion that the current Ncalt training package did not provide them with the relevant skills and knowledge to combat cyber related crimes. The online training pages should also provide a more detailed powers and policy input so those officers are fully aware of all the relevant current laws and procedures when dealing with cyber related crimes. It is important that case studies are also included within the new training package, as case studies would provide the users a better and more in-depth knowledge and understanding of the problem.

The online Ncalt training package should also provide officers with the relevant knowledge and skill set to be able to effectively respond to a cybercrime incident and provide a good service to the victims of cyber related crimes. The training package could also provide a section on crime prevention advice and officers would be able to provide guidance and advice to victims of crimes in safeguarding their ICT systems for future attacks.

When a crime is reported it will be computerised on a Crime Record Information System (CRIS). Such system will hold all details of the investigations and officers' actions on the matter. A random selected sampling process should be carried out in order to quality check the standard of the investigation. Currently there are no monitoring processes for the training of cybercrime officers; usually they are informed via an email through their supervisors that they need to complete the Ncalt mandatory training package. There is no process in place to monitor if officers have completed their Ncalt training package. There is also no feedback option available for officers to raise concerns if the training provided is of relevance or there needs to be further improvements to the package or further training updates are required. It is important to create a dedicated cybercrime unit for each 32 London Police Boroughs in order to review and monitor the standard of investigation, awareness and education of officers. The dedicated unit will also be able to raise cybersecurity awareness among police officers in order for them to protect and safeguard the information held within MPS ICT systems. This would also help to measure officers training needs or effectiveness of the training they have had in tacking cybercrime.

In addition, a monthly feedback form will be requested from each participant for a period of 6 months to identify if the training they have had assisted them in tackling cybercrime and how the package material could be developed to assist future participants to be more knowledgeable and adequately training when investigating cyber related crimes.

Currently the Ncalt online training package has few basic knowledge check questions that needs to be completed by the participants. However, it could be argued that with such a complex subject and with a high volume of information, newly crime methods and legislations to be learned, it is important to measure officers understanding and awareness of cybercrime through a more detailed knowledge check question test. This would be a method to measure officers understanding of the subject matter.

It is extremely pivotal for the MPS to constantly monitor and review the recommended online training package in order for the context to be relevant and up-to-date with the recent methods, powers and policies, tools and deployed tactics so that officers are fully equipped and skilled to tackle such complex investigations.

The police forces across the country have been faced with an increase in cybercrime over the recent years. The alarming factor is that the crime stats clearly illustrates that cybercrime has taken over physical crime within the UK. The chances of one to falling short of being a victim of cybercrime in the UK than robbery is 20 times more and 10 times more likely to be victims of cybercrime than theft (Ford 2016). Cybercrime comes in many different types therefore the police service within the UK need to be up-to-date through training, education and awareness in order to be able to combat such a complex crime.

Cybercrime will be one of the UK's crime priorities due to the scale of damage that is poses. The police forces must work in partnership to raise awareness of the police staff and of the community through training, education and awareness to protect themselves and the society to the best of their ability in preventing, detecting, locating and bring those responsible to justices. As cybercriminals based their beliefs that the law enforcement struggles to identify and bring those responsible to justice.

A fast growing multi-technological environment generates a high number of security risks for the MPS, if not addressed could result in both intelligence and information leakage, which would lead to significant consequences for the MPS. In order for the MPS to be operational effective, it is extremely important for the MPS staff to be adequately trained and educated in order to safeguard their Information Communication Technology (ICT) systems from possible attacks, breaches and hackers. The staff should also be made aware that deliberate or misuse of the MPS ICT systems would result in the public's loss of confidence in the MPS's and their ability to safeguard personal and highly sensitive information and or intelligence. This would further lead the MPS employees either investigated for criminal proceeding and or the MPS being subject to legal sanctions. The questionnaire that was disturbed among 250 frontline officers clear highlighted the issue this as an issues where a total of 83.1% of officers believed that the current MPS training and education was not sufficient to be able to keep their computer systems and devices safe.

As indicated by the Home Office (2010 The Cybercrime Strategy Report) the threat of cybercrime will never go away or get weaker as business become more cyber security aware but in fact such threats are continuously evolving and becoming more and more complex by nature. New methods involving high technology crimes are being committed, which were not in existence a few years ago and thus this supports the fact that police officers need to be contentiously updated with recent methods and high-tech cybercrime to be able to tackle and to prosecute those responsible.

As demonstrated within the main body of this chapter the MPS needs to value, protect and process its information and intelligence with confidentiality, integrity and legality, as this would have a direct impact on the public's trust and confidence in the police. If such basic level of information protection is breached this could have a detrimental effect on the MPS service deliver, public confidence and the organisations reputational values. Therefore, it is extremely important that the right level of training and education be provided to front line officers to be able to protect and safeguard the information on their ICT systems.

There needs to be a wider partner agency and government approach in order to tackle this high rising technical crime. The police services are usually a step behind the criminals and virtually as soon as a new technology appears, the criminals will commission that within their attacks. Furthermore, such an approach around information sharing and collaborative working would provide the MPS staff with adequate level of training and education to be able to investigate cybercrime and to safeguard their information systems from potential future breaches or cyber attacks.

5 Conclusions

As the scale of dependency on technology and the internet is growing rapidly in the UK and worldwide so does the complexity of cyber criminals in increasing their attacks on individuals, businesses and government organisations.

It is a clear that there is a need for the MPS to recognise the necessity to improve its cybercrime awareness, learning and training of staff in order for the MPS to be able to tackle such high rising and complex crime.

This chapter's aim was to understand the level of training MPS provides currently and provide a number of recommendations in order to improve the level of training and awareness among front line police officers. These recommendations include a more user-friendly interactive electronic cybercrime package with mandatory class based training inputs days for the all officers will yield better understanding and raise awareness among police officers.

It is also recommended to create a dedicated cybercrime unit for each 32 London Police Boroughs so that monitor the standard of investigation, awareness and education of officers can be reviewed more effectively. The dedicated unit will also be able to raise cyber security awareness among police officers in order for them to protect and safeguard the information held within MPS ICT systems in light of the GDPR.

References

Action Fraud. (2017). https://www.actionfraud.police.uk/news/the-threat-of-pbx-dial-through-fraud-apr17.
Bell, D. (2004). *Cyberculture: The key concepts*. Psychology Press.
Bidgoli, M., & Grossklags, J. (2016). End user cybercrime reporting: What we know and what we can do to improve it. In *Cybercrime and Computer Forensic (ICCCF), IEEE International Conference on* (1–6). IEEE.
Brown, C. (2015). Investigating and prosecuting Cyber Crime: Forensic dependencies and barriers to justice. *International Journal of Cyber Criminology, 9*(1).
Clough, J. (2015). *Principles of cybercrime*. Cambridge: Cambridge University Press.
Ford, R. (2016). Fraud doubles the number of crimes. *The Times*.
GOV. UK. (2015). *2010 to 2015 government policy: cyber security*. https://www.gov.uk/government/publications/2010-to-2015-government-policy-cyber-security/2010-to-2015-government-policy-cyber-security.
HMIC. (2015). *Real lives, real crimes. A study of digital crime and policing*. Available at: www.justiceinspectorates.gov.uk/hmic. Accessed on 21 Feb 2018.
Home Office, March. (2010). *The Cybercrime strategy report*. The Secretary of State for the Home Department.
Kesar, S. (2011). Is cybercrime one of the weakest links in electronic government. *Journal of International Commercial Law and Technology, 6*.
Langan, S. (2017). Despite major cyberattacks, businesses have been slow to react. *LSE Business Review*. Available at: https://www.cnbc.com/2017/02/07/cybercrime-costs-the-global-economy-450-billion-ceo.html. Accessed 18 Jan 2018.
Loveday, B. (2017). Still plodding along? The police response to the changing profile of crime in England and Wales. *International Journal of Police Science & Management, 19*(2), 101–109.
McGuire, M., & Dowling, S. (2013). Cyber crime: A review of the evidence. *Summary of key findings and implications. Home Office Research report, 75*.
Metropolitan Police Service. (2016). *Year end crime statistics 2015/2016*. Available at: https://www.met.police.uk/stats-and-data/year-end-crime-statistics. Accessed 21 Sept 2017.
Metropolitan Police Service. (2017). Available at: www.met.police.uk. Accessed 13 Sept 2017.
National Crime Agency. (2016). *NCA Strategic Cyber Industry Group*. http://www.nationalcrimeagency.gov.uk/publications/709-cyber-crime-assessment-2016/file. Accessed 6 Jan 2018.
Office for National Statistics. (2016). *Statistics bulletin 'Crime in England and Wales –year ending March 2016'*.
Swire, P. (2009). No cop on the beat: Underenforcement in e-commerce and cybercrime. *Journal on Telecommunications and High Technology Law, 7*, 107.
Wall, D. (2007). *Cybercrime: The transformation of crime in the information age* (Vol. 4). Cambridge: Polity.

Combating Cyber Victimisation: Cybercrime Prevention

Abdelrahman Abdalla Al-Ali, Amer Nimrat, and Chafika Benzaid

1 Introduction

The global penetration of networked communications has exposed different areas of society to the threats of cybercrimes. These levels of society include, but are not limited to, nations and communities. Today, individual organisations and governments are significantly more likely to be victimised through the use of information and communications technologies than experience conventional forms of victimisation (UNODC 2013).

All members of society are subjected to victimisation in different ways and often these practices are online versions of traditional forms of crimes (Baxter 2014). People can be harassed by unsolicited digital communications or interactions that threaten, are inappropriate or defame. They can also be subjected to the repeated use of digital media to bully through the sending of abusive or humiliating comments or pictures, and the personal identification of victims (Lipton 2011). Victimisation can further encompass cyberstalking, where an individual's digital and social media footprint is followed and used to and harass and intimidate (Roberts 2008). Victims may also suffer the theft of their identity and be subject to malware and viruses (Baxter 2014).

A. A. Al-Ali (✉)
University of East London, London, UK
e-mail: u1626185@uel.ac.uk

A. Nimrat
University of Gloucestershire, UK
e-mail: a.nimrat@glos.ac.uk

C. Benzaid
University of Sciences and Technology Houari Boumediene, Algeria
e-mail: cbenzaid@usthb.dz

© Springer Nature Switzerland AG 2018
H. Jahankhani (ed.), *Cyber Criminology*, Advanced Sciences and Technologies for Security Applications, https://doi.org/10.1007/978-3-319-97181-0_16

Cybercrime against business, organisations and government institutions has major financial and economic impacts. Attacks targeting computer systems, servers and data through hacking and spreading of viruses and malware programmes are major forms of cyber victimisation from an organisational perspective (Trim and Lee 2015). On an individual level people are victimised both psychologically and financially. Psychological impacts represent a significant and pervasive problem for cyber victims and can result in depression, fear and anxiety, emotional trauma, and even suicide (Dredge et al. 2014; Bonanno and Hymel 2013; Cénat et al. 2014). Individuals may further suffer the direct loss of savings and assets in addition to indirect financial impacts such as loss of employment and impaired credit ratings (Roberts 2008). The financial cost to organisations and businesses from cyber victimisation can be high. Firms may be hacked, defrauded, extorted, and have their financial and intellectual assets stolen. Globally, the cost of cyber victimisation to firms is an average $7.7 million annually (CNNMoney 2015). Governments can also be victimised for the purpose of gaining financial or political advantage. Increasingly government information and systems are the target of sophisticated and severe attacks representing a broader systemic threat with significant consequences for national security and infrastructure (KPMG 2017; Agustina 2015).

Efforts towards cybercrime prevention for individuals are acknowledged to have frequently focused on technology and protection of computers and devices. This is argued to diverge strongly from mainstream models of crime prevention, which principally focus on the human factor in crime. Criminal theories are therefore argued to have not yet integrated the fast-paced development of the Internet and associated cybercrime (Jahankhani 2013). Jahankhani and Askerniya (2012 cited in Akhgar and Yates 2013) introduced a grid model with the purpose of ordering cybercrime prevention strategies into four different classifications. Fig. 1 shows on the x-axis the level of tech-savviness, different individual levels of risk on the y-axis and the cognitive developmental stages of individuals on the z-axis. The idea/theme axis indicates the prevention initiative objective. The model integrates social aspects as a key factor within crime prevention strategies and emphasises education and awareness as a critical element in crime reduction (Jahankhani 2013).

As shown in the model (Fig. 1) the level of tech-savviness of individual users is the first dimension targeted by cybercrime prevention interventions. The activities included in the model are based on the most prevalent individual activities conducted on the Internet. With the aim of reducing individual risk and enhancing protective aspects, interventions centre on enhancing user education, awareness, and training in relation to the particular skills required for the different activities listed (Jahankhani 2013). Jahankhani (2013) highlights that interventions should be applied at the different individual developmental stages to help reduce their exposure to cybercrime.

The next grid dimension concerns the users' risk levels and the degree to which interventions are required, focusing on user knowledge levels, training and awareness. Low risk is associated with users who have considerable knowledge of technology and online exposure risks. Medium risk pertains to those with insufficient knowledge of online exposure risks who may have above average levels

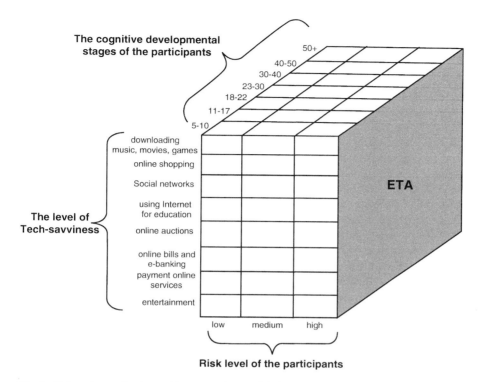

Fig. 1 Cybercrime reduction and/or prevention model (Source: Jahankhani and Askerniya 2012 cited in Akhgar and Yates 2013 p. 264)

of cyber victimisation in terms of computer viruses or falling victim to identity theft and online financial fraud. Such users are identified as having current knowledge of computer and device security but lack the more in-depth knowledge needed to change their behaviour online. High risk is related to users who use the Internet extensively with limited regard for risk exposure (Jahankhani 2013).

The third grid dimension considers the cognitive developmental stages of users. It is argued that diverse impacts from risk and protective factors exist at different stages of development. User ages are therefore found to be significant in the different interventions required (Jahankhani 2013).

The final dimension refers to the main objective of the prevention programme. Jahankhani (2013) argues that the most successful approach towards reducing and preventing cyber victimisation is to enhance individual behavioural skills and cognitive development through the development of a range of awareness, education and training interventions focused on online exposure risks and online behaviours.

Liechti and Sumi (2002) define Internet awareness in terms of awareness of other users and the maintenance of knowledge in relation to the activities and situation of others. It is argued that a general idea of what is occurring or that there is something that is occurring is already valuable knowledge.

2 Legal Perspectives

2.1 Balancing Freedom and Protection

A review of the literature identifies a recurring theme across different countries concerning civil liberties and the challenge of balancing restricting and punishing certain behaviours with freedom of expression. One of the main problems with the introduction of specific laws to cover cybercrime emerges from the notion that cyberspace is not a physical space owned by anyone. The Internet is a manmade device created to allow better connectivity among people (Marczak and Coyne 2010). "No one fully monitors or censors information entered to servers interconnected around the world" (Barker 2002, p. 85). National and political boundaries do not exist in cyberspace and this reality has compounded the problem of how and where jurisdiction can be established. The proponents of the Internet state it could not be and should not be regulated because of its openness and international nature (Netanel 2000). Therefore, not only legally does it become a problem to convict an individual who engages in cybercrime across different jurisdictions; consideration must be given to whether regulation of the Internet is merited. There is a fine balance here between protection of cyberspace and maintaining the openness and freedom of cyberspace (Netanel 2000).

The issue of balancing freedom of expression with protection has assumed paramount importance and generates significant complexities in legislating against cyber harassment, bullying and other forms of online abuse. In the UK the issue of online harassment in the political arena has been widely debated in the media and government, as MP's voting on sensitive issues have become the subject of online abuse, harassment and stalking. There are major tensions and fierce debate regarding defining clear boundaries and legislating for prevention, which by some has been viewed as encroaching on civil freedoms and the right to protest and campaign (Emm 2009).

The tension between security and civil liberty is exemplified by fierce protests from campaigners on both sides. Campaigners for protection have pressed for legislation for greater protection against stalkers, 'trolls', and online bullies (Edwards 2012).

It has been argued that laws to combat cyber victimisation and abuse need to consider concerns related to the First Amendment, designed to safeguard free speech in the USA and the entitlement of individuals to accept speech free without government interference. Lipton (2011) acknowledges the sensitivity and problematic nature of introducing laws that limit speech, emphasising that this is more challenging for communications that do not conform to traditional types previously subject to legislation. Lipton argues that in the real world, legislation has been able to successfully criminalise many of the wrongs now committed online and therefore it is acceptable that judges continue to distinguish between protected and unlawful speech in the online context (Lipton 2011).

2.2 Legal Approaches

In terms of the legal approach two key issues consistently arise in the literature. Firstly, whether specific new legislation is required or whether amendment of existing legislation is sufficient. The second issue concerns the choice between a criminal or civil approach to combating cyber victimisation.

Fukuchi highlights that to overcome many of the challenges in legislating against cyber victimisation authorities have generally adopted two main approaches. One approach involves the modification of existing legislation on stalking or harassment through inserting references to digital initiation of contact. Therefore, in these jurisdictions there are no specific laws targeting cyber victimisation practices; however, actions which constitute these are proscribed (Hazelwood and Koon-Magnin 2013).

While new laws have been drafted to combat specific forms of cybercrime, research demonstrates that existing legislation enacted before the existence of cyberspace is still relied on significantly. In the UK there are plans by the government to amend existing criminal legislation to target forms of cyber harassment specifically relating to 'trolls', who sexually harass and subject their victims to verbal abuse on the Internet or mobile communication. This is another example of a legal approach that seeks to build on existing legislation to extend protection for victims from harassment and abuse through texts messaging. In addition, in the UK the Computer Misuse Act in 2008 was amended to increase penalties for hacking and to facilitate the extradition of hackers under existing treaties (p.15). Across all countries existing legislation is being amended to combat specific acts (Emm 2009). However, there is a lack of research in relation to the effectiveness of this approach. In some areas, the limitations of existing legislation are being recognised, for instance in relation to protecting victims of cybercrime. In Australia there have been calls to reform outdated legislation (Baxter 2014).

In some cases, new legislation is required to deal with specific new forms of cybercrime. For example, the alternative approach has been to introduce new laws clearly defining and proscribing cyber stalking and cyber harassment. Frequently, in these jurisdictions distinct laws identify traditional and online forms of harassment and stalking (Hazelwood and Koon-Magnin 2013).

In the UK 'revenge porn' has been identified as one area requiring attention, combined for calls for specific laws in relation to "cyberbullying". Further progress has been made in the United States where some states are introducing specific legislation. The State of California passed legislation on "revenge porn" in 2013 making it a criminal offence to engage in "disorderly conduct", to take and then circulate with the intention to cause grave psychological harm, private and intimate photos and videos (Agate and Ledward 2013). However, this California law has been highly criticised as it only applies to offenders who actually take the photos they then distribute, and do so to intentionally cause serious emotional distress. As up to 80% of the photos used in revenge porn are "selfies", according to a survey by the Cyber Civil Rights Initiative, and the law does not concern itself with hackers and redistributors who copy and republish the images, only a minority of victims are protected. On the other hand, the Cyber Civil Rights Initiative sees it as a positive initial move (Agate and Ledward 2013). This case exemplifies the tensions between protection and civil freedom that relate to other forms of cyber victimisation.

New legislation in Canada is being enacted to provide protection for citizens against cybercrimes. Draft legislation currently being discussed in Canada and named the "Protecting Canadians from Online Crime Act", includes provisions to make the non-consensual online publishing of intimate pictures a crime, as well as aiming to cover other acts not currently legally considered crimes (Agate and Ledward 2013).

Research also emphasises the role of new specific legislation. In a wide-ranging review of US state cyberstalking legislation by Goodno (2007), five significant dissimilarities between offline and online stalking were recognised. Cyberstalking differs in that the stalker could be anywhere in the world, is able to remain anonymous, may impersonate another's identity to stalk the victim, can use third parties to contact the victim, while anyone with Internet access may be messaged by communications which are immediately present and cannot be deleted. It is thus concluded that to address cyberstalking, unique laws are required which can ameliorate the focus on physical stalking in traditional legislation (Hazelwood and Koon-Magnin 2013). This would have the effect of making the classification of both online and offline bullying clearer by revising and combining current legislation, which was a key recommendation (Agate and Ledward 2013).

The question of whether criminal legislation is an appropriate approach to combating cyber victimisation is a further theme identified in the literature, combined with questions relating to the role of civil law. A key issue in the debate concerning cybercrime legislation is whether certain issues can and should be addressed with civil law rather than a criminal approach. One concern raised in relation to a civil law approach is the lower burden of proof that is implied compared to criminal courts, which potentially increases the chance of a miscarriage of justice (Emm 2009).

While the aim of criminal law is to discourage and penalise criminal behaviour, civil law is directed towards the provision of remedies that compensate the victim. Criminal law is therefore acknowledged as an essential element in any regulatory regime where the focus is on discouraging and punishing misbehaviour (Lipton 2011). Lipton (2011) argues that as a result of its significance within regulatory approaches, criminal legislation needs to be more effectively harmonised and directed specifically towards addressing the most widespread online abuses.

Where criminalisation is adopted certain factors are noted to be beneficial for inclusion. In relation to online abuse, this means establishing a reasonable standard in terms of the victim's state of mind. It is argued that criminal liability should be incurred when a victim reasonably fears for their safety or security, thus protecting communications and speech that may be unpleasant or cause emotional distress but are mostly harmless (Lipton 2011).

The choice between these approaches is not a straightforward one and is underpinned by numerous complexities. At the time of writing, the United Kingdom for example, had not determined whether the civil or criminal legal route is appropriate to address cyberbullying. More widely, problematic Internet use has increased in both civil and criminal legal proceedings for all countries (Recupero et al. 2006). Different laws exist across countries that cover a variety of serious actions, for example downloading child pornography or sexual solicitation of minors, cyberstalking and committing technological crimes (Recupero 2008). As discussed, different educational regulatory frameworks and preventative plans that cover cyberbullying have been developed. Campbell et al. argue that creating a specific criminal law for cyberbullying in the United Kingdom may not be the way forward, especially as it may mean criminalising immature youths who may not be aware of the potential impact of their actions. Current civil law as well as criminal law seem to be appropriate to tackle serious forms of cyberbullying and the introduction of preventative methods (as with traditional bullying) may prove to be more effective.

2.3 Authorisation Requirements

One of the critical ways of assessing the effectiveness of laws dealing with cyber victimisation in UAE is to review the authorisation requirements. In the UAE offenders can take given measures to complicate the process of investigation. On the other hand, there have been incidences where police officers and law enforcement authorities have been using software that can enable anonymous communication as well as identification of complicated data, especially when offenders are using public Internet terminals or open wireless networks. This is where an authorisation requirement becomes effective; according to Giordano (2004), an authorisation requirement is a restriction that law enforcement authorities can impose to deter manufacturers or offenders from creating or developing software that makes it hard to identify them or collect their data. This approach seems to have been adopted from Article 7 of Italian Decree 144, which the UAE has converted into law to deter instances of cyber victimisation. Consequently, the individual in question is required by UAE law to request identification from her/his customers before giving them access to the use of Internet related services. Since a private person who sets up a wireless access point is in general not covered by this obligation, monitoring can quite easily be circumvented if offenders make use of unprotected private networks to hide their identity (ICT 2014).

There may be shortcomings in the extent to which the authorisation requirement is applied in the UAE. Such concerns relate to whether the extent of improvement in terms of investigation can justify the restriction of access to the Internet and as such can be extended to anonymous communication services. There is realisation that free access to Internet is currently recognised as a vital aspect of the right of free access to information, especially those rights that have been protected by the legal framework in UAE. Therefore, in as much as countries have adopted authorisation requirements to limit instances of cyber victimisation, registration obligations as applied currently seems to limit efforts aimed at ending cybercrime in totality. As Giordano (2004) highlights, the current registration obligation is seen to be interfering with the right to operate Internet related services without such authorisation. The UK in particular has been affected by the adoption of such a requirement, following the 2005 Joint Declaration of the United Nation (UN) Special Rapporteur on Freedom of Opinion and Expression as well as the OSCE Representative on Freedom of the Media and the OAS Special Rapporteur on Freedom of Expression (OSCE 2005). While the adoption of this regulation could be a significant means of reducing instances of cyber victimisation in the UAE, it is also likely that the move would impact the use of the Internet, insofar as users may then fear that their Internet usage would always be monitored and investigated. Even when offenders are aware that their actions are legal or in good faith, it can still influence their interaction and usage. On the same note, offenders who would like to prevent identification can easily circumvent the identification procedures. Offenders can use prepaid phone cards that are bought in a foreign country and which do not require identification to access the Internet. There have been similar concerns regarding the legislation targeting anonymous communication services. In the wake of this, there is already a debate on whether similar instruments of encryption technology should be adopted for anonymous communication technology as well as services (Forte 2002). Other than the conflict between protecting privacy and ensuring the ability to investigate offences, the arguments arising against the practicability of the many legal approaches to addressing the challenge of encryption (especially enforceability) can be applicable equally to anonymous communication.

2.3.1 Regulation

Regulatory perspectives represent a further area of debate that provide insights to alternative or mutually supportive measures for combating cyber victimisation, including emphasis on self-regulation and shifting the burden of regulation to Internet Service Providers (ISPs).

Izak (2013) stresses that the majority of challenges within current regulatory approaches to cyber victimisation can be resolved by placing the burden of accountability for harassing behaviour and comments onto ISPs. Cyber harassment is noted to have significant parallels with copyright law and the potential usefulness of the Digital Millennium Copyright Act as a model for enforcing ISP liability is emphasised. According to Izak (2013), the principal failing of current cyber harassment law is the inability to take down offending content even once it has been identified. This recommendation implies that an ISP will be accountable for the harassing or defamatory comments of their users unless they move rapidly to remove or disable access to it once they have been notified (2013).

In contrast, Lee (2011) advocates implementing regulations that enforce take-down policy. Requirements enforcing a strict take-down policy are suggested by Lee (2011) to essentially include rapid response by Internet portals to objections in relation to insulting or defamatory content. This would involve temporary account deletion account for 30 days unless illegal content is not found; however, this is frequently not applied effectively, to the detriment of victims. A further impediment for victims is highlighted in the current Korean policies on take-down, which shifts the burden of evidence in terms of the content at issue to the victim. This is suggested not to be fair or in the victim's interests (Lee 2011). Lee (2011) advocates several proposals for formulating legal responses to cyber victimisation. Firstly, that online defamation be decriminalised in favour of an emphasis on public remedies. Secondly, strict definition and precision of expression in relation to identifying what constitutes online defamation is further advised. Also, Lee emphasises strict self-regulation.

The New Zealand Law Commission notes the essential importance of not locating legal and policy responses to cyber victimisation within conservative or defensive outlooks towards new technology. Its three-tiered approach proposed to address the issue of cyber victimisation includes self-regulation as well as legal responses and solutions. The first tier focuses on the notion of user empowerment, stressing user education about cyber rights and responsibilities and the provision of technical knowledge, allowing their exercise by users. The next tier involves increasingly adopted self-regulation systems to control bad behaviours and maintain standards, which frequently include contract 'terms of use' identifying unacceptable behaviours and procedures for dealing with breaches. The final tier is the regulatory structure of statutory and common law identifying the frontiers of acceptable communication for individuals that are applicable across all channels.

2.3.2 Jurisdiction

Jurisdiction emerges as a major challenge to combating cyber victimisation. Cyber victimisation occurs within the cyberspace, which exists without national boundaries or jurisdictions and provides great scope for the victim and perpetrator to be physically located in different countries (Levin et al. 2007). However, the municipal nature of the law influences a national and inward focus that creates jurisdictional complexities (Emm 2009). The legal system in the UAE for instance, even in relation to its neighbouring countries, would need to be significantly flexible and incorporate provisions to counter jurisdictional challenges relating to such matters as rule procedures, execution of legal decisions, and penalties.

An element of understanding cyber victimisation in UAE is that the act links different countries together naturally in as much as offenders and victims may be located in UAE (Frith and Frith 2003). Cases of victimisation that have targeted UAE and those that have been committed from within UAE were borderless and may not have acknowledged time or place elements. This challenges understanding of the demographics of cyber victimisation and demands strategies that can be used to understand the dispersion of international coordination and investigation to trace victims and offenders.

2.3.3 International Co-operation

The role of international co-operation in addressing and regulating cyber victimisation, given its transnational nature, is consistently stressed in the literature. Agate and Ledward (2013) argue that the continually evolving nature of cybercrime and cyber victimisation will potentially give rise to new legislation and regulations to combat new issues and emphasise the importance of international co-operation to share lessons and solutions (Agate and Ledward 2013).

There is considerable emphasis on harmonisation of legal approaches and collaboration across international institutions and governments to share expertise and practices and to develop a co-ordinated response. The implication for the UAE and for this research study is to ensure a framework that embraces this philosophy at each stage of the legal reform process. A global approach that reflects a high level of co-ordination towards legislation is viewed as vital to avoid the risk of disconnected and ineffective measures that fail to effectively protect citizens from becoming victims of cybercrime (Chawki 2005).

Major legal approaches to cybercrime around the world can provide valuable insight to the development of a legal approach in the UAE. In the Anglosphere of English-speaking countries (Australia, Canada, New Zealand, United Kingdom and the United States) discussions aimed at harmonisation commenced in 2011. Across these countries co-ordinated action and a global approach where possible is noted to address jurisdictional issues, and evidence the introduction of new cyber related legislation where needed (Emm 2009).

3 Theoretical Framework

The key themes reviewed in this chapter provide critical insights into each of three dimensions: the individual context, the cybercrime context, and the social control context. At the individual level the theory identifies the role of the individual and its relationship to vulnerability or exposure to cybercrime. Meanwhile social control theory provides a perspective on the role of society in preventing and combating cybercrime to reduce cyber victimisation. The cybercrime context establishes critical insight from the perpetuators' perspective in understanding the structures, motivations and resources that drive cybercrime. In order to examine the tenets of cyber victimisation, a highly contextual perspective is required that acknowledges the relationship between three key facets that have emerged in this literature review.

Kshetri's (2009) analysis of cybercrime supports this perspective in emphasising the interrelation and dynamic between three key dimensions outlined in. This shows that characteristics and key processes of law enforcement agencies, cyber criminals and cybercrime victims perpetuate and drive the growth of cybercrime. In terms of law enforcement agencies, the failure to keep pace with cybercrime technologies, a lack of experience and abilities in solving cybercrime, and in particular a lack of collaboration with the private sector and cooperation at the global

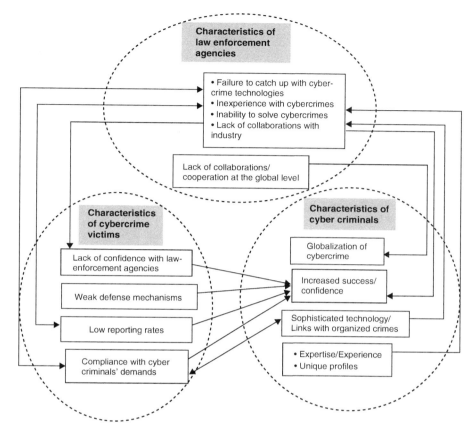

Fig. 2 Characteristics of key dimensions driving the cybercrime dynamic. (Source: Kshetri 2009, p. 39)

level contributes to cybercrime growth. Cybercriminal characteristics, including the globalisation of cybercrime, growing crime success and confidence, rising technological sophistication and linkages with other types of crime, cybercriminal expertise and experience and their unique abilities are revealed as further significant drivers perpetuating cybercrime. Finally, characteristics of cybercrime victims are proposed to contribute to driving the cybercrime dynamic. These include a lack of confidence in law enforcement, weak measures of self-protection, minimal reporting rates and a tendency to comply with the demands of cybercriminals (Kshetri 2010) (Fig. 2).

This model underscores the need for an integrated understanding of combating cybercrime by integrating the individual victim context with the cybercriminal behaviour and social context in terms of law enforcement. This literature review points to the significance of theories such as routine activity and self-control in explaining reasons why people are likely to engage in a given aspect of cyber

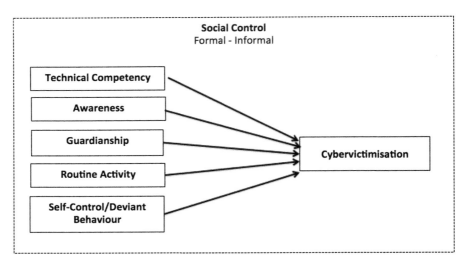

Fig. 3 Theoretical framework

victimisation, and why individuals perpetuate cybercrime. These theories focus on individual level features or characteristics or the situational context to increase chances of cyber victimisation occurring. The literature review points to a number of themes underpinning cyber victimisation.

As shown in Fig. 3, different themes such as technical competency, guardianship and self-control all have a role to play in cyber victimisation. This theoretical framework provides a basis for investigating the context of cyber victimisation and the effect of the characteristics of online individuals cybercrime victimisation. For instance, technical competency relates to a user's technical knowledge and abilities, which can significantly influence their behaviour, potential for being victimised and their ability to protect themselves. User awareness is considered to be a critical element of capable guardianship, holding strong significance for the reduction of cyber victimisation (Jahankhani and Al-Nemrat 2011, 2010). Guardianship is underlined as a key factor for minimising cyber victimisation (Jansen and Leukfeldt 2016) and relates to self-protection and the protective measures that users undertake to make themselves less vulnerable to cyber victimisation, linked to the level of technical skills and awareness users have to support a reduction of victimisation. The literature further points to routine activity as a key factor in cyber victimisation, relating to the extent to which a user's routine activities contribute to vulnerability and making themselves a suitable target for a motivated offender (Cohen and Felson 1979). The review has highlighted the influence of self-control and deviant behaviours within cybercrime: low self-control among other users may influence participation in and execution of deviant online behaviours and drive cyber victimisation (Gercke 2007).

4 Conclusion

This chapter has presented a review of key themes in the literature relevant to the focus of this research. Cybercrime has been defined and examined, identifying the significance and nature of this phenomena and providing an overview of the key forms of cyberattacks and the key technologies that constitute a broad array of systems, infrastructures and tools that are facilitators of cybercrime. A perspective has been explored in terms of cybercriminals and examining classifications of cybercriminal groups, structures, motivations and the nature of crime and the dark web. Space transition theory provides critical insights into the behaviours of criminals. In Sect. 2.4 the focus is placed on understanding the impact of cybercrime in terms of the different forms of cyber victimisation and examines the diverse range of ways in Key theory has been presented that is critical in understanding the role of the individual's behaviour and attitude and its relationship to cybercrime. Self-control theory and deviant behaviour and routine activity theory are discussed in relation to cyber victimisation and user behaviour enabling or mitigating cybercrime. A review of social control theory has provided insights into types of informal and formal social control mechanisms that can be applied to combat cybercrime and reduce and prevent cyber victimisation.

The themes reviewed critically underpin the theoretical framework presented in section that provides a theoretical basis for this research based on the factors and dimensions identified in the literature review.

References

Agate, J., & Ledward, J. (2013). Social media: How the net is closing in on cyber bullies. *Entertainment Law Review, 24*(8), 263–268.

Agustina, J. R. (2015). Understanding cyber victimisation: Digital architectures and the disinhibition effect. *International Journal of Cyber Criminology, 9*(1), 35.

Akhgar, B., & Yates, S. (Eds.). (2013). *Strategic intelligence management: National security imperatives and information and communications technologies*. Waltham: Butterworth-Heinemann.

Barker, R. A. (2012). An examination of organizational ethics. *Human Relations, 55*(9), 1097–1116.

Baxter, A. (2014). *Improving responses to cyber victimisation in South Australia*. http://www.victimsa.org/files/cybercrime-report-2014.pdf.

Bonanno, R. A., & Hymel, S. (2013). Cyber bullying and internalizing difficulties: Above and beyond the impact of traditional forms of bullying. *Journal of Youth and Adolescence, 42*(5), 685–697.

Cénat, J. M., Hébert, M., Blais, M., Lavoie, F., Guerrier, M., & Derivois, D. (2014). Cyberbullying, psychological distress and self-esteem among youth in Quebec schools. *Journal of Affective Disorders, 169*, 7–9. https://doi.org/10.1016/j.jad.2014.07.019.

Chawki, M. (2005). A critical look at the regulation of cybercrime. *The ICFAI Journal of Cyberlaw, IV*(4), 1–56.

CNNMoney. (2015). Cybercrime costs the average U.S. firm $15 million a year. *CNN*. [Online]. Available at: http://money.cnn.com/2015/10/08/technology/cybercrime-cost-business/index.html. Accessed 3 Apr 2017.

Cohen, L., & Felson, M. (1979). Social change and crime rate trends: A routine activity approach. *American Sociological Review, 44*, 588–608.

Dredge, R., Gleeson, J., & de la Piedad Garcia, X. (2014). Cyberbullying in social networking sites: An adolescent victim's perspective. *Computers in Human Behavior, 36*(0), 13–20.

Edwards, L. (2012). Defining the 'object' of public relations research: A new starting point. *Public Relations Inquiry, 1*(1), 7–30.

Emm, D. (2009). *Cybercrime and the law: A review of UK computer crime legislation* [Online]. Available at: http://www.securelist.com/en/analysis/204792064/Cybercrime_and_the_law_a_review_of_UK_computer_crime_legislation. Accessed 3 Apr 2017.

Forte, D. (2002). Analyzing the difficulties in backtracing onion router traffic. *International Journal of Digital Evidence, 1*(3), 1–7.

Frith, U., & Frith, C. D. (2003). Development and neuropsychology of mentalizing. Philosophical Transactions of the Royal Society. *Biological Sciences, 358*, 459–473.

Gercke, M. (2007). *Cyberterrorism. How terrorists use the internet* (p. 62). Computer und Recht.

Giordano, S. M. (2004). Electronic evidence and the law. *Information Systems Frontiers, 6*(2), 161–174.

Goodno, N. H. (2007). CS, a new crime: Evaluating the effectiveness of current state and federal laws. *Missouri Law Review, 72*, 125–197.

Hazelwood, S. D., & Koon-Magnin, S. (2013). Cyber stalking and cyber harassment legislation in the United States: A qualitative analysis. *International Journal of Cyber Criminology, 7*(2), 155.

ICT Regulation Toolkit, Privacy and Data Retention Policies in Selected Countries [Online]. Available at: www.ictregulationtoolkit.org/en/PracticeNote.aspx?id=2026. Accessed 18 Sept 2014.

Izak, J. (2013). Cyberharassment Is a true danger: Is the solution found in copyright law? Doctoral dissertation, Michigan State University.

Jahankhani, H. (2013). Developing a model to reduce and/or prevent cybercrime victimization among the user individuals. In: B. Akhgar & S. Yates (Eds.), *Strategic intelligence management: National security imperatives and information and communications technologies*. Butterworth-Heinemann.

Jahankhani, H., & Al-Nemrat, A. (2010). Examination of cyber-criminal behaviour. *International Journal of Information Science and Management, 2010*, 41–48.

Jahankhani, H., & Al-Nemrat, A. (2011). Cybercrime profiling and trend analysis. *Intelligence Management*, 181–197.

Jahankhani and Askerniya, I. (2012). How best to protect the user-individuals in Moscow from cyber crime attacks. [Online]. Available at: https://pdfs.semanticscholar.org/c0bc/ff81bad5282abfe1a4e2d4ab2d58a3b3471b.pdf. Accessed 31 March 2017.

Jansen, J., & Leukfeldt, R. (2016). Phishing and malware attacks on online banking customers in the Netherlands: A qualitative analysis of factors leading to victimization. *International Journal of Cyber Criminology, 10*(1), 79.

Kshetri, N. (2009). Positive externality, increasing returns, and the rise in cybercrimes. *Communications of the ACM, 52*(12), 141–144.

KPMG. (2017). Cybercrime survey report: Insights and perspectives [Online]. Available at: https://assets.kpmg.com/content/dam/kpmg/in/pdf/2017/12/Cyber-Crime-Survey.pdf. Accessed 7 April 2015.

Lee, H. K. (2011). Cultural consumer and copyright: A case study of anime fansubbing. *Creative Industries Journal, 3*(3), 237–252.

Levin, A. M., Dato-on, M. C., & Manolis, C. (2007). Deterring illegal downloading: The effects of threat appeals, past behavior, subjective norms, and attributions of harm. *Journal of Consumer Behaviour, 6*(2-3), 111–122.

Liechti, O., & Sumi, Y. (2002). Editorial: Awareness and the WWW. *International Journal of Human-Computer Studies, 56*(1), 1–5.

Lipton, J. D. (2011). Combating cyber-victimization. *Berkeley Technology Law Journal, 26*(2), 1103–1155.

Marczak, M., & Coyne, I. (2010). Cyberbullying at school: Good practice and legal aspects in the United Kingdom. *Australian Journal of Guidance & Counselling, 20*(2), 182–193.

Netanel, N. W. (2000). Cyberspace self-governance: A skeptical view from liberal democratic theory. *California Law Review, 88*(2), 395–498.

OSCE. (2005). Joint declaration by the UN special rapporteur on Freedom of opinion and expression, the OSCE representative on Freedom of the media and the OAS special rapporteur on Freedom of expression [Online]. Available at: https://www.osce.org/fom/27455?download=true. Accessed 31 March 2017.

Recupero, R. P., Harms, S. E., & Noble, M. J. (2006). Googling suicide: Surfing for suicide information on the Internet. *The Journal of Clinical Psychiatry, 69*(6), 878–888.

Roberts L. (2008). Cyber victimisationin Australia: Extent, impact on individuals and responses. *Tasmanian Institute of Law Enforcement Studies,* (6), 1–12.

Trim, P. R. J., & Lee, Y. I. (2015). Issues that managers need to consider when undertaking research relating to the cyber environment. In P. R. J. Trim & H. Y. Youm (Eds.), *Korea-uk initiatives in cyber security research: Government, University and Industry collaboration* (pp. 66–79). Republic of Korea: British Embassy Seoul.

UNODC. (2013). *Cybercrime*. Retrieved from United Nations office on drugs and crime website: https://www.unodc.org/documents/data-and-analysis/tocta/10.Cybercrime.pdf.

Information Security Landscape in Vietnam: Insights from Two Research Surveys

Mathews Nkhoma, Duy Dang Pham Thien, Tram Le Hoai, and Clara Nkhoma

1 Background

The advancement of information technology brings both benefits and risks to organisations. For instance, modern technology such as analytics for storing and analysing a massive volume of data allows companies to rapidly expand their databases of customer information and create values from big data (Erevelles et al. 2016). Additionally, contemporary practices such as cloud computing and 'Bring Your Own Device' provide employees with greater flexibility and access to organisational information assets stored on devices that are not managed by the companies' IT departments (Morrow 2012). As organisations rely more on novel technology and practices, they become more exposed to the increasing number of cyber-threats leading to data breaches.

In fact, data breaches can be costly to organisations. In 2016, the total average damage caused by data breaches was 4 million US dollars (Ponemon Institute LLC 2017). Although the cost of data breaches decreased in 2017 to 3.62 million US dollars, it remained a major damage to organisations. Schneier (2011) suggested that the level of InfoSec investment should be equivalent to the value of information assets that can be potentially lost from data breaches. Moreover, Von Solms and Von Solms (2004) discussed that InfoSec is not only a technological issue. The cause of InfoSec incidents not only comes from inadequate technological protection but also from the deliberate or careless misbehaviours of internal stakeholders, which are often referred to as the insider threats (Safa et al. 2018).

M. Nkhoma (✉) · T. Le Hoai · C. Nkhoma
Business & Management, RMIT University, Vietnam, HCMC, Vietnam
e-mail: mathews.nkhoma@rmit.edu.vn

D. Dang Pham Thien
Science & Technology, RMIT University, Vietnam, HCMC, Vietnam
e-mail: duy.dangphamthien@rmit.edu.vn

The undesirable InfoSec behaviours of the insiders can be classified as intentional or unintentional (Crossler et al. 2013). Additionally, Stanton et al. (2005) added technical expertise as another dimension that further categorised the insiders' InfoSec behaviours. For example, insiders having malicious intention and a high level of technical skills can cause destruction such as hacking into the company's information systems to steal confidential data. On the other hand, insiders having high-level technical expertise but less malicious intention would be more likely to perform dangerous tinkering such as opening a wireless gateway for external usage (Stanton et al. 2005). The potential threats coming from the insiders were regarded as even greater than those performed by external agents, due to the insiders' easy access to the confidential assets and their knowledge of the organisations (Colwill 2009). In fact, the damage caused by the insiders' malicious behaviours, especially those carried out by insiders who are recruited by competitors and trained to excerpt sensitive data, is considered to be the most severe (Roy Sarkar 2010).

In this book chapter, we contribute a review of the organisational dimensions that have impacts on the InfoSec environment. By combining the findings derived from two nationwide surveys and a case study conducted with InfoSec experts in Vietnam, a developing country in Southeast Asia, we analyse those dimensions and discuss the challenges and opportunities for InfoSec improvements in such a context. Then, we provide recommendations for InfoSec practitioners and managers.

The two research surveys and a qualitative study were conducted in Vietnam, a developing country with an increasing penetration of the Internet. As of 2016, the Internet penetration in Vietnam climbed to 52% and ranked 13th in the world (Internet Live Stats 2016). Nevertheless, the information security has not grown at the same pace. Although the majority of Vietnamese companies has implemented technological solutions such as anti-virus or firewall, only 19% of them have an information system management (Vietnam MIC 2017). In 2016 and 2017, there was multiple security incidents in the country that causes severed damages. The most talked-about incident is the attack on Vietnam Airline's customers database causing the leak of 410,000 VIP members data (Tuoi Tre 2016a). Around the same time, the two major airports Tan Son Nhat and Noi Bai information system was hacked to display inappropriate messages about the relationship between Viet Nam and the Philippines (Tuoi Tre 2016b). This particular attack, as admitted by the Ministry of Communication, resulted in a damage that was five times more severe than the previous attack in 2015 (Vietnam MIC 2017). In addition to all the major incidents, Vietnam was listed among the top ten countries with the most computers compromised by banking trojan in 2016 (Symantec 2017).

2 Framework of Analysis

The effective protection of information assets requires a multidisciplinary approach, which demands all organisational dimensions to be taken into consideration when designing and implementing InfoSec measures. A comprehensive list of dimension in Von Solms (2001) proposed a list of 13 dimensions of the InfoSec research discipline, namely strategic/corporate governance, governance/organisational, policy, best practice, ethical, certification, legal, insurance, personnel/human, awareness, technical, measurement/metrics and audit. Consistent with these dimensions, Da Veiga and Eloff (2007) discussed six organisational domains of InfoSec, which consists of leadership and governance, security management and organisation, security policies, security program management, user security management, technology protection and operations. Dzazali et al. (2009) discussed in their study the dimensions of corporate governance, organisational policy, best practice, ethical, compliance, legal, personnel/human, awareness, technology, measurement/metrics and audit. Additionally, budget and economics were considered as part of the organisational InfoSec's dimensions (Silic and Back 2014). More recently, Safa (2017) presented another set of five aspects affecting organisational InfoSec, namely technological, human, managerial, education and awareness, and social and cultural. In line with the discussions of these studies, we elaborate on the four key dimensions of organisational InfoSec that have been consistently emphasised by the extant literature. These key dimensions are (1) technology, (2) employees, (3) management and (4) legal and compliance.

2.1 Technology

Historically, information security is perceived to be a technological issue. Research in information security field prior to 2007 places a primary attention to technological aspects while overlooking other dimensions (Siponen and Oinas-kukkonen 2007). The focus on technology in designing an effective information security defence system elevates the role of technology. As much as it claims the trophy of protecting information assets, technology takes the blame for data breaches when they occur (Dodds and Hague 2004). Although it is no longer only about technology in information security, the application of technology to build a strong denfense system remains vital. Across organisations, a major scope of the IT department is to ensure technological part of information security is up-to-date.

Technological aspects of information technology cover all methods that prevent and tackle security breaches by software, hardware or a combination of both, classified as proactive and reactive technology. The preventive technology that organisations rely on is referred to as the proactive information security technology (Venter and Eloff 2003). Different from proactive technology that is utilised before the breach occurs, reactive technology is triggered as soon as the breach is detected (Venter and Eloff 2003).

The table below introduces some of the popular threats that can be prevented by security software presented by Safa (2017).

Threats/incidents	Definition	Technological solution
Adware	Programs that monitor Internet users' online activities in order to initiate pop-up advertising or other targeted marketing activities	Anti-adware, anti-spyware
Keyloggers	Programs that capture and record Internet users' every keystroke, including personal information and passwords	Anti-logger, anti-virus, anti-malware
Trojans	Malicious programs that appear as harmless or desirable applications, but are designed to cause loss or theft of computer data, or even to destroy the system	Firewalls. anti-virus

2.2 Employees

People have direct access to information assets and network in the organisation. Despite the existence of software, hardware and practice to prevent security threats, it all depends on users to execute. Therefore, protecting information assets has always been about the technologies, process and the users. Although a numerous literature studying malicious insiders as a risk to InfoSec, it has also been recognised that inadvertent actors make up a significant share of incidents leading to security attack. For example, InfoSec report throughout the year has shown that employees intentional or unintentional InfoSec mishandling are amongst the top cause of attacks (Kaseya 2013). Employees sometimes do not aware of the risk of sharing the same account and password or leaving their devices unattended.

On the other hand, employees can help organisations secure their environment for information assets by complying with the security and procedures. Performing InfoSec behaviours according to the policies and procedures depend on a number of factors: benefits of compliance, cost of compliance and cost of noncompliance (Bulgurcu et al. 2017). The evaluation of benefits and cost of compliance/ noncompliance is influenced by the employees' perception of risks and barrier to secure the information assets (Dzazali et al. 2009). Such perception, as well as the awareness of InfoSec, can be improved through education and awareness program.

Information security skills can be improved over time. However, it is not an intuitive process that people can acquire the skills naturally. For most people, a training program must be in place to guide them through the process. According to (Puhakainen and Siponen 2010), there are a number of criteria in a training program that aid the effectiveness of the program:

– Training method should enable employees' cognitive processing of information
– Learning tasks should be personally relevant to employees daily tasks
– Training should be according to employees' previous knowledge of InfoSec
– Training should be made into a continuous communication rather than a one-off

The increase in InfoSec awareness comes together with the enrichment of security knowledge and application. The positive outcomes of the training program has been evidenced in a number of studies. Eminağaoğlu et al. (2009) found three ways the awareness training program yield impact on employees behaviours: (1) decreasing InfoSec bad practices, (2) increasing the InfoSec good practices, (3) involving employees in InfoSec controls and mechanism, (4) changing the attitude of InfoSec compliance (from reluctant to willing to comply). The empirical evidence of the relationship between awareness and actions toward InfoSec was presented in (Choi et al. 2014), in which the positive influence of managerial awareness on managerial actions, and of managerial actions on the overall security performance of the company was concluded.

2.3 Management

Chang and Ho (2015) argue that InfoSec is a business issue, so top managers should be involved. Information security management spans across three layers: strategic, tactical and operational (Narain Singh et al. 2014). The strategic role of managing security is to ensure that the goals of this task align with the business objectives. In order to do that, planning must take place but it is not always an easy task. The challenge of security planning is when there is a conflict of security management goal and business objective. Dzazali et al. (2009) illustrated the point by introducing the dilemma the Immigration Department of Malaysia faced. While enabling online services allowing citizens to access their own database, which subsequently improves the efficiency of public service, the process exposes citizens personal information to the risks of data breaches. In those cases, the policy and procedures- the outcome of the planning process- need to be in place to minimise the risks. Complimenting the policy and procedures is a guideline- the tactical aspect of security management. Thanks to the tactical element of security management, it supports the fast pace of organisational and technological change, allowing the organisation stakeholders to prevent the associated threats while maintaining the momentum (Caralli and Wilson 2004). The last element of security management relates to the measurement of security. Eloff and Eloff (2005) suggest that by having key performance indicators audit, organisation can evaluate the improvement of the security architectures and compare themselves against other organisation; thus make necessary changes.

2.4 Legal and Compliance

The compliance component of information security covers both internal compliance with the company information security policy and the company compliance with the government laws and regulations. A report from Ernst and Young (2007) revealed that compliance with regulation is one of the top drivers of information practice in organisations (cited in Breaux et al. 2009). Over the decades, there has been multiple legal laws and regulations concerning the confidentiality, privacy and data integrity. Examples include the EU privacy directive, GLMR Bill in the USA. Most recently is the launch of General Data Protection Privacy in the EU, which will affect firms that are handling personal information of EU citizens (McKinsey 2017). Although failing to comply to the laws and regulations yield legal and financial risks to organisations, not all companies move fast enough to reflect the change in the legal requirements. The change in information security practice is complicated due to the human aspect involved and oftentimes delayed due to the company norm and culture (Smith et al. 2010).

3 Research Design

The objective of this chapter is to draw a information security landscape in Vietnamese small and medium businesses. To achieve this objective, we used a multilevel methodological approach. At the first stage of the project, we surveyed 504 IT/information security experts who were working for SMEs across industries. Only respondents who were in managerial position and had the power to make decisions regarding information security were qualified to answer the online questionnaire. Each of the respondents represented one Vietnamese company. Questions included in this phase were to understand information security from the managerial and organisational perspectives. Three main parts of the questionnaire included: (1) industry, data storage and information security threats, (2) organisational investment in information security and (3) information security training across industries.

At the second stage of the project, we sought to understand information security practice from the end users' perspective, in this case – the employees'. Surveys were randomly distributed to employees in both managerial and non-managerial positions, asking them to describe their level of security expertise and common practices of handling the security of their personal devices. Four hundred respondents across 27 industries in Vietnam agreed to answer and completed our surveys.

At the last stage of the project, we conducted 23 in-depth interviews with information security experts and end-users. To recruit the respondents, we sent invitations through social media and professional forums. Qualified respondents must pass the screening criteria, which required them to have working experience in information security field (for experts). Another criteria respondents need to meet was that their company must have security policy and require security compliance at work (for end-users). One hour of the interview centered around the topic of information security support and compliance in Vietnamese organisations. The interview were one hour long and carried in either Vietnamese or English, depending on which language the respondent felt comfortable with. The profiles of the respondents are stated in Table 1.

Table 1 Respondents profiles

End-user (U) Security manager/expert (E)	Occupation	Industry
U1–6	Counter teller	Banking
U7–8	Accountant	
U9–12	University lecturers	Education
U13–14	Admin staff	
U15–16	Marketing executive	Oil distribution
EX1	IT Auditor/consultant	Financial
EX2	IT manager	IT services
EX3	Security consultant	Banking
EX4	Security officer	IT services
EX5	Deputy IT director	Banking
EX6	Data security manager	Engineering
EX7	IT director	Education

4 Findings

4.1 Sensitive Data

Our research conducted in Vietnam in 2013 revealed that 53.4% of 504 surveyed organisations considered at least one out of five types of data as sensitive and critical for their business, including data about their customers, business partners and suppliers, finance and products. Customer data were defined to cover personal details and historical transactions between the customers and the companies. Similarly, data about business partners and suppliers contain names, key contact persons and records of past transactions. Financial data include figures about budgets and cash flows of the companies, while product data cover aspects related to the design and manufacturing of the products.

As shown in Fig. 1, 19.8% of organisations reported to treat all four types of data as equally important, followed by two (14.3%) and three types of data (12.5%).

As shown in Fig. 2, 59.13% of the surveyed companies reported to store sensitive data about their customers, followed by data about business partners and suppliers (48.81%), finance (47.02%), products (42.46%) and other types (1.79%). The statistics in Fig. 2 also indicate that there were more firms perceiving data about customers and business partners and suppliers as sensitive, whereas financial and product data were identified by fewer firms as their sensitive data.

Consumers information were popular across all industries, especially B2C enterprises. The leading industries which stored a large amount of consumers data were travelling, tourism, restaurant (95%) and retails (76%). On the other hand, partners and suppliers information were commonly held by B2B companies in industries such as journalism (75%), trading, import and export (62%). One

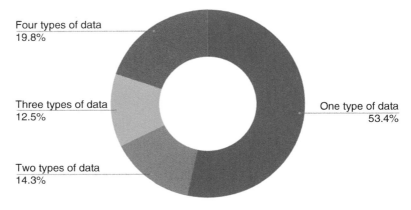

Fig. 1 Number of sensitive data types

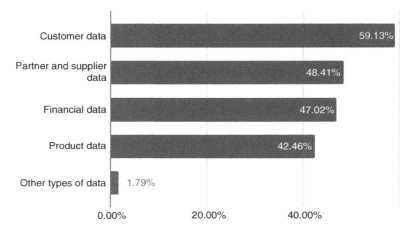

Fig. 2 Types of sensitive data

particular case was advertising agencies, which majority of them stored both types of information: consumers data (82%) and partners information (63%) due to its nature of business. Advertising agencies' stakeholders includes both business clients and consumers. Their clients, who pay for the service, are other businesses, while the target audiences of their service (advertisements) are consumers.

Most of the banking & finance companies reported to store financial data (73%). Coming in the second place was companies in the construction company (65%). Since financial information is the core of businesses in banking & finance industry, the scope of financial data hold by companies in this industry might be much wider than those operating in construction industry, which financial data they store might only be the clients' budget.

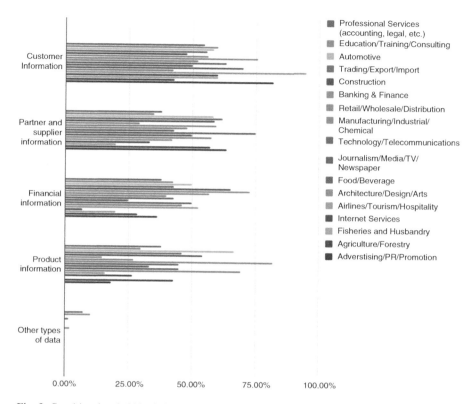

Fig. 3 Sensitive data held by industries

Product information becomes popular, and perhaps more sensitive to industries in which companies compete against each other on the aspects of the products. Those includes companies in manufacturing (82%), art and design (69%) and automotive (67%) (Fig. 3).

4.2 Targeted Attacks and Data Breaches in Vietnam

It's worth looking back to the previous year to see how the targeted attackss organisations trend. In 2012, 22% of the interviewed companies claimed that they had been targeted for information security attack. The rate climbs to 27% in 2013 resulting in a 23% increase compared to the previous year. The trend in Vietnam contradicted to the global stagnant trend but on average, the risk for targeted attacks in small and medium Vietnamese organisations was lower than the global level. In 2012, 31% of the small and medium businesses suffered from the risk of targeted attacks. The rate slightly dropped by 30% in 2013 (Symantec, 2013) (Fig. 4).

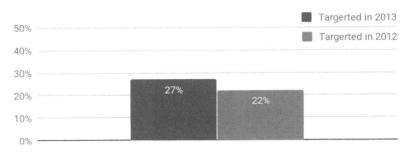

Fig. 4 Targeted attacked in Vietnamese SMEs

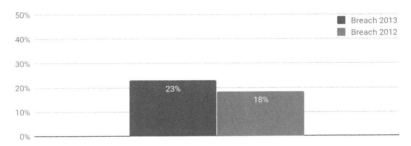

Fig. 5 Data breached incidents in Vietnamese SMEs

2013 globally was described as the year of mega breach not only due to the humongous amount of incidents but also the large scale of breaches during this year. At the global level, the total number of breaches grew by 62% compared to 2012. There were eight major breaches which leaked over ten million identities, while in 2012 there was only one (Symantec, 2013). PwC in the UK reported up to 87% small organisations with security breaches incidents in 2013, an increase from 76% the previous year (PwC, 2013). The rate of breaches in Vietnam was not as high as in a developed country like the UK but shared the same upward trend. Twenty-three percent of the interviewed Vietnamese companies reported the breach of InfoSec in 2013, while that of 2012 was only 18% (Fig. 5).

4.3 Managerial Aspect

Getting top management support to implement a proper information security system is vital. Our in-depth interviews with security experts in Vietnamese companies shared a common viewpoint about the role of top management in this process. With the approval of the top manager, it is easier to pull necessary human and financial resource to build up an effective system.

> *Top management's buy-in is extremely important for InfoSec implementation. (EX3)*
>
> *The second stage [designing the InfoSec implementation] is very important. The key activity in this stage is to present your implementation plan to the board of management, and you must convince them that the proposed InfoSec measures are crucial for the company and receive their support. (EX6)*

However, the security experts we interviewed did not satisfy with the support they received from the managers. They mentioned the low security awareness of the top management, who believed that technical implementation such as software, hardware and network security are enough. The people component was tremendously overlooked; therefore, education and training do not occur often. In general, the interviewed security experts agreed that top managers in their company understate the benefit of having a strong information security system, which then results in a lack of resources and support for security system.

> *Large enterprises in Vietnam do have InfoSec departments that are dedicated to take care of InfoSec issues, but they mainly focus on the hardware, software, or network security ... they have not yet realised the importance of the people and process components of InfoSec management. That's why they are not very supportive when it comes to training and enforcing procedures. (EX5)*
>
> *Most top management of companies in Vietnam has not yet developed a mindset that sees InfoSec as important. It is understandable, since the companies in Vietnam still remain at the level of thinking about how to survive, rather than how to improve. (EX2)*

The findings from the qualitative phase explained the low budget for InfoSec we discovered in the prior phase. The budget for InfoSec varies depending on the company size and which industry it is operating in. For small size businesses, the investment for InfoSec ranging from $100,000 to $500,000 per year and typically accounts for 3–4% of the IT budget (Filkins 2016). Results from our survey in Vietnam in 2013 indicated a much lower budget for InfoSec investment with 86% of the small companies spent less than $5000. Although the budget for InfoSec in Vietnam was low, it does not necessarily mean that companies undervalue the security defence. It might reflect the low investment in IT infrastructure for Vietnamese companies in general (Fig. 6).

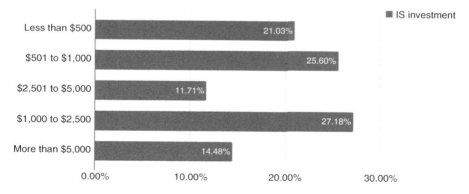

Fig. 6 InfoSec investment

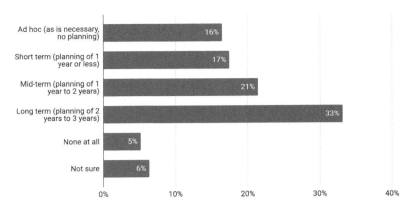

Fig. 7 InfoSec planning

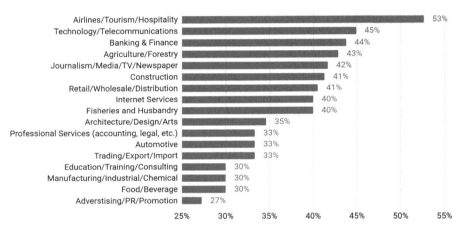

Fig. 8 Training occurence by industries

Planning for InfoSec in Vietnam does occur in the organisations. The majority of the company (33%) relied on a long-term plan while 38% of them had either short-term or mid-term plan. Sixteen percent of the interviewed company took counterpart actions whenever it is necessary without any plan (Fig. 7).

The majority of Vietnamese companies in 2013 did not frequently provide training for employees. Our survey revealed only 37% of the organisations frequently trained their employees on this topic. The industry with a significantly more company provided training relatively to other industry was travelling and tourism with 57%. Following that was tech organisations (45%), banking and finance companies (44%). Advertising companies, despite most of them storing both clients' data and consumers information, rarely organised any training for their employees (Fig. 8).

4.4 Employees and Compliance

In another survey conducted in 2015, results indicated that employees were self-reported to have skills in InfoSec. Sixty-five percent of the respondents believed their skills were intermediate, 14% of them had advanced skills, 6% were expert in InfoSec and the rest just began to develop this skill set. Although the majority of the respondents claimed to be intermediate users or higher, most of them did not demonstrate behaviours that were helpful for minimising the risk of information security breach. For example, 68% of the respondents did not lock the screen nor sign out of their working devices, 67% still clicked through suspicious websites or emails. When working with emails, most of them neglected the examination of email senders, link or attachment. More than half of the respondents reported to open the emails, link or attachment without knowing if it came from genuine senders.

Security experts and employees in Vietnamese companies all agreed that the cost of information security compliance is inevitable.

> *There is no way that information security is comfortable, it is simply a trade-off of being secure and other things. (E2, IT Manager)*

The cost of compliance for employees is referred to as time-consuming and too complicated to follow. Experts believed that such cost is a hurdle to increase the employees' compliance. That explains our results from the quantitative survey, in which the majority of the respondents still failed to comply with the good security practice.

> *For me as a user I understand that security compliance is only complying with organisation's requirements. However, I do not see the problem–why I need to do that. For the users, I find it wastes too much time. (U9, University lecturer)*

> *The security task is time-consuming. Some require just a couple of minutes of time while others virtually take away my time. (U3, Counter teller)*

Another paradox from the surveys is that despite self-reported intermediate skills, employees did not exercise simple practices that can prevent the information leakage. From the in-depth interview, we found out that employees did not recognise which skills they were lacking until they were asked to perform complicated security procedures. Intermediate skills that the majority of the respondents claimed they possessed might be an overstatement, as one IT director mentioned in the in-depth interview (Figs. 9, and 10):

> *Some end-users can set up own WI-FI network at work or home. There is no password, security protection on the WI-FI. Anyone can access their own network. People think they know but actually, they don't. (E7, IT director)*

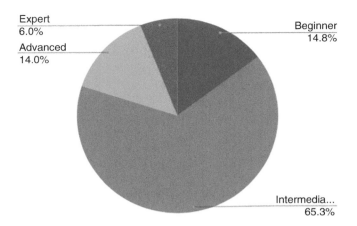

Fig. 9 Perceived InfoSec skills

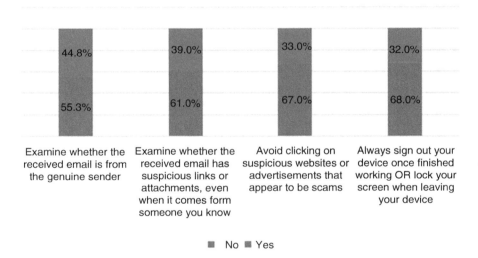

Fig. 10 InfoSec behaviours

5 Discussion

Information security is a multi-discipline matter. A holistic examination of InfoSec requires reviewing categories including technological aspect, human aspect- social and culture included, education and awareness and managerial aspect (Safa et al. 2018). Taken together, the reviews in this paper reveal current InfoSec state in small and medium businesses in Vietnam. The technological aspect was omitted due to the limitation in data collection.

Like large enterprises, small and medium businesses in Vietnam deal with a lot of data and suffer from the risk of being targeted. Most of the companies are working with at least one type of sensitive and critical data for their business, primarily consumer data. The type of sensitive data companies store depends on the industries which they are operating in. There is an increasing amount of businesses as victims of targeted attackss and data breaches over year, however not as much as in a developed country.

Information security investment is much less than the global average. There are three potential reasons for this. Filkins (2016) suggested that the investment for InfoSec typically accounts for 4–6% of the total IT budget. If IT budget in Vietnamese companies is small, that explains the tight budget spent on information security. Second, the value of the data stored in organisations is not significant or is underestimated. Industry experts believed that a good InfoSec investment should reflect the cost of data loss caused by a security breach (Schneier 2011). The managerial security awareness might be at a low level, which will then result in managerial non-action regarding improving security infrastructure in the companies (Choi et al. 2014). The InfoSec skills of employees are perceived to be at the intermediate level. However, their behaviours in some stated scenarios in the interview mostly does not follow the best practice. Giving the low security training rate, the awareness about the risk and threats of misbehaviour might as well be low. Alternatively, while working on their devices, employees might weight the benefits of noncompliance over the cost of noncompliance, leading to the noncompliance to best practices (Bulgurcu et al. 2010).

References

Breaux, T. D., Antón, A. I., & Spafford, E. H. (2009). A distributed requirements management framework for legal compliance and accountability. *Computers & Security, 28*(1–2), 8–17. https://doi.org/10.1016/j.cose.2008.08.001.

Bulgurcu, B., Cabusoglu, H., & Benbasat, I. (2010). Information security policy compliance: An emperical study of rationality- based belief's and information security awareness. *MIS Quarterly, 34*(3), 523–548.

Bulgurcu, B., Cavusoglu, H., & Benbasat, I. (2017). Information security policy compliance : An empirical study of rationality-based beliefs and information security awareness. *MIS Quarterly, 34*(3), 523–548.

Caralli, R., & Wilson, W. (2004) *The challenges of security management*. Pittsburgh: CERT, Software Engineering …. Available at: http://www.ready.gov/sites/default/files/documents/files/challenges_of_security_management[1].pdf.

Chang, S. E., & Ho, C. B. (2015). Organisational factors to the effectiveness of implementing information security management. *Industrial Management & Data Systems, 106*(3), 345–361. https://doi.org/10.1108/02635570810844124.

Choi, N., Kim, D., Goo, J., & Whitmore, A. (2014). Knowing is doing: An empirical validation of the relationship between managerial information security awareness and action. *Information Management & Computer Security, 16*(5).

Colwill, C. (2009). Human factors in information security: The insider threat – Who can you trust these days? *Information Security Technical Report*. Elsevier Ltd, *14*(4), 186–196. https://doi.org/10.1016/j.istr.2010.04.004.

Crossler, R. E., Johnston, A. C., Lowry, P. B., Hu, Q., Warkentin, M., & Baskerville, R. (2013). Future directions for behavioral information security research. *Computers & Security*. Elsevier Ltd, *32*, 90–101. https://doi.org/10.1016/j.cose.2012.09.010.

Da Veiga, A., & Eloff, J. H. P. (2007). An information security governance framework. *Information Systems Management, 24*(4), 361–372. https://doi.org/10.1080/10580530701586136.

Dodds, R., & Hague, I. (2004). Information security – More than an IT issue. *Chartered Accountants Journal, 83*(11), 56.

Dzazali, S., Sulaiman, A., & Zolait, A. H. (2009). Information security landscape and maturity level: Case study of Malaysian Public Service (MPS) organisations. *Government Information Quarterly*. Elsevier Inc., *26*(4), 584–593. https://doi.org/10.1016/j.giq.2009.04.004.

Eloff, J. H. P., & Eloff, M. M. (2005). Information security architecture. *Computer Fraud & Security, 2005*(11), 10–16. https://doi.org/10.1016/S1361-3723(05)70275-X.

Eminağaoğlu, M., Uçar, E., & Eren, Ş. (2009). The positive outcomes of information security awareness training in companies – A case study. *Information Security Technical Report, 14*(4), 223–229. https://doi.org/10.1016/j.istr.2010.05.002.

Erevelles, S., Fukawa, N., & Swayne, L. (2016). Big Data consumer analytics and the transformation of marketing. *Journal of Business Research*. Elsevier Inc., *69*(2), 897–904. https://doi.org/10.1016/j.jbusres.2015.07.001.

Ernst & Young. (2007). *Tenth annual global information security survey: Achieving a balance of risk and performance.*

Filkins, B. (2016). *IT security spending trends.*

Internet Live Stats. (2016). *Internet users by country (2016)*. Available at: http://www.internetlivestats.com/internet-users-by-country/.

Kaseya. (2013). *The top seven causes of major security breaches.*

McKinsey. (2017). *Tackling GDPR compliance before time runs out*. Available at: https://www.mckinsey.com/business-functions/risk/our-insights/tackling-gdpr-compliance-before-time-runs-out.

Morrow, B. (2012). BYOD security challenges: Control and protect your most sensitive data. *Network Security, 2*(12), 5–8. https://doi.org/10.1016/S1353-4858(12)70111-3 Elsevier Ltd.

Narain Singh, A., Gupta, M. P., & Ojha, A. (2014). Identifying factors of "organisational information security management". *Journal of Enterprise Information Management, 27*(5), 644–667. https://doi.org/10.1108/JEIM-07-2013-0052.

Ponemon Institute LLC. (2017). *2017 cost of data breach study* (pp. 1–34). Available at: https://www-01.ibm.com/common/ssi/cgi-bin/ssialias?htmlfid=SEL03130WWEN&.

Puhakainen, P., & Siponen, M. (2010). 'Improving employees' compliance through information systems security training: An action research study. *MIS Quarterly, 34*(4), 757–778.

Roy Sarkar, K. (2010). Assessing insider threats to information security using technical, behavioural and organisational measures. *Information Security Technical Report*. Elsevier Ltd, *15*(3), 112–133. https://doi.org/10.1016/j.istr.2010.11.002.

Safa, N. S. (2017). The information security landscape in the supply chain. *Computer Fraud and Security*. Elsevier Ltd, *2017*(6), 16–20. https://doi.org/10.1016/S1361-3723(17)30053-2.

Safa, N. S., Maple, C., Watson, T., & Von Solms, R. (2018). Motivation and opportunity based model to reduce information security insider threats in organisations. *Journal of Information Security and Applications*. Elsevier Ltd, *40*, 247–257. https://doi.org/10.1016/j.jisa.2017.11.001.

Schneier, B. (2011) *Secrets and lies: Digital security in a networked world.* Wiley.

Silic, M., & Back, A. (2014). Shadow IT – A view from behind the curtain. *Computers & Security*. Elsevier Ltd, *45*, 274–283. https://doi.org/10.1016/j.cose.2014.06.007.

Siponen, M. T., & Oinas-kukkonen, H. (2007). A review of information security issues and respective contributions. *The Data Base for Advances in Information Systems, 38*(1), 60–80. https://doi.org/10.1145/1216218.1216224.

Smith, S., Winchester, D., Bunker, D., Jamieson, R., Jamieson, R., & Bunker, D. (2010). Circuits of power: A study of mandated compliance to an information systems security "De Jure" standard in a government organisation. *MIS Quarterly, 34*(3), 463–486.

Stanton, J. M., Stam, K. R., Mastrangelo, P., & Jolton, J. (2005). Analysis of end user security behaviors. *Computers & Security, 24*(2), 124–133.
Symantec. (2017) *2017 internet security threat report.*
Tuoi Tre. (2016a) *Banks on the defence following Vietnam Airlines data breach.* Available at: https://tuoitrenews.vn/business/36271/banks-on-the-defence-following-vietnam-airlines-data-breach.
Tuoi Tre. (2016b) *Cyberattacks on Vietnam airports were well-planned: Association.*
Venter, H. S., & Eloff, J. H. P. (2003). A taxonomy for information security technologies. *Computers & Security, 22*(4), 299–307. https://doi.org/10.1016/S0167-4048(03)00406-1.
Vietnam MIC. (2017) *White book of Viet Nam information and communication technology 2017.*
Von Solms, B. (2001). Information security – A multidimensional discipline. *Computers & Security, 20*(6), 504–508. https://doi.org/10.1016/S0167-4048(01)00608-3.
Von Solms, B., & Von Solms, R. (2004). The 10 deadly sins of information security management. *Computers & Security, 23*(5), 371–376. https://doi.org/10.1016/j.cose.2004.05.002.

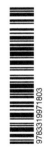